Emerging Low-Dimensional Materials

Emerging Low-Dimensional Materials

Editors

Bo Chen
Rutao Wang
Nana Wang

MDPI • Basel • Beijing • Wuhan • Barcelona • Belgrade • Manchester • Tokyo • Cluj • Tianjin

Editors

Bo Chen
City University of Hong Kong
Hong Kong
China

Rutao Wang
Shandong University
Jinan, Shandong
China

Nana Wang
University of Wollongong
Wollongong, NSW
Australia

Editorial Office
MDPI
St. Alban-Anlage 66
4052 Basel, Switzerland

This is a reprint of articles from the Special Issue published online in the open access journal *Crystals* (ISSN 2073-4352) (available at: https://www.mdpi.com/journal/crystals/special_issues/Emerging_Low_Dimens_Materials).

For citation purposes, cite each article independently as indicated on the article page online and as indicated below:

LastName, A.A.; LastName, B.B.; LastName, C.C. Article Title. *Journal Name* **Year**, *Volume Number*, Page Range.

ISBN 978-3-0365-6828-7 (Hbk)
ISBN 978-3-0365-6829-4 (PDF)

© 2023 by the authors. Articles in this book are Open Access and distributed under the Creative Commons Attribution (CC BY) license, which allows users to download, copy and build upon published articles, as long as the author and publisher are properly credited, which ensures maximum dissemination and a wider impact of our publications.

The book as a whole is distributed by MDPI under the terms and conditions of the Creative Commons license CC BY-NC-ND.

Contents

About the Editors . . . . **vii**

Preface to "Emerging Low-Dimensional Materials" . . . . **ix**

Bo Chen, Rutao Wang and Nana Wang
Emerging Low-Dimensional Materials (Volume I)
Reprinted from: *Crystals* 2023, 13, 166, doi:10.3390/cryst13020166 1

Mingguang Ma, Yunxia Wei, Jie Chen and Qiong Shang
Electrochemical In Situ Fabrication of Titanium Dioxide Nanotubes on a Titanium Wire as a Fiber Coating for Solid-Phase Microextraction of Polycyclic Aromatic Hydrocarbons
Reprinted from: *Crystals* 2021, 11, 1384, doi:10.3390/cryst11111384 5

Chuancang Zhou, Feipeng Zhang and Hongyu Wu
Boosting pH-Universal Hydrogen Evolution of FeP/CC by Anchoring Trace Platinum
Reprinted from: *Crystals* 2022, 12, 37, doi:10.3390/cryst12010037 19

Dong Chen and Zhongren Nan
Layer-by-Layer Assembly of Polyelectrolytes on Urchin-like MnO_2 for Extraction of Zn^{2+}, Cu^{2+} and Pb^{2+} from Alkaline Solutions
Reprinted from: *Crystals* 2022, 12, 358, doi:10.3390/cryst12030358 29

Shao-Bo Guo, Wei-Bin Zhang, Ze-Qin Yang, Xu Bao, Lun Zhang, Yao-Wen Guo, et al.
The Preparation and Electrochemical Pseudocapacitive Performance of Mutual Nickel Phosphide Heterostructures
Reprinted from: *Crystals* 2022, 12, 469, doi:10.3390/cryst12040469 41

Yanjun Zhai, Shuli Zhou, Linlin Guo, Xiaole Xin, Suyuan Zeng, Konggang Qu, Nana Wang and Xianxi Zhang
Zeolitic Imidazolate Framework 67-Derived Ce-Doped CoP@N-Doped Carbon Hollow Polyhedron as High-Performance Anodes for Lithium-Ion Batteries
Reprinted from: *Crystals* 2022, 12, 533, doi:10.3390/cryst12040533 57

Liangfeng Niu, Shoujie Guo, Wei Liang, Limin Song, Burong Song, Qianlong Zhang and Lijun Wu
In Situ Electrochemical Derivation of Sodium-Tin Alloy as Sodium-Ion Energy Storage Devices Anode with Overall Electrochemical Characteristics
Reprinted from: *Crystals* 2022, 12, 575, doi:10.3390/cryst12050575 69

Kyungjin Im, Dong-Hyoup Seo and Hyunwook Song
Bias-Voltage Dependence of Tunneling Decay Coefficient and Barrier Height in Arylalkane Molecular Junctions with Graphene Contacts as a Protecting Interlayer
Reprinted from: *Crystals* 2022, 12, 767, doi:10.3390/cryst12060767 79

Seock-Hyeon Hong, Dong-Hyoup Seo and Hyunwook Song
Demonstration of Molecular Tunneling Junctions Based on Vertically Stacked Graphene Heterostructures
Reprinted from: *Crystals* 2022, 12, 787, doi:10.3390/cryst12060787 89

Dong-Hyoup Seo, Kyungjin Im and Hyunwook Song
High-Temperature Electronic Transport Properties of PEDOT:PSS Top-Contact Molecular Junctions with Oligophenylene Dithiols
Reprinted from: *Crystals* 2022, 12, 962, doi:10.3390/cryst12070962 99

Baoshou Shen, Rong Hao, Yuting Huang, Zhongming Guo and Xiaoli Zhu
Research Progress on MXene-Based Flexible Supercapacitors: A Review
Reprinted from: *Crystals* **2022**, *12*, 1099, doi:10.3390/cryst12081099 . . . . **107**

Guo-Ping Shen, Ruo-Yao Fan, Bin Dong and Bo Chen
Ferrocene Formic Acid Surface Modified Ni(OH)$_2$ for Highly Efficient Alkaline Oxygen Evolution
Reprinted from: *Crystals* **2022**, *12*, 1404, doi:10.3390/cryst12101404 . . . . **145**

About the Editors

Bo Chen

Bo Chen received his B.S. degree, M.S. degree, and Ph.D. degree from the Lanzhou University (2009), Shandong University (2012), and Nanyang Technological University (2017, under the supervision of Prof. Hua Zhang), respectively. Then, he worked as a research fellow at the Nanyang Technological University before moving to the City University of Hong Kong in 2020. He is currently a research associate at the City University of Hong Kong. His current research interests include the synthesis, characterization, and applications of nanomaterials with unconventional phases.

Rutao Wang

Rutao Wang is a Professor of the School of Materials Science and Engineering, Shandong University. He received his Ph.D. degree in Materials Science from the Lanzhou Institute of Chemical Physics, Chinese Academy of Sciences (2015). During 2015–2018, he joined Professor Li Zhang's group as a Research Fellow in the Department of Mechanical and Automation Engineering, at the Chinese University of Hong Kong. Now, his research focuses on developing advanced materials for supercapacitor and all solid-state batteries.

Nana Wang

Nana Wang is a research fellow at the Institute for Superconducting and Electronic Materials at UOW. She is a recipient of the 2020 Discovery Early Career Researcher Award. Her research interests focus on liquid secondary metal-ion batteries, designing and fabricating new electrode materials with desired properties, and understanding the fundamental science of the interfaces, charge transport, and reactions in and of the materials that have advanced characterization techniques.

Preface to "Emerging Low-Dimensional Materials"

Technologies for renewable energy and green environments have received extensive attention in the past few decades. Recently, low-dimensional materials, such as zero-dimensional (0D), one-dimensional (1D), and two-dimensional (2D) materials, have been intensively investigated due to their unique catalytic, mechanical, electronic, and optical properties as well as their various applications. Great efforts have been devoted to studying their synthesis strategies, unique properties, chemical reaction processes, and potential applications. Nevertheless, challenges still exist. It is therefore urgent and significant to appreciate new advances and to review recent progresses in novel low-dimensional materials.

Bo Chen, Rutao Wang, and Nana Wang
Editors

Editorial

Emerging Low-Dimensional Materials (Volume I)

Bo Chen [1,*], Rutao Wang [2,*] and Nana Wang [3,*]

1. Department of Chemistry, City University of Hong Kong, Hong Kong SAR, China
2. School of Materials Science and Engineering, Shandong University, Jinan 250100, China
3. Institute for Superconducting and Electronic Materials, University of Wollongong, Wollongong, NSW 2522, Australia
* Correspondence: bchen005@e.ntu.edu.sg (B.C.); rtwang@sdu.edu.cn (R.W.); nanaw@uow.edu.au (N.W.)

1. Introduction

We recently published the first volume of the Special Issue "Emerging Low-Dimensional Materials". Impressively, we have a great collection of 11 outstanding articles that have been published. You are welcome to access these articles without any charge via the following link: https://www.mdpi.com/journal/crystals/special_issues/Emerging_Low_Dimens_Materials (accessed on 30 December 2022).

In this Special Issue, Jianping Long and coworkers report high-performance nickel-phosphide-based hybrids prepared in one step used as supercapacitor electrodes without a polymer binder [1]. They showed good electrochemical performance due to their good catalytic activity, excellent conductivity, increased catalytic active sites, and improved ion transmission caused by the interface effect.

MXenes, as emerging 2D materials, have been widely applied in various fields due to their unique properties. A review article focusing on MXene-based materials used as electrodes for flexible supercapacitors has been timely summarized by Zhu's group [2]. The recent progress in fabrication methods of MXene-based flexible electrodes, their corresponding electrochemical performance, and future possibilities are discussed in detail.

Besides supercapacitors, there are two research articles focusing on anodes for Li/Na-ion batteries. Zhai and her coworkers designed cerium-doped cobalt phosphide@nitrogen-doped carbon (Ce-doped CoP@NC), which possesses a hollow polyhedron architecture, using Zeolitic Imidazolate Framework 67 as a template [3]. The Ce-doped CoP@NC exhibited excellent electrochemical performance as an anode in Li-ion batteries due to Ce doping's structural merits, the carbon network, and the well-designed hollow polyhedron.

Wu et al. successfully synthesized a sodium–tin alloy anode, which was in situ electrochemically formed via a straightforward design of Sn foil integrated with a Na ring [4]. The fluffy, porous structure of a Na–Sn alloy anode can effectively alleviate the change in the volume of Sn metal during cycling, and can also inhibit the formation of sodium dendrites. More importantly, the consumption of Na ions due to the repeated formation of SEI film can be instantaneously complemented, thus increasing the Coulombic efficiency.

For hydrogen/oxygen evolution, Chen et al. designed $Ni(OH)_2$/nickel-foam-based electrocatalysts with a tiny amount of ferrocene formic acid (FFA) (FFA-$Ni(OH)_2$/NF) via electrochemical activation and surface atom modulation [5]. FFA-$Ni(OH)_2$/NF exhibited outstanding OER performances in an alkaline solution, benefiting from the synergistic effects of Fe-Ni heteroatoms and the strong electron interaction.

Regarding improving electrocatalytic performances for hydrogen evolution reactions, Hongyu Wu and coworkers fabricated 3D FeP-Pt film on carbon cloth (3D FeP-Pt/CC) using hydrothermal methods together with phosphating as well as electro-deposition [6]. Three-dimensional FeP-Pt/CC demonstrated superior electrocatalytic performances for hydrogen evolution reactions at all pHs, with excellent long-term stability and remarkable durability.

Zhongren Nan's group at Lanzhou University prepared three-dimensional urchin-like MnO_2@poly (sodium 4-styrene sulfonate) (PSS)/poly (diallyl dimethylammonium

Citation: Chen, B.; Wang, R.; Wang, N. Emerging Low-Dimensional Materials (Volume I). *Crystals* **2023**, *13*, 166. https://doi.org/10.3390/cryst13020166

Received: 30 December 2022
Accepted: 9 January 2023
Published: 18 January 2023

Copyright: © 2023 by the authors. Licensee MDPI, Basel, Switzerland. This article is an open access article distributed under the terms and conditions of the Creative Commons Attribution (CC BY) license (https://creativecommons.org/licenses/by/4.0/).

chloride) (PDDA)/PSS particles (MnO$_2$@PSS/PDDA/PSS) through the layer-by-layer assembly strategy [7]. MnO$_2$@PSS/PDDA/PSS demonstrated high efficiency in removing Zn^{2+} from an aqueous solution at a pH of 13. Moreover, it was effective in removing Pb^{2+} and Cu^{2+} from slightly alkaline water.

Qiong Shang's group synthesized unique titanium dioxide nanotube (TiO$_2$NTs)-coated fiber through the anodization of Ti wire in the electrolyte [8]. Moreover, the extraction mechanism of polycyclic aromatic hydrocarbons by TiO$_2$NT fibers was explained in detail.

Hyunwook Song's group studied a molecular junction with two graphene contacts incorporated with self-assembled arylalkane monolayers [9]. Their results demonstrated reliable and stable molecular junctions with graphene contacts as well as intrinsic charge transport characteristics. Meanwhile, the potential application of the voltage-induced barrier-lowering approximation to the graphene-based molecular junction was justified.

Based on their last work, Hyunwook Song and coworkers further fabricated vertical molecular tunneling junctions with graphene heterostructures, where arylalkane molecules can act as charge transport barriers [10]. Various characterization techniques and an intact statistical analysis were adopted in this research.

Another paper related to molecular junctions is reported by Seo et al. [11]. The high-temperature electronic transport activities of spin-coated PEDOT:PSS junctions based on self-assembled oligophenylene dithiol monolayers were investigated.

In conclusion, this Special Issue presents recent progress on emerging low-dimensional materials and could encourage future investigations into them.

After finishing the first volume successfully, we think that there is still plenty of room for research on low-dimensional materials. Therefore, we have decided to announce the second volume of this Special Issue on low-dimensional materials. You are welcome to submit any high-quality manuscripts related to low-dimensional materials to the second volume. Please find more details at the following link: https://www.mdpi.com/journal/crystals/special_issues/HN197LMB87 (accessed on 30 December 2022).

Author Contributions: Conceptualization, writing—original draft preparation, review and editing, B.C., R.W. and N.W. All authors have read and agreed to the published version of the manuscript.

Acknowledgments: The contributions of all of the authors are gratefully acknowledged. We would like to express our gratitude to the editorial team of *Crystals*.

Conflicts of Interest: The authors declare no conflict of interest.

References

1. Guo, S.-B.; Zhang, W.-B.; Yang, Z.-Q.; Bao, X.; Zhang, L.; Guo, Y.-W.; Han, X.-W.; Long, J. The Preparation and Electrochemical PseudocapacitivePerformance of Mutual Nickel Phosphide Heterostructures. *Crystals* **2022**, *12*, 469. [CrossRef]
2. Shen, B.; Hao, R.; Huang, Y.; Guo, Z.; Zhu, X. Research Progress on MXene-Based Flexible Supercapacitors: A Review. *Crystals* **2022**, *12*, 1099. [CrossRef]
3. Zhai, Y.; Zhou, S.; Guo, L.; Xin, X.; Zeng, S.; Qu, K.; Wang, N.; Zhang, Z. Zeolitic Imidazolate Framework 67-Derived Ce-Doped CoP@N-Doped Carbon Hollow Polyhedron as High-Performance Anodes for Lithium-Ion Batteries. *Crystals* **2022**, *12*, 533. [CrossRef]
4. Niu, L.; Guo, S.; Liang, W.; Song, L.; Song, B.; Zhang, Q.; Wu, L. In Situ Electrochemical Derivation of Sodium-Tin Alloy as Sodium-Ion Energy Storage Devices Anode with Overall Electrochemical Characteristics. *Crystals* **2022**, *12*, 575. [CrossRef]
5. Shen, G.-P.; Fan, R.-Y.; Dong, B.; Chen, B. Ferrocene Formic Acid Surface Modified Ni(OH)$_2$ for Highly Efficient Alkaline Oxygen Evolution. *Crystals* **2022**, *12*, 1404. [CrossRef]
6. Zhou, C.; Zhang, F.; Wu, H. Boosting pH-Universal Hydrogen Evolution of FeP/CC by Anchoring Trace Platinum. *Crystals* **2022**, *12*, 37. [CrossRef]
7. Chen, D.; Nan, Z. Layer-by-Layer Assembly of Polyelectrolytes on Urchin-like MnO$_2$ for Extraction of Zn^{2+}, Cu^{2+} and Pb^{2+} from Alkaline Solutions. *Crystals* **2022**, *12*, 358. [CrossRef]
8. Ma, M.; Wei, Y.; Shang, Q. Electrochemical In Situ Fabrication of Titanium Dioxide Nanotubes on a Titanium Wire as a Fiber Coating for Solid-Phase Microextraction of Polycyclic Aromatic Hydrocarbons. *Crystals* **2021**, *11*, 1384. [CrossRef]
9. Im, K.; Seo, D.-H.; Song, H. Bias-Voltage Dependence of Tunneling Decay Coefficient and Barrier Height in Arylalkane Molecular Junctions with Graphene Contacts as a Protecting Interlayer. *Crystals* **2022**, *12*, 767. [CrossRef]

10. Hong, S.-H.; Seo, D.-H.; Song, H. Demonstration of Molecular Tunneling Junctions Based on Vertically Stacked Graphene Heterostructures. *Crystals* **2022**, *12*, 787. [CrossRef]
11. Seo, D.-H.; Im, K.; Song, H. High-Temperature Electronic Transport Properties of PEDOT:PSS Top-Contact Molecular Junctions with Oligophenylene Dithiols. *Crystals* **2022**, *12*, 962. [CrossRef]

Disclaimer/Publisher's Note: The statements, opinions and data contained in all publications are solely those of the individual author(s) and contributor(s) and not of MDPI and/or the editor(s). MDPI and/or the editor(s) disclaim responsibility for any injury to people or property resulting from any ideas, methods, instructions or products referred to in the content.

Article

Electrochemical In Situ Fabrication of Titanium Dioxide Nanotubes on a Titanium Wire as a Fiber Coating for Solid-Phase Microextraction of Polycyclic Aromatic Hydrocarbons

Mingguang Ma, Yunxia Wei *, Jie Chen and Qiong Shang

College of Chemical Engineering, Lan Zhou City University, Lanzhou 730070, China; mamg001@lzcu.edu.cn (M.M.); shuyeal@lzcu.edu.cn (J.C.); sq@lzcu.edu.cn (Q.S.)
* Correspondence: weiyx07@lzu.edu.cn

Abstract: A novel titanium dioxide nanotube (TiO_2NTS) coated fiber for solid-phase microextraction (SPME) was prepared by in situ anodization of titanium wire in electrolyte containing ethylene glycol and ammonium fluoride (NH_4F). The effects of different electrolyte solutions (NH_4F and ethylene glycol) and oxidation voltages on the formation and size of TiO_2NTs was studied. It was obtained from the experiment that TiO_2NTs arrays were arranged with a wall thickness of 25 nm and the diameter of 100 nm pores in ethylene glycol and water (v/v, 1:1) containing NH_4F of 0.5% (w/v) with a voltage of 20 V at 25 °C for 30 min. The TiO_2NTs were used as solid-phase microextraction fiber coatings coupled with high-performance liquid chromatography (HPLC) in sensitive determination of polycyclic aromatic hydrocarbons (PAHs) in spiked real samples water. Under the optimized SPME conditions, the calibration curve has good linearity in the range of 0.20–500 $\mu g \cdot L^{-1}$, and the correlation coefficient (R^2) is between 0.9980 and 0.9991. Relative standard deviations (RSDs) of 3.5–4.7% (n = 5) for single fiber repeatability and of 5.2% to 7.9% for fiber-to-fiber reproducibility (n = 3) was obtained. The limits of detection (LOD) (S/N = 3) and limits of quantification (LOQ) (S/N = 10) of PAHs were 0.03–0.05 $\mu g \cdot L^{-1}$ and 0.12–0.18 $\mu g \cdot L^{-1}$. The developed method was applied to the preconcentration and determination of trace PAHs in spiked real samples of water with good recoveries from 78.6% to 119% and RSDs from 4.3 to 8.9%, respectively.

Keywords: solid-phase microextraction; titanium dioxide nanotube; high-performance liquid chromatography; polycyclic aromatic hydrocarbons

Citation: Ma, M.; Wei, Y.; Chen, J.; Shang, Q. Electrochemical In Situ Fabrication of Titanium Dioxide Nanotubes on a Titanium Wire as a Fiber Coating for Solid-Phase Microextraction of Polycyclic Aromatic Hydrocarbons. *Crystals* 2021, 11, 1384. https://doi.org/10.3390/cryst11111384

Academic Editors: Rajratan Basu and Vladislav V. Kharton

Received: 14 October 2021
Accepted: 10 November 2021
Published: 12 November 2021

Publisher's Note: MDPI stays neutral with regard to jurisdictional claims in published maps and institutional affiliations.

Copyright: © 2021 by the authors. Licensee MDPI, Basel, Switzerland. This article is an open access article distributed under the terms and conditions of the Creative Commons Attribution (CC BY) license (https://creativecommons.org/licenses/by/4.0/).

1. Introduction

Solid-phase microextraction (SPME) has attracted extensive attention due to its high sensitivity, rapidity, simplicity and being free of solvents [1,2]. In principle, the technique is based on the distribution of the target analyte between the sample matrix and the thin extraction coating deposited on the fine solid fibers. Currently, microscale fused silica rods are used as the matrix for commercial SPME fibers. As a result, commercial SPME coatings are generally characterized by a low operating temperature, instability, less selectivity and swelling in organic solvents [3,4]. In addition, the SPME coating variety of the product is single, and the application is limited. To overcome the above problems, an important development in the SPME fiber preparation technology is to obtain a coating with high mechanical strength and good chemical stability by using a metal as a matrix to improve the sensitivity and selectivity of the coating to the target compound [5–9]. Metal nanofibers, in particular, have attracted the attention of many researchers due to their unique physical chemistry properties, including a large specific surface area, good chemical and thermal stability, and favorable adsorption performance [10,11]. These metal fibers with high mechanical strength and nanostructured coatings exhibit higher extraction capacity, faster extraction rate and better extraction selectivity for the target analytes. Due to the chemical inertness of the metal fiber surface, researchers proposed novel strategies to prepare novel metal SPME fibers [12–14].

Since the Gong group reported the formation of titanium dioxide porous structure by anodizing titanium foils in fluorine-containing aqueous electrolyte, various attempts were made to prepare TiO_2 nanostructures by anodization [15,16]. At present, the preparation, characterization, and application of TiO_2NTs have attracted wide attention. The importance of TiO_2 is growing as a major component of photocatalysis and photovoltaic devices, as well as applications in sensors, biomedical coatings, preservatives or antioxidants [17–20].

The preparation methods include hydrothermal synthesis, template synthesis and anodization. TiO_2NTs arrays not only have larger surfaces, but also possess some special properties, including unique electronic and chemical properties. The TiO_2NTs prepared by anodization are perpendicularly orientated and well-aligned structures with controllable operation, good repeatability and a simple oxidation preparation process. In the application of titanium dioxide nanotubes, the dimensions of the nanotubes, such as the outer diameter and length, has a profound influence on its ability to produce an excellent performance. Different electrolytes will affect the morphology of coating, which will affect the extraction effect of TiO_2NTs. PAHs are a kind of widely distributed organic pollutant which has carcinogenic, teratogenic and mutagenic effects, especially those containing four or more aromatic rings [21,22]. Many modern analytical techniques have been developed and applied to the determination of PAHs in environment water [23,24]. Since the concentration levels of PAHs are present in the aquatic environment at trace levels in complex matrices, this makes accurate determination difficult. Sample pretreatment in trace analysis is the bottleneck and the main source of error in the analysis process. Therefore, PAHs need to be enriched before instrumental analysis and determination.

At present, techniques for the enrichment of PAHs from environmental samples include solvent extraction [25,26], solid-phase extraction [27,28], SPME [29,30], pressurized liquid extraction [31] and supercritical fluid extraction [32]. The combination of SPME and HPLC is the most commonly used technology, which has the characteristics of a simple preparation method, high enrichment efficiency and less time consumption.

The shape of nano TiO_2 produced by the in situ oxidation of titanium wire is closely related to the electrolyte. In this paper, we investigated the effects of fluoride concentration and the amount of ethylene glycol on the formation and dimensions of the TiO_2NTs. The as-fabricated TiO_2NTs coating was employed to extract polycyclic aromatic hydrocarbons (PAHs) with different ring numbers in combination with high-performance liquid chromatography-UV detection (HPLC-UV). Meanwhile, the extraction mechanism of PAHs by TiO_2 nanotube fibers was discussed. It was found that the TiO_2NTs coating exhibited high extraction capability and good selectivity for some PAHs from the water phase. Under the optimized SPME conditions, the fiber was used to measure PAHs in spiked real samples and from which we draw a satisfactory result.

2. Experimental

2.1. Chemicals and Reagents

A roll of Ti wire of 0.25 mm diameter with high purity was obtained from Alfa Aesar (Ward Hill, MA, USA). A 0.45 μm micropore membrane of polyvinylidene fluoride was supplied by Xingya Purifying Material Factory (Shanghai, China). Ammonium fluoride (NH_4F) and ethylene glycol were obtained from Sinopharm Chemical Reagent Co., Ltd. (Shanghai, China). The HPLC-grade methanol was purchased from Yuwang Chemical Company (Shandong, China). Naphthalene (Nap), phenanthrene (Phe), fluorene (Flu), anthracene (Ant), fluoranthene (Flt), and pyrene (Pyr) were obtained from the Aldrich (Steinheim, Germany). Individual standard stock solutions were prepared in methanol at a concentration of 100 mg·L^{-1} and stored at 4 °C in the refrigerator. All other reagents were of analytical grade.

2.2. Instruments

Anodization of the Ti wire was performed with a precise WY-3D power supply (Nanjing, China). A Waters 600E multi-solvent delivery system (Milford, MA, USA) equipped with Waters 2487 dual λ absorbance detector and a Waters Sunfire C18 chromatographic column (150 mm × 4.6 mm, 5 μm) was used as all separations. A N2000 workstation (Zhejiang University, Hangzhou, China) was used for the acquisition of data. The mobile phase was methanol/water and the wavelength of UV detection was set at 254 nm. A desorption chamber was used in a commercially available Supelco SPME-HPLC interface (Bellefonte, PA, USA). Scanning electron microscope micrographs of TiO_2 fibers were obtained on a field emission scanning electron microscope (Zeiss, Oberkochen, Germany) equipped with an energy dispersive X-ray (EDX) spectrometer. X-ray diffraction (XRD) was performed on a D8 Advance diffractometer (Bruker, Germany). Ultrapure water was obtained from the Sudreli-system (Chongqing, China). The DF-101S Magnetic stirrer (Zhengzhou, China) was used to heat and stir.

2.3. Preparation of SPME Fiber

The Ti wire was cut into lengths of 6 cm then thoroughly washed with ethanol, acetone and distilled water in sequence prior to the anodization. TiO_2NTs were prepared by a potentiostatic anodization method in a two-electrode electrochemical cell. In a typical process, the commercial Ti wire (99.9%) was used as a working electrode, and a Pt rod as the counter electrode. Ti wire was anodized in different NH_4F concentrations and the amount of ethylene glycol at a controlled potential of 20 V for 30 min at room temperature. After the anodization, the fiber was washed with deionized water and then dried in air.

2.4. The Procedure of SPME-HPLC

To carry out the extraction, 15 mL of the standard solution or sample solution was added into a 20 mL glass vial equipped with 1 cm magnetic stirrer bar inside and a Teflon septum. The TiO_2NTs/Ti fiber was exposed to the heated and stirred sample solution for extraction. Subsequently, the fiber was pulled out and immersed into an SPME-HPLC interface using the static mode for desorption with mobile phase. After desorption, a six-port valve was switched from the load to inject position, the mobile phase was passed through the interface and PAHs were introduced into the analytical column at a flow rate of 1 mL·min^{-1} for analysis. The mixtures of methanol and water of 90/10 (v/v) were employed as mobile phases for HPLC analyses of PAHs at a flow rate of 1 mL·min^{-1}. Corresponding chromatographic signals was monitored at 254 nm. For the next extraction, the fiber was immersed into methanol and ultrapure water to eliminate possible carry-over for 10 min and 5 min, respectively.

3. Results and Discussion

3.1. The Effect of NH_4F on TiO_2NTs Coating

3.1.1. Aqueous Solution System

According to the oxidation mechanism proposed by Macak et al. [33–35], in the process of preparing TiO_2NTs by in situ anodization of the Ti matrix, the anodic oxidation rate and chemical etching rate jointly determine the morphology and structure of TiO_2NTs array. The growth and formation rate of TiO_2NTs array is controlled by anodic oxidation rate, while the dissolution rate of TiO_2NTs array is controlled by chemical etching rate. The composition and concentration of electrolyte have an important effect on the structure of TiO_2NTs array. The formation of TiO_2NTs was investigated by adding different concentrations of NH_4F in aqueous electrolyte. In aqueous solution of NH_4F, the surface of Ti wire can be easily oxidized to TiO_2. With the increase in NH_4F concentration, the chemical etching rate is much higher than the formation rate of nanotube array. As shown in Figure 1, when the NH_4F concentration increases to 2.5%, flower-shaped rather than tubular TiO_2 is generated.

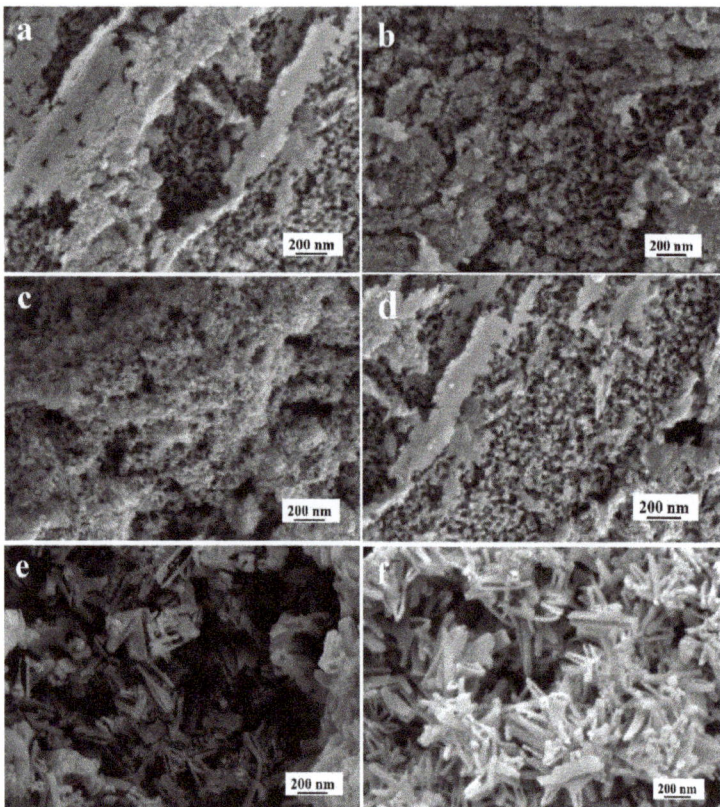

Figure 1. SEM images of Ti wires anodized in aqueous solution of NH$_4$F. Anodic conditions: time, 30 min; voltage, 20 V; temperature, 25 °C. Concentration of NH$_4$F (ω%): 0.1% (**a**), 0.25% (**b**), 0.5% (**c**), 1.0% (**d**), 2.5% (**e**), 5.0% (**f**).

3.1.2. Organic Electrolyte Solution System

In order to reduce the chemical etching rate of TiO$_2$NTs grown in situ by anodized Ti wire, organic solvent electrolyte was introduced. By adding the surface-active agent to change its crystal structure during the process of making TiO$_2$NTs, reduce nano TiO$_2$ bonding and improve its dispersion. In the organic electrolyte, the etching rate of TiO$_2$NTs by F$^-$ is small, and the anodized oxidation rate is greater than the chemical etching rate within the reaction time, so as to obtain regular TiO$_2$NTs array. The chemical etching rate increases with the increase in F$^-$ concentration, while the formation rate of TiO$_2$NTs array is the opposite. Therefore, by adjusting and optimizing the concentration of NH$_4$F, the dynamic balance of anodic oxidation and chemical etching in the growth of TiO$_2$NTs array can be controlled to achieve the controllable growth of TiO$_2$NTs array structure.

Figure 2 is the SEM diagram of Ti anodic oxidation with the same ethylene glycol concentration and different NH$_4$F concentration. As can be seen from Figure 2a,b, the concentration of NH$_4$F is low and the corresponding chemical etching rate is slow, and the lamellar layer covers the nanotubes, resulting in the failure to form continuous and regular nanotubes. The nanotubular morphology is preserved for the fiber coating with the narrower inner diameter at NH$_4$F concentration of 0.5–1.0% (ω%) (Figure 2c,d). However, the uniform frameworks of the TiO$_2$NTs structure were rapidly deteriorated at a higher concentration of NH$_4$F (Figure 2e,f). The main reason for this problem is that the chemical etching rate is much greater than the oxidation rate, the migration rate of Ti^{4+} increases, the structure of array TiO$_2$NTs is damaged, and finally, flake TiO$_2$ is formed.

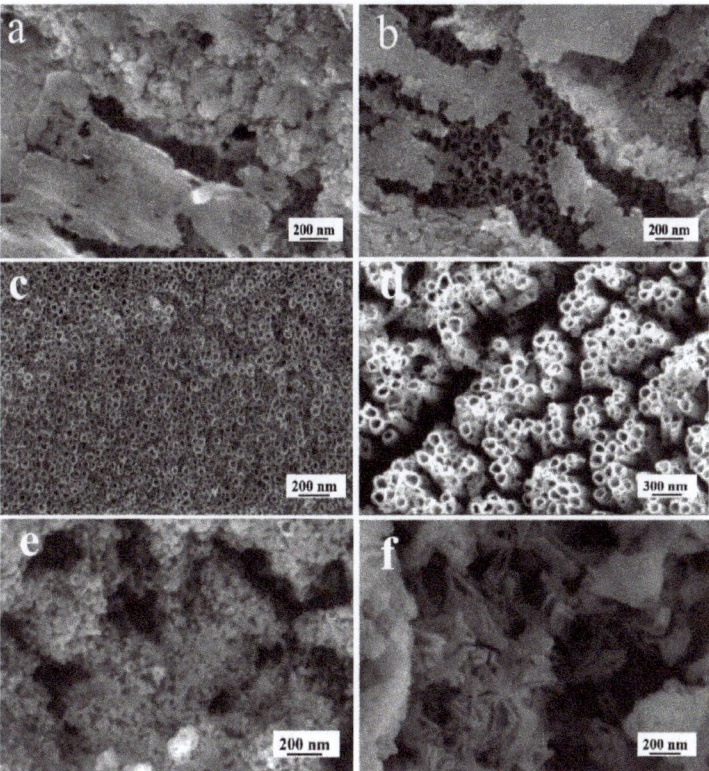

Figure 2. SEM images of Ti wires anodized in ethylene glycol containing NH$_4$F. Anodic conditions: time, 30 min; voltage, 20 V; temperature, 25 °C; electrolyte composition, ethylene glycol concentration (50%, v/v). Concentration of NH$_4$F (ω%): 0.1% (**a**), 0.25% (**b**), 0.5% (**c**), 1.0% (**d**), 2.5% (**e**), 5.0% (**f**).

3.2. The Effect of Ethylene Glycol

The excellent properties of TiO$_2$NTs/Ti fibers largely depend on the pore size of nanotubes. In the electrolyte with water as solvent, the TiO$_2$NTs obtained by anodic oxidation are shorter (Figure 1a–d). In order to increase the length of TiO$_2$NTs, NH$_4$F was used as an electrolyte and the concentration was 0.5%, and the concentration of ethylene glycol was 20%, 50%, 60% and 80% to obtain the TiO$_2$NTs/Ti fibers, respectively.

As depicted in Figure 3, considerable changes in the surface structures are observed after electrolysis in different electrolytes at 20 V constant voltage for 30 min. Ethylene glycol is an organic solvent electrolyte, which is less acidic than the water-soluble electrolyte, and the etching rate of TiO$_2$NTs by F$^-$ decreases. In a longer reaction time, the anodic oxidation rate is greater than the chemical etching rate, thus obtaining a longer TiO$_2$NTs array. However, as the concentration of ethylene glycol continued to increase, the nanotube lengthened and its irregularity increased, so that the nanotube was superimposed, and its mechanical strength decreased. Considering the length and strength of the nanotubes, the optimal electrolyte composition is 0.5% (ω%) NH$_4$F and 50% (v/v) ethylene glycol.

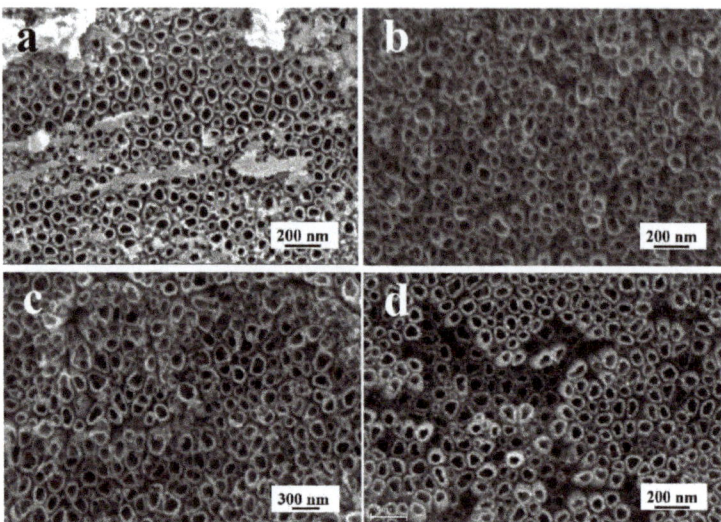

Figure 3. SEM images of Ti wire anodized in ethylene glycol solution of NH$_4$F. Electrochemical anodization condition: time, 30 min; voltage, 20 V; temperature, 25 °C; electrolyte composition, NH$_4$F concentration 0.50% (w%). The concentration of ethylene glycol (v/v): 20% (**a**), 50% (**b**), 60% (**c**), 80% (**d**).

3.3. The Effect of Voltages

Different anodic voltages produce different morphologies of titanium dioxide coating. As can be seen from Figure 4, a passive film of titanium dioxide is rapidly formed on the surface of titanium wire during oxidation (Figure 4a,b). TiO$_2$ was provided with poorer conductivity and a high voltage is required to maintain the reaction. With the increase in voltage, the reaction speed is accelerated, and nanotubes are formed on the surface of titanium wire (Figure 4b,c). However, high voltage can lead to the dissociation of water and the breakdown of the nanotube structure due to the formation of gas (Figure 4e,f).

3.4. The Characterization of the TiO$_2$NTs

SEM was used to characterize the surface of the Ti wire before and after anodizing. As shown in Figure 5, highly ordered TiO$_2$NTs arrays (Figure 5d) were generated on the surface of the anodized Ti wire with an average aperture of about 100 nm and a wall thickness of about 25 nm.

The surface elemental analysis of the Ti wire before and after oxidation was performed by energy dispersive X-ray spectroscopy. As shown in Figure 6a, the presence of weak O peak in the EDS spectrum illustrates the formation of a very thin passivation layer at the surface of the commercial Ti wire. Figure 6b shows peaks corresponding to the presence of Ti and O, and their mass composition was close to the composition molar ratio of TiO$_2$. The EDS analysis demonstrates the drastic increase in O content due to the formation of TiO$_2$NTs coating. From the SEM image and EDS spectrum analysis results, the TiO$_2$NTs are tightly embedded into the Ti wire substrate.

In the X-ray diffraction (XRD) image of TiO$_2$NTs, diffraction peaks at 25.2° and 27.3° are the characteristic diffraction peaks of 101 lattice plane of anatase and 110 lattice planes of rutile (see Figure S1 in the Supplementary Materials). As can be seen from Figure S1, the crystalline structures of TiO$_2$NTs obtained by in situ anodization of Ti wire was almost all anatase type.

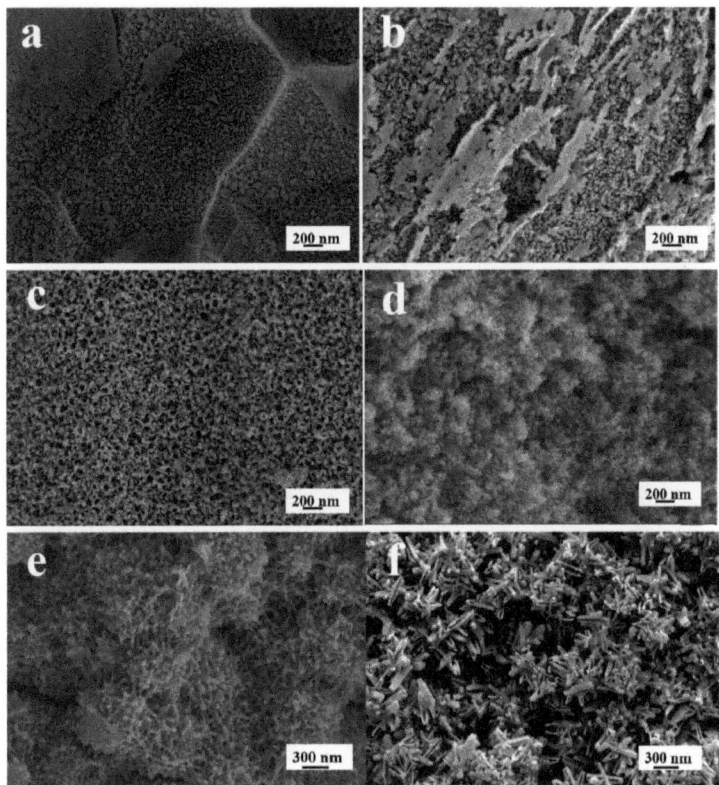

Figure 4. SEM images of in-situ titanium oxide wires at different voltages (**a**) 10 V; (**b**) 15V; (**c**) 20 V; (**d**) 25 V; (**e**) 30 V; (**f**) 25 V.

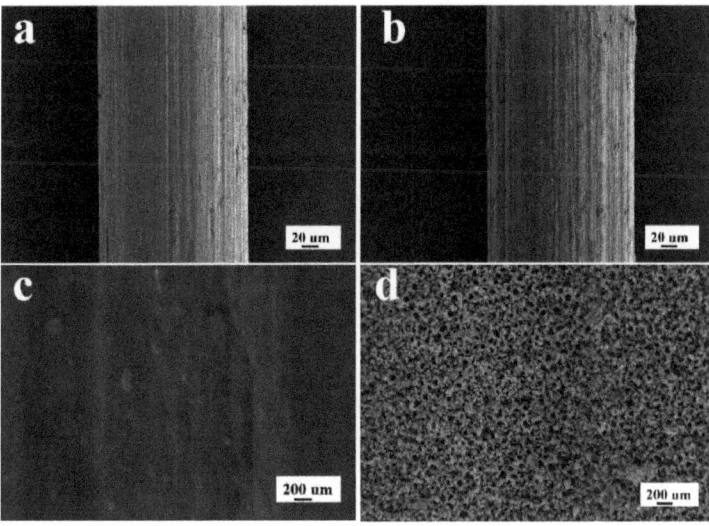

Figure 5. SEM images of the bare Ti wire (**a**) × 500, (**c**) ×50,000) and the Ti-TiO$_2$NTs fiber (**b**) ×500, (**d**) × 50,000).

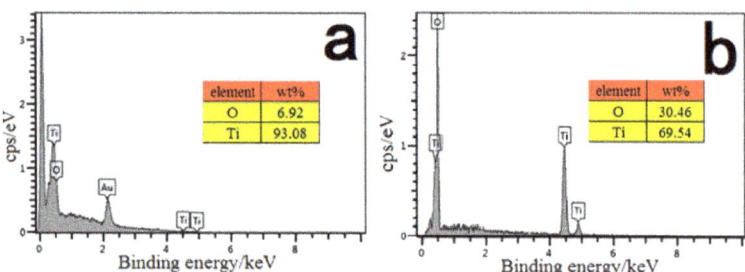

Figure 6. EDS spectra of the Ti wire (**a**) and the TiO$_2$NTs/Ti fiber (**b**).

3.5. The Extraction Mechanism of TiO$_2$NTs Fiber

It was found that the fiber had excellent extraction performance for PAHs. This result may be attributed to the inherent physicochemical nature of TiO$_2$NTs coatings. On the one hand, the as-fabricated fiber have special nanostructures, such as larger surface area, more open access points and better durability. These characteristics are most desirable for highly efficient SPME. On the other hand, the special anion-π orbital (electron donor-acceptor) interaction between TiO$_2$ and PAHs can take actions and result in the better extraction efficiencies. Therefore, the TiO$_2$NT-coated fiber just provides a hopeful approach to specifically extract and analyze trace amount of PAHs from complex environment water samples.

3.6. The Optimization of SPME Conditions for PAHs

The as-prepared TiO$_2$NTs/Ti fiber coupled with HPLC was used for the SPME of PAHs mixtures from water samples. To achieve the optimal extraction, the effects of extracting parameters, such as the extraction time, desorption time, temperature, stirring and ionic strength of sample solutions were optimized at the concentration level of 50 μg·L^{-1}.

3.6.1. Extraction and Desorption Time

In general, the extraction time depends on the analyte distribution equilibrium time between the fiber and water samples. Long extraction time is advantageous to reach the best equilibrium translation [36]. If the extraction time is short and the content of polycyclic aromatic hydrocarbons enriched in the fiber coating is too small, the sensitivity of the determination method will be affected. On the contrary, the extension of extraction time leads to the increase of determination time, and the method will lose its timeliness. Within 10–60 min, the effect of time on extraction efficiency was investigated. The amount of PAHs in the extraction coating increased with the extension of extraction time (expressed by peak area) before reaching equilibrium. As shown in Figure 7a, the extraction equilibrium was almost obtained within 50 min except for Ant. Considering only a slight increase in the peak areas of Ant after 50 min, 50 min is a reasonable compromise time between a good peak area and an acceptable extraction time for all PAHs.

In the SPME process, two steps are very important; one is the adsorption of the analyte on the surface of the fiber coating, and the other is the desorption of the analyte from the fiber coating in mobile phase. For target PAHs, the peak area reached constant maximum within 6 min (Figure 7b). Thus, 50 min extraction and 6 min desorption were employed in subsequent experiments.

3.6.2. Extraction Temperature

It is generally accepted that temperature is very important for SPME because of its potential influence. On the one hand, an elevated temperature can enhance the migration rate of molecules and so quicken the extraction rate. On the other hand, it would decrease the partitioning coefficient of the analyte between the fiber coating and the sample solution [37]. Therefore, the selection of a proper temperature is necessary for the extraction process. In our experiments, the effects of the temperature on the extraction efficiency of analytes were studied in a range of 25–70 °C. The results were shown in Figure 7c, the chromatographic peak area reached the maximum value of most compounds at 50 °C, so 50 °C was chosen for subsequent experiments.

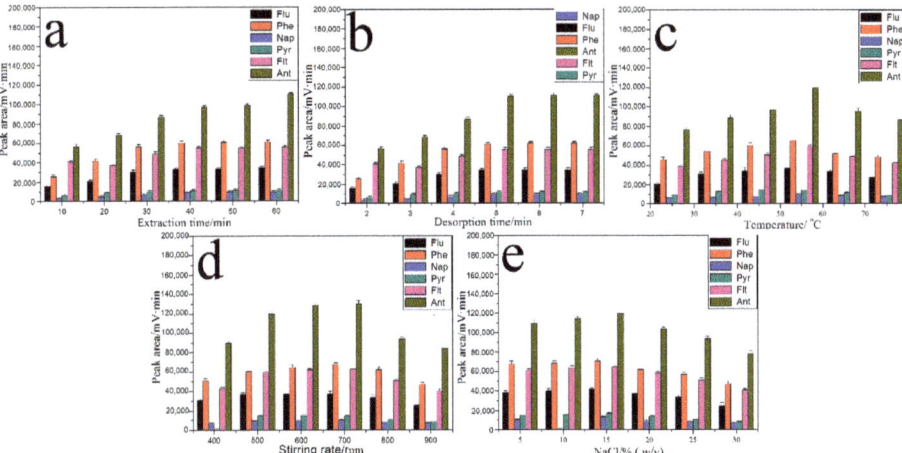

Figure 7. Effect of extraction time (**a**), desorption time (**b**), temperature (**c**), stirring rate, (**d**), and ionic strength (**e**) on extraction efficiency.

3.6.3. Stirring Speed

The extraction efficiency is notably affected by stirring speed, the solution agitation can reduce the equilibrium time by accelerating the diffusion of analytes from the samples to the fiber [38]. The stirring speed ranging from 200 to 1200 rpm was investigated. As shown in Figure 7d, the extraction efficiency increased with the stirring speed. However, a stirring rate above 700 rpm would lead to lower extraction efficiency. This may be due to strong fluid shear stress which made analytes absorbed in coating surface wash down easily. Therefore, the magnetic agitator was fixed at 700 rpm during the extraction.

3.6.4. Ionic Strength

It is known that the addition of a salt (NaCl) into a solution might either help with the extraction by the "salt out effect" or deteriorate the extraction due to the competitive adsorption of Na^+ and Cl^- [39]. The extraction efficiency as a function of salt concentration from 0% to 30% (30% is the saturated solubility of NaCl) was studied and is shown in Figure 7e. It is found that with the increase in salt concentration, chromatographic peak areas for the most analytes increase firstly up to 15% (w/v) salt concentration, indicating the "salt out effect" plays a dominant role at this stage and then the peak areas decrease because of the competitive adsorption. Therefore, a concentration of 15% (w/v) was selected as the optimized salt concentration.

3.7. Analytical Performance

Under the optimized conditions, figures of merit, including linear range, precision in terms of reproducibility and repeatability quantification (RSD%) and limits of detection (LOD). were evaluated. As shown in Table 1, good linearities are achieved in the range of 0.20–500 µg·L^{-1} for all target analytes with satisfactory correlation coefficients. The limits of detection (LOD) (S/N = 3) and limits of quantitation (LOQ) (S/N = 10) ranged from 0.03 to 0.05 µg·L^{-1} and from 0.12 to 0.18 µg/L for PAHs, respectively. The single fiber repeatability for five replicate extractions of PAHs at the spiking level of 50 µg·L^{-1} varied from 3.5% to 4.7%. The fiber-to-fiber reproducibility for three parallel fibers fabricated in different batches ranged from 5.2% to 7.9%. These data clearly indicate that satisfactory accuracy, good precision and high sensitivity were achieved with the proposed SPME-HPLC procedure.

Table 1. Analytical parameters of the proposed method with the TiO_2NTs/Ti fiber.

PAHs	Linearity ($\mu g \cdot L^{-1}$)	R^2	Recovery [a] %	RSD (%) [a]		LODs/($\mu g \cdot L^{-1}$)	LOQs/($\mu g \cdot L^{-1}$)
				Single Fiber (n = 5)	Fiber-to-Fiber (n = 3)		
Nap	0.2–500	0.9989	90.4	4.7	7.6	0.04	0.15
Flu	0.2–500	0.9985	107	3.8	6.8	0.05	0.17
Phe	0.2–500	0.9992	118	4.2	5.2	0.04	0.13
Ant	0.2–500	0.9980	92.4	4.0	7.4	0.03	0.12
Flt	0.2–500	0.9987	87.4	4.3	7.9	0.04	0.14
Pyr	0.2–500	0.9991	92.8	3.5	6.3	0.05	0.18

[a] Calculated at the concentration level of 50 $\mu g \cdot L^{-1}$.

3.8. Spiked Real Samples Analysis

To demonstrate the applicability and reliability of the proposed method, the presented method using the TiO_2NTs/Ti fiber was applied to the preconcentration and determination of target PAHs in tap water, river water, rainwater and wastewater. All water samples were filtered through 0.45 μm micropore membranes and then adjusted to pH 7.0 with phosphate buffer. The results of three replicate analyses are shown in Table 2. No PAHs was found in tap water but were detected in other water samples. In addition, for the sake of demonstrating the applicability and reliability, the recoveries of the target compounds are also determined. The median recoveries ranged from 78.6% to 119.0% with the RSD values from 4.3% to 8.9%.

The data in Table 2 indicated that the proposed method could be used in the analysis of spiked real samples. Figure 8 shows typical chromatograms of direct HPLC and SPME-HPLC for PAHs in wastewater. These data indicate that the proposed method is suitable for the extraction, enrichment and determination of target PAHs in different environmental water samples with a minor matrix effect.

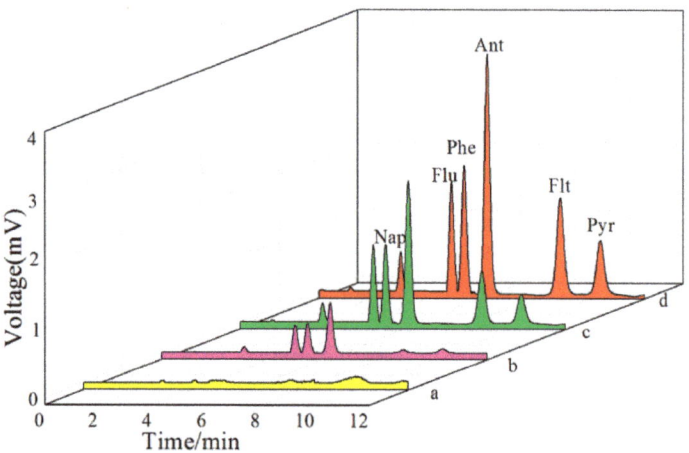

Figure 8. Chromatograms of direct HPLC and SPME-HPLC for PAHs in wastewater. Direct injection (a), SPME-HPLC with TiO_2NTs/Ti fiber (b), SPME-HPLC with TiO_2NTs/Ti fiber spiked wastewater at 5 $\mu g \cdot L^{-1}$ (c) and at 10 $\mu g \cdot L^{-1}$ (d).

Table 2. Analytical results of PAHs in different environmental water samples.

Water Samples	PAHs	Original/(μg·L^{-1})	Spiked with 5 μg·L^{-1}			Spiked with 10 μg·L^{-1}		
			Detected/(μg·L^{-1})	Recovery/%	RSD/%	Detected/(μg·L^{-1})	Recovery/%	RSD/%
Tap water	Nap	ND [a]	4.85	97.0	7.1	9.12	91.2	7.9
	Flu	ND	5.38	108	5.6	11.3	113	6.7
	Phe	ND	4.22	84.4	6.3	11.72	117.2	7.1
	Ant	ND	4.06	81.2	5.4	8.73	87.3	5.8
	Flt	ND	5.95	119	6.8	11.68	116.8	7.3
	Pyr	ND	4.41	88.2	4.3	9.05	90.5	5.5
River water	Nap	3.69	9.26	106.6	7.8	12.27	89.6	8.7
	Flu	2.57	6.83	90.2	4.9	11.85	94.3	5.8
	Phe	3.43	7.85	93.1	6.2	15.21	113.3	6.9
	Ant	1.12	4.81	78.6	5.9	10.16	91.4	6.1
	Flt	1.34	7.02	110.7	6.5	12.05	106.3	7.3
	Pyr	0.82	6.78	116.5	4.4	11.24	103.9	5.1
Wastewater	Nap	1.97	6.23	89.4	8.4	13.02	108.8	8.9
	Flu	3.12	8.65	106.5	6.2	12.36	94.2	7.6
	Phe	2.62	6.87	90.2	7.9	11.24	89.1	8.3
	Ant	2.34	6.63	90.3	7.3	11.14	90.3	8.5
	Flt	0.36	6.34	118.3	8.1	11.76	113.5	8.7
	Pyr	0.89	6.53	110.9	5.2	11.98	111.0	6.3
Rain water	Nap	1.82	7.17	105.1	7.9	11.37	96.2	8.6
	Flu	ND	4.89	97.8	5.3	8.52	85.2	6.1
	Phe	0.88	5.07	86.2	7.3	11.76	108.1	8.2
	Ant	3.75	8.28	94.6	5.8	15.22	110.7	6.7
	Flt	1.54	6.94	106.1	6.9	10.99	95.23	7.6
	Pyr	ND	5.07	101.4	4.6	10.78	107.8	5.1

[a] ND, Not detected or lower than LOD.

3.9. The Stability of the TiO$_2$NTs/Ti Fiber

The stability of the metal-based fibers greatly depends on their coating procedures, coating structures and surface properties [40]. In our study, TiO$_2$NTs coating was grown in situ on the Ti fiber substrate with the characteristics of good mechanical properties and easy acquirement, flexible and non-fragile. The fabricated fiber was immersed in 0.01 mol·L^{-1} NaOH and 0.01 mol·L^{-1} H$_2$SO$_4$ for 36 h, respectively. As a matter of fact, negligible morphological changes were observed from its SEM image. In order to examine its reusability, the TiO$_2$NTs/Ti fiber was soaked in buffer solution and methanol for 15 min in sequence to imitate the extraction process in the SPME procedure. The obtained results revealed that the fabricated fiber could be used at least 250 times for extraction and desorption of PAHs. In this case, the acceptable average recovery (87.5–107.2%) and RSDs (4.53–8.04%) were still achieved for five replicate analyses of target PAHs at the level of 50 μg·L^{-1}. Clearly, the physical and chemical stability demonstrate that the TiO$_2$NTs/Ti fiber will find its practical applications in environmental water samples.

3.10. A Comparison of the Developed Method with Other Methods

To further evaluate the analytical performance of the TiO$_2$NTs/Ti fiber, several typical analytical parameters such as extraction time, linear range, LOD, relative RSD and recovery rate of this method were compared with those of previously reported SPME method for the preconcentration and determination of PAHs [41–46]. Some statistics are presented in Table 3, this method offers relatively short extraction time, lower LODs than the methods [41,42,44], comparable RSDs to the methods [41,42,44–46], and satisfactory recoveries of the proposed method is comparable or superior to the other reported microextraction methods for the determination of PAHs. However, compared with the reported methods, this method has the advantages of simple production, relatively low cost of preparation and good reproducibility.

Table 3. Comparison of the proposed method with other reported methods.

Methods [a]	Analytes	Time (min)	Linear Ranges ($\mu g \cdot L^{-1}$)	LOD ($\mu g \cdot L^{-1}$)	RSD (%)	Recovery (%)	Refs
PDMS-SPME-GC-MS	Nap, Flu, Phe, Ant, Flt, Pyr	90	0.1–100	0.03–0.24	<19	69–105	[41]
AuNPs-SPME-GC-FID	Nap, Flu, Ant, Flt	50	0.05–300	0.025–0.25	2.49–7.90	78.4–119.9	[42]
PDMS/DVB-SPME-HPLC-UV	Nap, Flu, Ant, Phe, Pyr	60	0.04–15	0.005–0.027	0.97–2.21	81.23–89.11	[43]
AuMPs-SPME-HPLC-UV	Nap, Flu, Ant, Phe, Pyr	50	0.20–500	0.016–0.22	2.03–11.7	86.0–112.9	[44]
ph-TiO$_2$NS-SPME-HPLC-UV	Nap, Flu, Phe	40	0.05–300	0.008–0.043	6.13–9.45	86.2–112	[45]
AuNPs-SPME-HPLC-UV	Nap, Phe	20	0.1–300	0.008–0.037	3.49–9.26	82.74–110.0	[46]
TiO$_2$NTs/Ti-SPME-HPLC-UV	Nap, Flu, Phe, Ant, Flt, Pyr	50	0.20–500	0.03–0.05	3.5–7.9	87.4–118	Present method

[a] PDMS, polydimethylsiloxane; MS, mass spectrometry; AuNPs, gold nanoparticles; DVB, divinylbenzene; AuMPs, gold microparticles; ph-TiO$_2$NS, phenyl-functionalization of titanium dioxide-nanosheets; AuNPs, Au nanoparticles.

4. Conclusions

In this paper, the TiO$_2$NT-coated fiber was in situ fabricated through the anodization of Ti wire substrates in electrolyte containing ethylene glycol and NH$_4$F. The TiO$_2$NTs coating was performed in a highly reproducible manner and the TiO$_2$NTs were embedded into the Ti wire substrate. Under the optimized preparation conditions, uniform pore size array TiO$_2$NTs were obtained. The TiO$_2$NTs/Ti fiber possessed high surface area, good mechanical strength and chemical stability. Coupled to HPLC, the prepared fiber was investigated using six PAHs. It offered a simple, rapid, sensitive and inexpensive pretreatment way for selective concentration and the determination of PAHs in real environmental water samples. The SPME-HPLC analytical method earned a good linear relation, wider linear ranges (0.20–500 $\mu g \cdot L^{-1}$), better reproducibility and accuracy (RSDs, 3.5–7.9%), and higher sensitivity (LODs, 0.03–0.05 $\mu g \cdot L^{-1}$) for target pollutants. In addition, this robust fibre can be used for more than 250 extraction and desorption cycles without the loss of the extraction capability. All these indicated that this novel SPME-HPLC-UV technique would be a good selection and have a great potential for the detection or quantification of trace PAHs in complex samples.

Supplementary Materials: The following are available online at https://www.mdpi.com/article/10.3390/cryst11111384/s1. Figure S1: XRD patterns of TiO$_2$NTs

Author Contributions: Conceptualization, M.M., Y.W., J.C. and Q.S.; Methodology, M.M. and Y.W.; Project administration: M.M. and Y.W.; investigation, M.M., Y.W., J.C. and Q.S.; Writing—original draft: M.M., Writing—review & editing: M.M. and Y.W. All authors have read and agreed to the published version of the manuscript.

Funding: This research work was financially supported by the Key Talent Project of Gansu Province (2021), Science and Technology Projects of Gansu Province (20JR10RA290), Innovation Fund for Higher Education of Gansu Province (2021B-282) and Horizontal development project (LZCU-KJ/2021-040).

Institutional Review Board Statement: Not applicable.

Informed Consent Statement: Not applicable.

Data Availability Statement: Not applicable.

Conflicts of Interest: The authors declare no conflict of interest.

References

1. Alpendurada, M.D.F. Solid-Phase Microextraction: A promising technique for sample preparation in environmental analysis. *J. Chromatogr. A* **2000**, *889*, 3–14. [CrossRef]
2. Risticevic, S.; Niri, V.H.; Vuckovic, D.; Pawliszyn, J. Recent developments in solid-phase microextraction. *Anal. Bioanal. Chem.* **2009**, *393*, 781–795. [CrossRef]
3. Bagheri, H.; Piri-Moghadam, H.; Naderi, M. Towards greater mechanical, thermal and chemical stability in solid-phase microextraction. *TrAC Trends Anal. Chem.* **2012**, *34*, 126–139. [CrossRef]
4. Sajid, M.; Nazal, M.K.; Rutkowska, M.; Szczepańska, N.; Namieśnik, J.; Płotka-Wasylka, J. Solid phase microextraction: Apparatus, sorbent materials, and application. *Crit. Rev. Anal. Chem.* **2018**, *49*, 271–288. [CrossRef] [PubMed]
5. Feng, J.J.; Qiu, H.D.; Liu, X.; Jiang, S.X. The development of SPME fibers with metal wires as supporting substrates. *TrAC Trends Anal. Chem.* **2014**, *44*, 44–58.
6. Sun, M.; Feng, J.J.; Bu, Y.N.; Wang, X.J. Graphene coating bonded onto stainless steel wire as a solid-phase microextraction fiber. *Talanta* **2015**, *134*, 200–205. [CrossRef] [PubMed]
7. Gholivand, M.B.; Piryaei, M.; Abolghasemi, M.M. Anodized aluminum wire as a solid-phase microextraction fiber for rapid determination of volatile constituents in medicinal plant. *Anal. Chim. Acta* **2011**, *701*, 1–5. [CrossRef]
8. Djozan, D.; Abdollahi, L. Anodized zinc wire as a solid-phase microextraction fiber. *Chromatographia* **2003**, *57*, 799–804. [CrossRef]
9. Wang, H.J.; Zhang, Y.D.; Zhang, M.; Zhen, Q.; Wang, X.M.; Du, X.Z. Gold nanoparticle modified NiTi composite nanosheet coating for efficient and selective solid phase microextraction of polycyclic aromatic hydrocarbons. *Anal. Methods* **2016**, *8*, 6064–6073. [CrossRef]
10. Feng, J.J.; Sun, M.; Liu, H.M.; Li, J.B.; Liu, X.; Jiang, S.X. Au nanoparticles as a novel coating for solid-phase microextraction. *J. Chromatogr. A* **2010**, *1217*, 8079–8086. [CrossRef]
11. Mehdinia, A.; Aziz-Zanjani, M.O. Recent advances in nanomaterials utilized infiber coatings for solid-phase microextraction. *TrAC Trends Anal. Chem.* **2013**, *42*, 205–215. [CrossRef]
12. Aziz-Zanjani, M.O.; Mehdinia, A. Electrochemically prepared solid-phase microextraction coatings—A review. *Anal. Chim. Acta* **2013**, *781*, 1–13. [CrossRef]
13. Płotka-Wasylka, J.; Szczepańska, N.; la Guardia, M.; Namieśnik, J. Modern trends in solid phase extraction: New sorbent media. *TrAC Trends Anal. Chem.* **2016**, *77*, 23–43. [CrossRef]
14. Płotka-Wasylka, J.; Szczepańska, N.; la Guardia, M.; Namieśnik, J. Miniaturized solid-phase extraction techniques. *TrAC Trends Anal. Chem.* **2015**, *73*, 19–38. [CrossRef]
15. Gong, D.W.; Grimes, C.A.; Varghese, O.; Hu, W.C.; Singh, R.S.; Chen, Z.; Dickey, E.C. Titanium oxide nanotube arrays prepared by anodic oxidation. *J. Mater. Res.* **2001**, *16*, 3331–3334. [CrossRef]
16. García-Valverde, M.T.; Lucena, R.; Cárdenas, S.; Valcárcel, M. Titanium-dioxide nanotubes as sorbents in (micro)extraction techniques. *TrAC Trends Anal. Chem.* **2014**, *62*, 37–45. [CrossRef]
17. Liu, X.; Liu, Z.Q.; Lu, J.L.; Wu, X.L. Electrodeposition preparation of Ag nanoparticles loaded TiO_2 nanotube arrays with enhanced photocatalytic performance. *Appl. Surf. Sci.* **2014**, *288*, 513–517. [CrossRef]
18. Cabanas-Polo, S.; Boccaccini, A.R. Electrophoretic deposition of nanoscale TiO_2: Technology and applications. *J. Eur. Ceram. Soc.* **2015**, *36*, 265–283. [CrossRef]
19. Kafshgari, M.H.; Goldmann, W.H. Insights into theranostic properties of Titanium dioxide for nanomedicine. *Nano-Micro. Lett.* **2020**, *12*, 106–140.
20. Xu, X.; Chen, B.; Hu, J.P.; Sun, B.W.; Liang, X.H.; Li, N.; Yang, S.Y.; Zhang, H.; Huang, W.; Yu, T. Heterostructured TiO_2 spheres with tunable interiors and shells toward improved packing density and pseudocapacitive sodium storage. *Adv. Mater.* **2019**, *31*, 1904589. [CrossRef]
21. Wang, C.Y.; Wang, Y.D.; Herath, H.M.S.K. Polycyclic aromatic hydrocarbons (PAHs) in biochar-Their formation, occurrence and analysis: A review. *Org. Geochem.* **2017**, *114*, 1–11. [CrossRef]
22. Pulleyblank, C.; Cipullo, S.; Campo, P.; Kelleher, B.; Coulonet, F. Analytical progress and challenges for the detection of oxygenated polycyclic aromatic hydrocarbon transformation products in aqueous and soil environmental matrices: A review. *Crit. Rev. Env. Sci. Technol.* **2019**, *49*, 357–409. [CrossRef]
23. Gorshkov, A.G.; Izosimova, O.N.; Kustova, O.V. Determination of priority polycyclic aromatic hydrocarbons in water at the trace level. *J. Anal. Chem.* **2019**, *74*, 771–777. [CrossRef]
24. Zhang, Q.Y.; Liu, P.; Li, S.L.; Zhang, X.J.; Chen, M.D. Progress in the analytical research methods of polycyclic aromatic hydrocarbons (PAHs). *J. Liq. Chromatogr. Relat. Technol.* **2020**, *43*, 425–444. [CrossRef]
25. Riddle, S.G.; Robert, M.A.; Jakober, C.A.; Hannigan, M.P.; Kleeman, M.J. Size distribution of trace organic species emitted from light-duty gasoline vehicles. *Environ. Sci. Technol.* **2007**, *41*, 7464–7471. [CrossRef] [PubMed]

26. Zencak, Z.; Klanova, J.; Holoubek, I.; Gustafsson, O. Source apportionment of atmospheric PAHs in the western balkans by natural abundance radiocarbon analysis. *Environ. Sci. Technol.* **2007**, *41*, 3850–3855. [CrossRef]
27. Zhou, Y.Y.; Yan, X.P.; Kim, K.N.; Wang, S.W.; Liu, M.G. Exploration of coordination polymer as sorbent for flow injection solid-phase extraction on-line coupled with high-performance liquid chromatography for determination of polycyclic aromatic hydrocarbons in environmental materials. *J. Chromatogr. A* **2006**, *1116*, 172–178. [CrossRef]
28. Li, K.; Li, H.F.; Liu, L.B.; Hashi, Y.; Maeda, T.; Lin, J.M. Solid-phase extraction with C30 bonded silica for analysis of polycyclic aromatic hydrocarbons in airborne particulate matters by gas chromatography-mass spectrometry. *J. Chromatogr. A* **2007**, *1154*, 74–80. [CrossRef]
29. Liu, Q.Z.; Xu, X.; Wang, L.; Lin, L.H.; Wang, D.H. Simultaneous determination of forty-two parent and halogenated polycyclic aromatic hydrocarbons using solid-phase extraction combined with gas chromatography-mass spectrometry in drinking water. *Ecotoxicol. Environ. Saf.* **2019**, *181*, 241–247. [CrossRef]
30. Zhang, R.; Wang, Z.; Wang, Z.Y.; Wang, X.M.; Du, X.Z. Tailoring the selectivity of titania nanowire arrays grown on titanium fibers by self-assembled modification of trichlorophenylsilane for solid-phase microextraction of polycyclic aromatic hydrocarbons. *Microchim. Acta* **2019**, *186*, 536. [CrossRef]
31. Burkhardt, M.R.; Zaugg, S.D.; Burbank, T.L.; Olson, M.C.; Iverson, J.L. Pressurized liquid extraction using water/isopropanol coupled with solid-phase extraction clean up for semivolatile organic compounds, polycyclic aromatic hydrocarbons (PAH), and alkylate PAH homolog groups in sediment. *Anal. Chim. Acta* **2005**, *549*, 104–116. [CrossRef]
32. Hartonen, K.; Bøwadt, S.; Dybdahl, H.P.; Nylund, K.; Sporring, S.; Lund, H.; Oreld, F. Nordic laboratory intercomparison of supercritical fluid extraction for the determination of total petroleum hydrocarbon, polychlorinated biphenyls and polycyclic aromatic hydrocarbons in soil. *J. Chromatogr. A* **2002**, *958*, 239–248. [CrossRef]
33. Macak, J.M.; Hildebrand, H.; Marten-Jahns, U.; Marten-Jahns, P. Mechanistic aspects and growth of large diameter self-organized TiO_2 nanotubes. *J. Electroanal. Chem.* **2008**, *621*, 254–266. [CrossRef]
34. Macak, J.M.; Gong, B.G.; Hueppe, M.; Schmuki, P. Filling of TiO_2 nanotubes by self-doping and electrodeposition. *Adv. Mater.* **2007**, *19*, 3027–3031. [CrossRef]
35. Valota, A.; LeClere, D.J.; Skeldon, P.; Curioni, M.; Hashimoto, T.; Berger, S.; Kunze, J.; Schmuki, P.; Thompson, G.E. Influence of water content on nanotubular anodic titania formed in fluoride/glycerol electrolytes. *Electrochim. Acta* **2009**, *54*, 4321–4327. [CrossRef]
36. Tian, Y.; Sun, M.; Wang, X.Q.; Luo, C.N.; Feng, J.J. A nanospherical metal–organic framework UiO-66 for solid-phase microextraction of polycyclic aromatic hydrocarbons. *Chromatographia* **2018**, *81*, 1053–1061. [CrossRef]
37. Fang, L.; Hou, L.X.; Zhang, Y.H.; Wang, Y.K.; Yan, G.H. Synthesis of highly hydrophobic rutile Titania-silica nanocomposites by an improved hydrolysis co-precipitation method. *Ceram. Int.* **2017**, *43*, 5592–5598. [CrossRef]
38. Huang, Y.N.; Chen, J.; Li, Z.; Wang, L.; Guan, M.; Qiu, H.D. Porous graphene-coated stainless-steel fiber for direct immersion solid-phase microextraction of polycyclic aromatic hydrocarbons. *Anal. Methods* **2019**, *11*, 213–218. [CrossRef]
39. Li, Y.; Ma, M.G.; Zhang, M.; Wang, X.M.; Du, X.Z. In situ anodic growth of rod-like TiO_2 coating on a Ti wire as a selective solid-phase microextraction fiber. *RSC Adv.* **2014**, *4*, 53820–53827. [CrossRef]
40. Wang, Z.Y.; Wang, F.F.; Zhang, R.; Wang, Z.; Du, X.Z. A new strategy for electrochemical fabrication of manganese dioxide coatings based on silica nanoparticles deposited on titanium fibers for selective and highly efficient solid-phase microextraction. *New J. Chem.* **2019**, *43*, 5055–5064. [CrossRef]
41. Doong, R.A.; Chang, S.M.; Sun, Y.C. Solid-phase microextraction for determining the distribution of sixteen US Environmental Protection Agency polycyclic aromatic hydrocarbons in watersamples. *J. Chromatogr. A* **2000**, *879*, 177–188. [CrossRef]
42. Feng, J.J.; Sun, M.; Liu, H.M.; Li, J.B.; Liu, X.; Jiang, S.X. A novel silver-coated solid- phase microextraction metal fiber based on electroless plating technique. *Anal. Chim. Acta* **2011**, *701*, 174–180. [CrossRef]
43. Lucio-Gutiérrez, R.; Salazar-Cavazos, J.M.D.; de Torres, N.H.W.; Castro-Ríos, R. Solid-phase microextraction followed by high-performance liquid chromato-graphy with fluorimetric and UV detection for the determination of polycyclic aromatic hydrocarbons in water. *Anal. Lett.* **2008**, *41*, 119–136. [CrossRef]
44. Liu, H.X.; Liu, L.; Li, Y.; Wang, X.M.; Du, X.Z. Preparation of a robust and sensitive gold-coated fiber for solid-phase microextraction of polycyclic aromatic hydrocarbons in environmental waters. *Anal. Lett.* **2014**, *47*, 1759–1771. [CrossRef]
45. Guo, M.; Song, W.L.; Wang, T.E.; Li, Y.; Wang, X.M.; Du, X.Z. Phenyl-functionalization of titanium dioxide-nanosheets coating fabricated on a titanium wire for selective solid-phase microextraction of polycyclic aromatic hydrocarbons from environment water samples. *Talanta* **2015**, *144*, 998–1006. [CrossRef] [PubMed]
46. Zhang, Y.D.; Yang, Y.X.; Li, Y.; Wang, X.M.; Du, X.Z. Growth of cedar-like Au nanoparticles coating on an etched stainless steel wire and its application for selective solid-phase microextraction. *Anal. Chim. Acta* **2015**, *876*, 55–62. [CrossRef]

Article

Boosting pH-Universal Hydrogen Evolution of FeP/CC by Anchoring Trace Platinum

Chuancang Zhou [1], Feipeng Zhang [2,3,*] and Hongyu Wu [1,*]

1. College of Engineering, Lishui University, Lishui 323000, China; 30130815@hncj.edu.cn
2. Henan Provincial Engineering Laboratory of Building-Photovoltaics, Institute of Physics, Henan University of Urban Construction, Pingdingshan 467036, China
3. School of Materials Sciences and Engineering, Shijiazhuang Tiedao University, Shijiazhuang 050043, China
* Correspondence: zhfp@emails.bjut.edu.cn (F.Z.); zhoucc@lsu.edu.cn (H.W.)

Abstract: To improve the electrocatalytic properties for hydrogen evolution reactions, strategies need to be adopted, such as increasing specific surface area and active site, as well as decreasing interface energy. Herein, we report the preparation of FeP on carbon cloth using a two-step process of hydrothermal and phosphating. Otherwise, to utilize the excellent catalytic performance of Pt and decrease consumption of Pt, the hyperdispersed Pt nanoparticles for the sake of modifying transition-metal phosphides film were designed and fabricated. Finally, 3D FeP-Pt/CC was successfully prepared by means of electro-deposition using three electrodes. The crystalline structure, surface morphology and elemental composition of the synthesized samples have been investigated by X-ray diffraction (XRD), X-ray photoelectron spectroscopy (XPS), scanning electron microscopy (SEM) and energy dispersive X-ray analysis (EDS). The XRD results show that the as-prepared products are of orthorhombic FeP structure, and EDS results indicate that there exist Pt elements in 3D FeP-Pt/CC. The electrocatalytic performances were evaluated by, such as linear scan voltammetry, tafel plots and electrochemical impedance spectroscopy on electrochemical workstations. These results show that the FeP-Pt/CC exhibit a current density of 10 mA·cm^{-2} at an over-potential of 58 mV for HER in 0.5 M H$_2$SO$_4$, which is very close to the values of 20% Pt/C which was previously reported. FeP-Pt/CC has excellent durability.

Keywords: FeP-Pt/CC; hydrogen evolution reaction; electrochemical impedance spectroscopy; durability

1. Introduction

With the increasing depletion of fossil fuels and tremendous environmental pollution, hydrogen energy is extensively concerned as it is a clean, highly efficient and renewable energy alternative [1–6]. Electrochemical water splitting is considered to be one of the most promising large-scale hydrogen production methods. However, the industrialization of water electrolysis is still limited due to the lack of cheap and efficient catalysts [7–10]. Among the numerous catalysts, transition-metal phosphides with high conductivity, chemical stability, special crystal structure and abundant valence state have been intensively focused. To improve the electrocatalytic performances, an important strategy is increasing the number of active sites by means of generating a larger specific surface area [11–14]. Besides, three dimension (3D) basal electrodes, for example, carbon cloth has excellent conductivity, mechanical properties and larger specific surface area [15]. At present, Pt metal is regarded to be the most efficient electrocatalyst for hydrogen evolution reactions [9]. However, the resource of Pt is scarce and expensive. In this report, the hyperdispersed Pt nanoparticles for the sake of modifying transition-metal phosphides film is designed and fabricated in order to utilize the catalytic performance of Pt and reduce Pt dosage at the same time. The 3D FeP-Pt film was grown on conductive carbon cloth (CC) by means of hydrothermal methods with high temperature phosphating and electro-deposition [5,7,8].

The results show FeP-Pt/CC has excellent electrocatalytic performances for HER at all pH and has excellent durability and long term stability.

2. Experimental

Firstly, the CC matrix with an area of 1 cm × 1 cm is hydrothermally treated in 68% HNO_3 solution at 120 °C, and then it is washed by deionized water and electrochemically oxidized for 30 mins in NaCl solution. The solution of Ferric nitrate, ammonium fluoride and carbamide was prepared in a reactor, and then the treated CC matrix was placed in the solution at 120 °C for 12 h. Iron compound film attached to carbon cloth obtained after cooled and washed. The iron compound film was then dried at 100 °C and subjected to sintering in succession at 300–350 °C for 2–3 h in a tube furnace with argon and sodium hypophosphite to obtain the FeP/CC. Finally, the Platinum nanoparticles were prepared on the FeP/CC by electrodeposition on FeP/CC, and then the 3D FeP-Pt/CC was obtained. It is electrodeposited for 2 h under the voltage of −0.6−−0.7 V in the solution of potassium chloroplatinate and a small amount of boric acid.

The X-ray diffraction (XRD) measurements were carried out by using a Dandong DX-2700B diffractometer with CuKa radiation (λ = 1.5418 Å). The scanning electron microscopy (SEM) images were obtained by scanning electron microscopy (VEGA3, TESCAN, Brno, South Moravia, Czech Republic). The X-ray photoelectron spectroscopes (XPS) were carried out (ESCALAB 250Xi, Al Ka, 150 W, Waltham, MA, USA) to examine the chemical composition and valence state of the as-prepared samples. The electrochemical measurements were performed with a three-electrode system in 0.5 M H_2SO_4, 1 M PBS(Phosphate buffer solution), 1 M KOH solution using a princeton electrochemical workstation. A saturated calomel electrode (SCE) and a graphite rod were used as the reference electrode and counter electrode, respectively in acidic and neutral solutions. Hg/HgO is used as reference electrodes in alkaline solution. The as-prepared FeP-Pt/CC was used as the working electrode. The linear scan voltammetry (LSV) measurements were carried at a scanning rate of 5 mVs^{-1}. The electrochemical impedance spectra (EIS) were obtained at −200 mV vs reference electrode with a frequency range from 0.1 Hz to 100,000 Hz, 5 mg 20% Pt/C catalyst, add 100 μL 5w% Nafion solution and 900 μL ethanol and mix with ultrasound for at least 30 min; 1 μL of solution suck each time and drop it onto the glassy carbon electrode (with a diameter of 3 mm), drop it again after drying, and until the loading capacity is 0.212 mg/cm^2. All the measurements were corrected using iR compensation.

3. Results and Discussion

The XRD patterns of the FeP-Pt/CC and FeP/CC are shown in Figure 1. It can be seen from Figure 1 that all the samples before and after electro-deposition (FeP/CC and FeP-Pt/CC) can be well indexed as the orthorhombic FeP (JCPDS No. 390809) phase. It is evidenced that the FeP was well prepared. It can be seen from the figure that after the electrodeposition of platinum, the crystallinity and purity of the film decreased. In addition, the diffraction peaks of Platinum shown in the XRD pattern of FeP-Pt/CC in Figure 1 at 39.77°, 46.26° indicate that there has Pt metal in FeP-Pt/CC.

The SEM and edx mapping images of the FeP-Pt/CC and FeP/CC films are shown in Figures 2 and 3. It can be seen from the scanning electron microscope that the surface grains of the film before electrodeposition are mainly composed of rods with a diameter of less than 1 micron, and massive grains with a size of several microns appear after electrodeposition. According to the EDX diagram, the molar ratio of Fe:P:Pt is about 16.45:18.9:4.87 and there are some impurities of NaCl. According to EDX mapping photos, Fe,P,Pt has similar element distribution, indicating the dispersion uniformity of Pt, and a small amount of oxidation points can be seen. It can also be seen from Figure 3 that there have not been Pt elements in the FeP/CC particles and the molar ratio of Fe:P is 18.58:25.76. It is evidenced that the FeP-Pt/CC was successfully obtained.

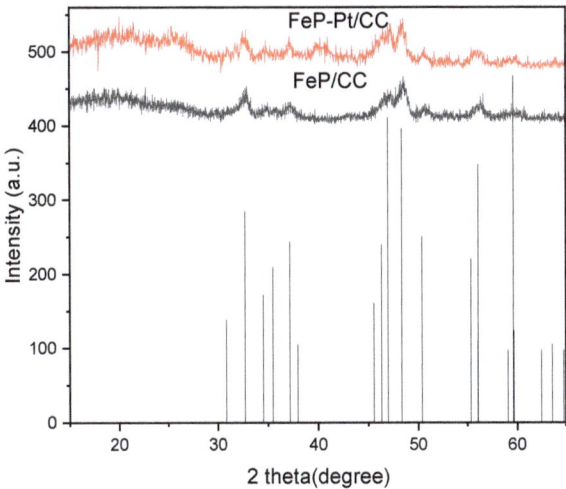

Figure 1. XRD patterns of the FeP/CC and FeP-Pt/CC.

In order to further verify the composition of the FeP-Pt/CC film surface, the X-ray photoelectron spectroscopy (XPS) survey is used. Figure 4 presents the XPS results of the FeP-Pt/CC. As shown in the figure, the total spectrum suggests the co-existence of Carbon, Oxygen, Ferrum, Phosphorus and Platinum element in the material. The Fe $2p_{3/2}$ corresponds to peak at 712 eV and the Fe$2p_{1/2}$ corresponds to peak at 726 eV. The P 2p corresponds to one peak at 134 eV and P 2s corresponds to the peak at 192 eV; 71.6 eV and 75.16 eV of the peak corresponds to zero valences of Pt. Besides, the C 1s and O 1s correspond to the peaks at 280 eV and 531.9 eV, respectively. The OKL1 and OKL2 correspond to peaks at 990 eV and 1000 eV. It is well evidenced that the FeP-Pt/CC was successfully fabricated with a small amount of precious Pt.

The electrocatalytic performances of FeP-Pt/CC and FeP/CC were investigated by a Princeton electrochemical workstation. All HER measurements were carried out at 25 °C. The results are given in Figure 5. It can be seen from Figure 5a that FeP-Pt/CC and FeP/CC possesses over-potentials of −58 mV and −110 mV reach the cathode current density of 10 mA·cm^{-2} in 0.5 M H$_2$SO$_4$, while 20% Pt/C has the smallest η_{10} (−36 mV). Their corresponding Tafel slope is 49.6, 70.6, 108.6 mV/dec. Tafel slopes suggest that the Volmer reaction is fast and the rate-limiting step is the Heyrovsky reaction. It shows that the addition of platinum greatly improves the hydrogen evolution performance of FeP. However, in the phosphate buffer solution, the hydrogen evolution performance is poor, η_{10} of FeP-Pt/CC, FeP/CC and Pt/C is −214, −187, −60 mV, respectively. The corresponding Tafel slope are 256, 510, 105.5 mV/dec. At a higher potential, platinum can improve the HER performance of FeP greatly, even more than 20% Pt/C. A neutral-effective electrocatalyst has apparently the best benefit of environmental benignity and very broad application prospects. Alkaline-efficient electrocatalysts are the most important and widely used technology in the industry. From Figure 5e, we can find that 20% Pt/C exhibits excellent electrocatalytic activity in 1 m KOH. η_{10} of FeP-Pt/CC and FeP/CC are −42.6, −44 mV, respectively. The corresponding Tafel slopes are 80, 60.5 mV/dec. LSV of FeP-Pt/CC has been very close to Pt/C catalyst. It shows that FeP-Pt/CC has a good application prospect.

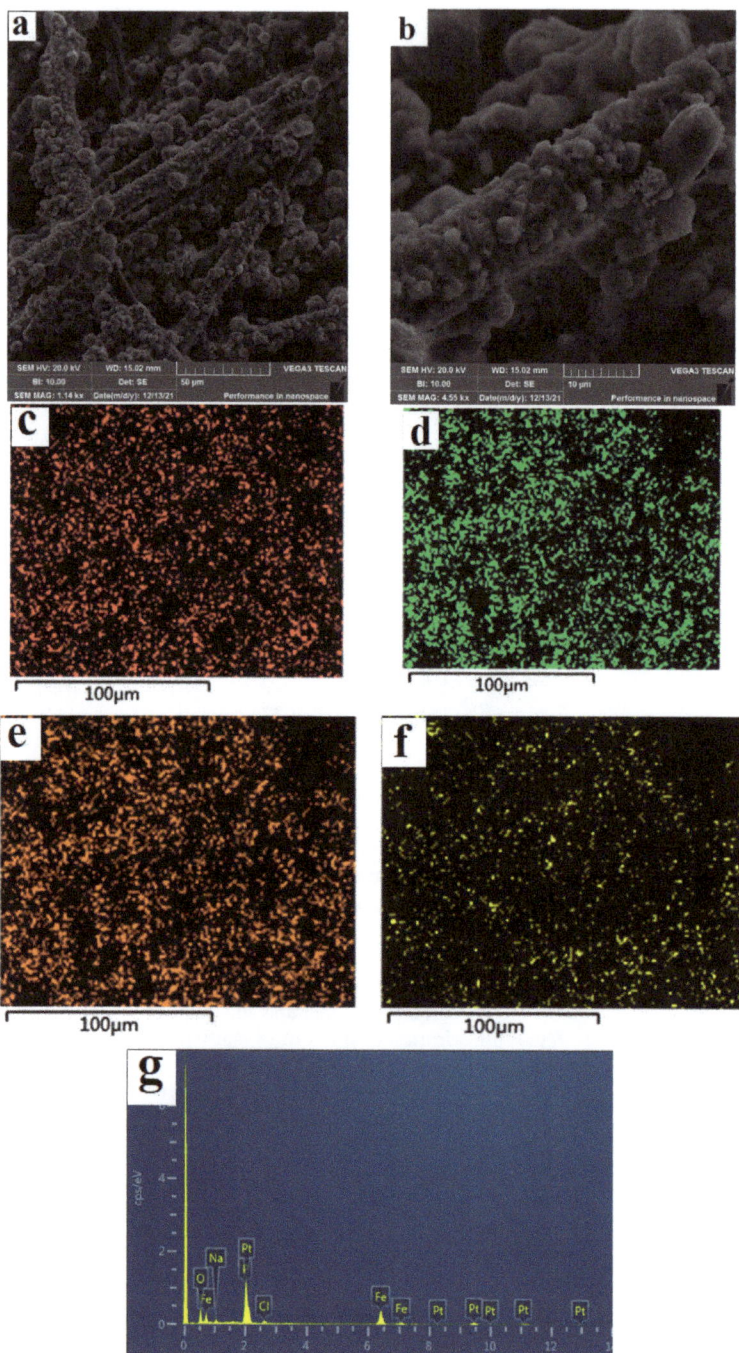

Figure 2. SEM images (**a,b**), EDX mapping (**c–f**) and EDX spectrum (**g**) of FeP-Pt/CC.

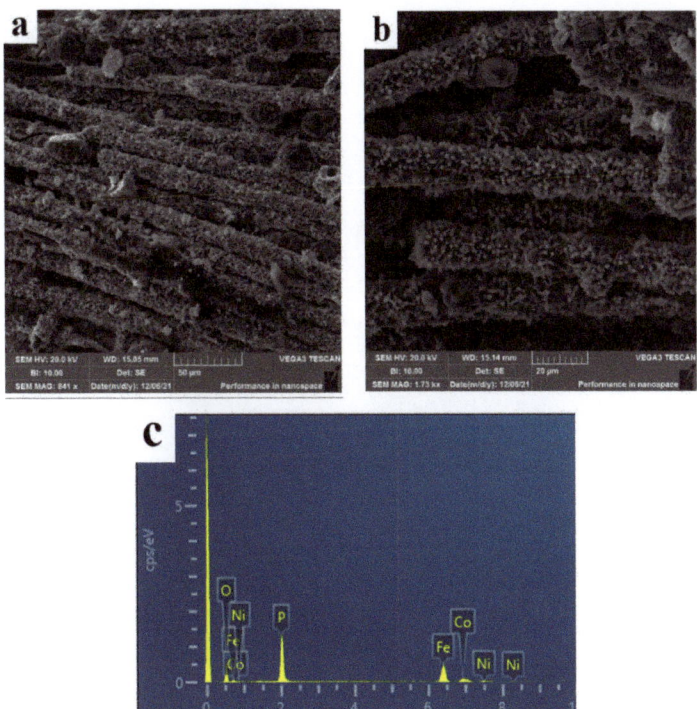

Figure 3. SEM images (**a**,**b**) and EDX spectrum (**c**) of FeP/CC.

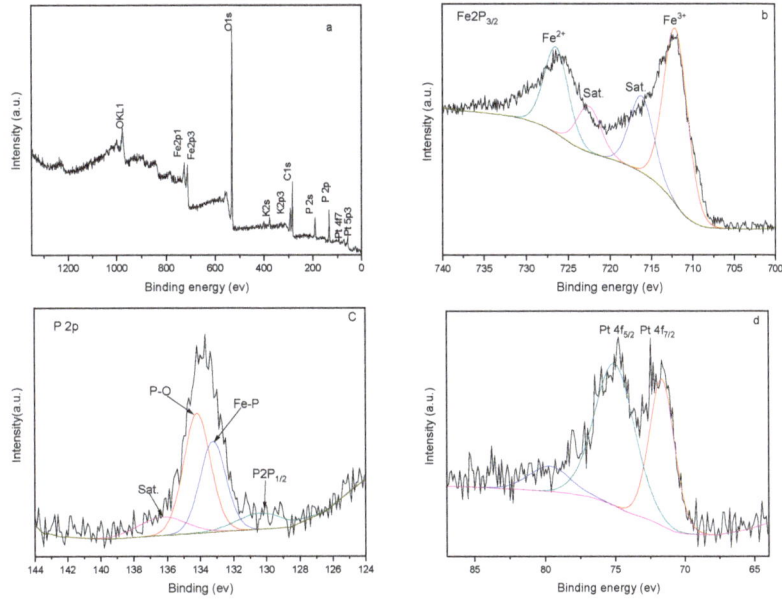

Figure 4. XPS spectra of FeP-Pt/CC (**a**) survey spectrum, (**b**) Fe 2p, (**c**) P 2p and (**d**) Pt 4f.

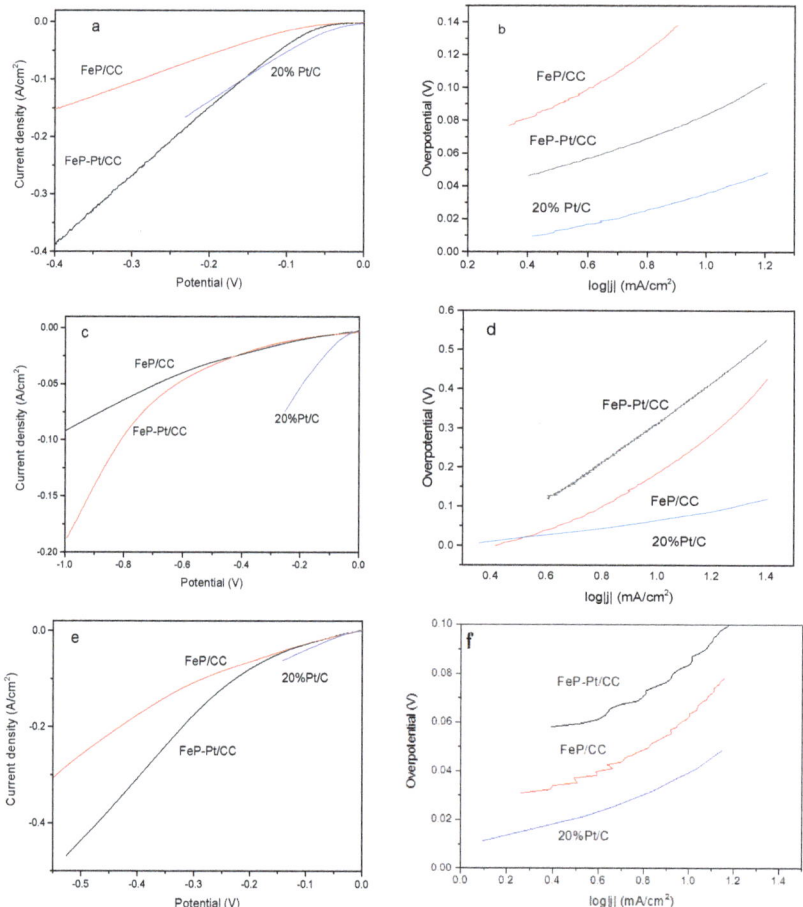

Figure 5. Electrochemical HER measurements. Linear sweep polarization curves obtained in 0.5 M H_2SO_4 (**a**), 1 M PBS (**c**), 1 M KOH (**e**). The corresponding Tafel slopes (**b,d,f**).

To further investigate the electrocatalytic performances of the materials, electrochemical impedance spectroscopy is also carried out. Figure 6 shows the results of Nyquist plots of FeP-Pt/CC and FeP/CC at −0.2 V in 0.5 M H_2SO_4. Table 1 shows the element values in the equivalent circuit of the AC impedance spectrum. It can be found from Figure 6 that the Nyquist plot shows a line with an angle of inclination of 45°, which suggests the phase angle of reactive ions concentration fluctuation on the electrode surface is 45 degrees lags behind the AC current. In addition, it can be seen from Table 1 that the R1 (reactive resistance) of FeP-Pt/CC is 0.75 Ω, which is very much less than that of FeP/CC (1.6 Ω). These features indicate that the electrode reaction is completely controlled by the diffusion step. It is estimated that the reason is Platinum particles reduce the reactive resistance and make the surface rougher.

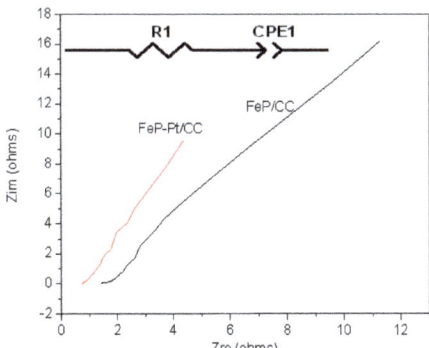

Figure 6. Nyquist plots of FeP-Pt/CC and FeP/CC at −0.2 V in 0.5 M H_2SO_4 with inset the equivalent electrical circuit.

Table 1. Element values in the equivalent circuit of AC impedance spectrum in 0.5 M H_2SO_4.

Element	FeP-Pt/CC	FeP/CC
R1/Ω	0.75157	1.603
CPE1-T/$\Omega^{-1} \cdot cm^{-2} \cdot s^{-n}$	0.15453	0.086979
CPE1-P/$\Omega^{-1} \cdot cm^{-2} \cdot s^{-n}$	0.74181	0.70683

Figure 7 shows the results of Nyquist plots of FeP-Pt/CC and FeP/CC at −0.2 V in 1 M KOH with an inset of the equivalent electrical circuit. It can be seen from Figure 7 that the inclined straight lines show an angle of nearly 45 degrees, this suggests that there is a thick and compact passivation film on the surface of the electrode, and the ion migration is greatly inhibited. These results indicate that a dense passivation film is easily formed on the surface of iron in a strongly alkaline solution. It can also be deduced that the circuit diagram is a series connection of resistor R1 and constant phase element (CPE). Here the constant phase element CPE has two values, CPE-P and CPE-T. Table 2 shows the element values in the equivalent circuit of the AC impedance spectrum. As is shown in Table 2, the CPE-P values of FeP-Pt/CC and FeP/CC are 0.7778 $\Omega^{-1} \cdot cm^{-2} \cdot s^{-n}$ and 0.7689 $\Omega^{-1} \cdot cm^{-2} \cdot s^{-n}$, respectively. It can be illustrated that the rough and porous electrode surface produces double-layer capacitance and there exists a dispersion effect on the electrode surface. The resistance should decrease because platinum has better conductivity than FeP after platinum plating. In addition, the increase of CPE value indicates the increase of capacitance effect and the increase of film thickness and roughness.

In addition, the durability and long term stability of materials were further investigated. Figure 8 shows the V-T curve of FeP-Pt/CC at 10 mA/cm^2 in 0.5 M H_2SO_4, 1 M PBS, 1 M KOH, respectively. Compared with other solutions, potential changes little in the 0.5 M H_2SO_4. In general, the potential changes little after 20 h in various solutions, which can be applied in practical production.

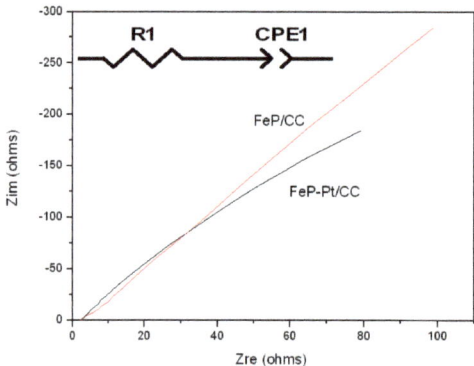

Figure 7. Nyquist plots of FeP-Pt/CC and FeP/CC at −0.2 V in 1 M KOH. The inset is the equivalent electrical circuit.

Table 2. Element values in the equivalent circuit of AC impedance spectrum in 1 M KOH.

Element	FeP-Pt/CC	FeP/CC
R1/Ω	1.827	2.184
CPE1-T ($\Omega^{-1} \cdot cm^{-2} \cdot s^{-n}$)	0.00795	0.00551
CPE1-P ($\Omega^{-1} \cdot cm^{-2} \cdot s^{-n}$)	0.7778	0.7689

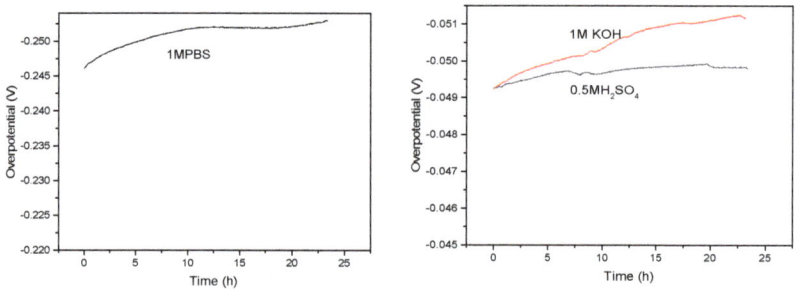

Figure 8. Time dependence of Overpotential for $NiCo_2P_x$ at 10 mA/cm^2 (in 0.5 M H_2SO_4, 1 M KOH and 1 M PBS).

4. Conclusions

FeP-Pt/CC has also been successfully prepared by means of hydrothermal-phosphatization and electrodeposition methods. The as-prepared FeP films have an orthorhombic structure. XPS and EDX tests proved the uniform distribution of Pt in FeP-Pt/CC, the addition of trace platinum can significantly improve the catalytic activity of FeP for hydrogen evolution. FeP-Pt/CC exhibit a current density of 10 mA·cm^{-2} at over-potential of 58 mV and 42.6 mV for HER in 0.5 M H_2SO_4 and 1 M KOH, respectively, this is very close to the values of 20% Pt/C V-t curves with 20 h in various solutions show FeP-Pt/CC has excellent durability and long term stability.

Author Contributions: Conceptualization, F.Z.; methodology, C.Z.; formal analysis, C.Z. and F.Z.; resources, H.W.; data curation, C.Z.; writing—original draft preparation, C.Z.; writing—review and editing, F.Z.; visualization, F.Z.; supervision, F.Z. and H.W.; project administration, H.W.; funding acquisition, H.W. All authors have read and agreed to the published version of the manuscript.

Funding: The authors would like to thank the support provided by National Natural Science Foundation of China under grant No.51572066.

Data Availability Statement: The datasets generated during and/or analyzed during the current study are available from the corresponding author on reasonable request.

Conflicts of Interest: Authors declare that they have no known competing financial interests or personal relationships that could have appeared to influence the work reported in this paper.

References

1. Niu, Z.; Qiu, C.; Jiang, J.; Ai, L. Hierarchical CoP–FeP Branched Heterostructures for Highly Efficient Electrocatalytic Water Splitting. *ACS Sustain. Chem. Eng.* **2018**, *7*, 2335. [CrossRef]
2. Anantharaj, S.; Ede, S.R.; Sakthikumar, K.; Karthick, K.; Mishra, S.; Kundu, S. Electrospun cobalt-ZIF micro-fibers for efficient water oxidation under unique pH conditions. *ACS Catalysis* **2016**, *6*, 8069. [CrossRef]
3. Liu, T.; Ma, X.; Liu, D.; Hao, S.; Du, G.; Ma, Y.; Asiri, A.M.; Sun, X.; Chen, L. Mn Doping of CoP Nanosheets Array: An Efficient Electrocatalyst for Hydrogen Evolution Reaction with Enhanced Activity at All pH Values. *ACS Catalysis* **2017**, *7*, 98. [CrossRef]
4. Guo, P.; Wu, Y.X.; Lau, W.M.; Liu, H. Porous CoP nanosheet arrays grown on nickel foam as an excellent and stable catalyst for hydrogen evolution reaction. *Int. J. Hydrog. Energy* **2017**, *42*, 26995. [CrossRef]
5. Yan, Y.; Thia, L.; Xia, B.Y.; Ge, X.; Liu, Z.; Fisher, A.; Wang, X. Construction of Efficient 3D Gas Evolution Electrocatalyst for Hydrogen Evolution: Porous FeP Nanowire Arrays on Graphene Sheets. *Adv. Sci.* **2015**, *2*, 1500120. [CrossRef] [PubMed]
6. Sun, Y.; Hang, L.; Shen, Q.; Zhang, T.; Li, H.; Zhang, X.; Lyue, X.; Li, Y. Mo doped Ni2P nanowire arrays: An efficient electrocatalyst for the hydrogen evolution reaction with enhanced activity at all pH values. *Nanoscale* **2017**, *9*, 16874. [CrossRef] [PubMed]
7. Zhang, J.; Liang, X.; Wang, X.; Zhuang, Z. CoP nanotubes formed by Kirkendall effect as efficient hydrogen evolution reaction electrocatalysts. *Mater. Lett.* **2017**, *202*, 146. [CrossRef]
8. Zhang, R.; Wang, X.; Yu, S. Ternary NiCo2Px Nanowires as pH-Universal Electrocatalysts for Highly Efficient Hydrogen Evolution Reaction. *Adv. Mater.* **2017**, *29*, 1605502. [CrossRef] [PubMed]
9. Jiang, K.; Liu, B.; Luo, M. Single platinum atoms embedded in nanoporous cobalt selenide as electrocatalyst for accelerating hydrogen evolution reaction. *Nat. Commun.* **2019**, *10*, 1743. [CrossRef] [PubMed]
10. Li, Y.; Cai, P.; Ci, S.; Wen, Z. Strongly Coupled 3D Nanohybrids with Ni2P/Carbon Nanosheets as pH-Universal Hydrogen Evolution Reaction Electrocatalysts. *ChemElectroChem* **2017**, *4*, 340. [CrossRef]
11. Feng, J.X.; Tong, S.Y.; Tong, Y.X.; Li, G.R. Pt-like Hydrogen Evolution Electrocatalysis on PANI/CoP Hybrid Nanowires by Weakening the Shackles of Hydrogen Ions on the Surfaces of Catalysts. *J. Am. Chem. Soc.* **2018**, *140*, 5118. [CrossRef] [PubMed]
12. Liu, Z.; Qi, J.; Liu, M.; Zhang, S.; Fan, Q.; Liu, H.; Liu, K.; Zheng, H.; Yin, Y.; Gao, C. Aqueous Synthesis of Ultrathin Platinum/Non-Noble Metal Alloy Nanowires for Enhanced Hydrogen Evolution Activity. *Angew. Chem.* **2018**, *130*, 11852. [CrossRef]
13. Ma, J.; Habrioux, A.; Alonso-Vante, N. Enhanced HER and ORR behavior on photodeposited Pt nanoparticles onto oxide–carbon composite. *J. Solid State Electrochem.* **2013**, *17*, 1913. [CrossRef]
14. Spori, C.; Kwan, J.T.H.; Bonakdarpour, A.; Wilkinson, D.; Strasser, P. Analysis of oxygen evolving catalyst coated membranes with different current collectors using a new modified rotating disk electrode technique. *Angew. Chem. Int. Ed. Eng.* **2017**, *56*, 5994.
15. Wang, Y.; Li, M.; Xu, L.; Tang, T.; Ali, Z.; Huang, X.; Hou, Y.; Zhang, S. Polar and conductive iron carbide@N-doped porous carbon nanosheets as a sulfur host for high performance lithium sulfur batteries. *Chem. Eng. J.* **2019**, *358*, 962. [CrossRef]

Article

Layer-by-Layer Assembly of Polyelectrolytes on Urchin-like MnO$_2$ for Extraction of Zn^{2+}, Cu^{2+} and Pb^{2+} from Alkaline Solutions

Dong Chen [1,2] and Zhongren Nan [1,*]

1 Gansu Key Laboratory for Environmental Pollution Prediction and Control, College of Earth and Environmental Sciences, Lanzhou University, Lanzhou 730000, China; chend16@lzu.edu.cn
2 School of Agriculture and Forestry Economics and Management, Lanzhou University of Finance and Economics, Lanzhou 730020, China
* Correspondence: nanzhongren@lzu.edu.cn; Tel.: +86-0931-891-2328

Abstract: Three-dimensional (3D) urchin-like MnO$_2$@poly (sodium 4-styrene sulfonate) (PSS)/poly (diallyl dimethylammonium chloride) (PDDA)/PSS particles were prepared via the layer-by-layer (LBL) assembly of polyelectrolytes for the extraction of Zn^{2+} from alkaline media. The adsorption performance of Zn^{2+} on MnO$_2$, MnO$_2$@PSS/PDDA/PSS, and MnO$_2$@(PSS/PDDA)$_3$/PSS was investigated in batch experiments. The adsorption of Zn^{2+} on MnO$_2$@PSS/PDDA/PSS has been studied under various conditions, such as initial Zn^{2+} concentration, adsorbent dosage, the solution's pH, and reaction time. The Zn^{2+} adsorption process is well represented by the pseudo-second-order kinetic model, and the equilibrium data fit the Freundlich isotherm well. MnO$_2$@PSS/PDDA/PSS also showed high efficiency for Pb^{2+} and Cu^{2+} removal from slightly alkaline water. Thus, our research provides a deep insight into the preparation of 3D manganese oxides with polyelectrolyte films for the extraction of heavy metal ions, such as Pb^{2+}, Cu^{2+}, and Zn^{2+}, from slightly alkaline wastewater.

Keywords: urchin-like MnO$_2$; LBL assembly; heavy metal ions; adsorption; polyelectrolytes

Citation: Chen, D.; Nan, Z. Layer-by-Layer Assembly of Polyelectrolytes on Urchin-like MnO$_2$ for Extraction of Zn^{2+}, Cu^{2+} and Pb^{2+} from Alkaline Solutions. *Crystals* 2022, 12, 358. https://doi.org/10.3390/cryst12030358

Academic Editors: Bo Chen, Rutao Wang and Nana Wang

Received: 19 January 2022
Accepted: 25 February 2022
Published: 8 March 2022

Publisher's Note: MDPI stays neutral with regard to jurisdictional claims in published maps and institutional affiliations.

Copyright: © 2022 by the authors. Licensee MDPI, Basel, Switzerland. This article is an open access article distributed under the terms and conditions of the Creative Commons Attribution (CC BY) license (https:// creativecommons.org/licenses/by/ 4.0/).

1. Introduction

The dissolved phase of heavy metals, such as Pb^{2+}, Cu^{2+}, and Zn^{2+}, in wastewater has become a matter of increasing concern due to their great transferability and bioavailability, as well as their severe cytotoxicity [1–3]. Lead (Pb^{2+}), mainly from petrol, paint, plumbing, pipes, car batteries, pigments, has a tolerable daily intake (TDI) value set at levels of <3.5 μg/kg body weight. An overdose of Pb^{2+} over a long period might cause irreversible damage to the central nervous system [4,5]. Copper (Cu^{2+}) and zinc (Zn^{2+}) are necessary for organism development in small quantities, but excessive exposure to them can lead to toxicity, disrupting the normal functions of cells and organs [6–8]. Thus, exploring efficient and cost-effective methods for heavy metal treatment is in demand, especially in developing countries.

Various materials, such as adsorbents [9,10], ion exchange resins [11,12], chemical precipitation agents [13,14], electrochemical anodes [15] and membranes [16,17] have been used to remove heavy metal ions [18]. Among them, adsorbents are widely used due to their highly cost-effective properties and easy operation [19]. Many studies have been published related to the removal of Zn^{2+} from acidic [20,21] to neutral wastewater; however, little information is available for Zn^{2+} adsorption in slightly alkaline water [22,23]. Manganese oxide (MnO$_2$) has been extensively reported as an efficient scavenger of many heavy metals, due to their unique physical and chemical properties, with the controllable tuning of structure [24–26], while it is still important to improve their stability and chemical activity. It has been reported that MnO$_2$ with a 3D urchin-like structure, with modest corrugating patterns, are considered to exhibit noticeable chemically stable and active properties, significantly differentiating them from particles with smooth surfaces [27–29]. Furthermore, the derivatizing process of urchin-like MnO_2 surfaces with macromolecular components also noticeably enhances their affinity for heavy metals. Layer-by-layer assembly (LBL) is one way to permit the molecular engineering of surfaces through the continuous depositing process of polyelectrolytes and numerous functional compounds [30,31].

In the present work, the urchin-like MnO_2, with outer diameters of 2 to 5 μm, was prepared in a hydrothermal process. Then, the polyelectrolytes, PSS and PDDA, were deposited sequentially via LBL assembly to form a strong, dense coating on urchin-like MnO_2 to form 3D adsorbent MnO_2@(PSS/PDDA/PSS). The adsorption properties of Zn^{2+} on MnO_2@(PSS/PDDA/PSS) were studied in batch experiments. Different experimental conditions affecting the uptake of Zn^{2+} were investigated, and the experimental data were fitted with various models to further understand the adsorption mechanisms.

2. Materials and Methodology

2.1. Reagents and Materials

$MnSO_4 \cdot H_2O$ and $(NH_4)_2S_2O_8$ were purchased from the Keda Reagent Factory (Shenyang, China). Nafion solution (5%) was obtained from the Yilong Energy Technology Co. Ltd. (Suzhou, China). Poly (sodium 4-styrene sulfonate) (PSS, Mw 70,000 g/mol) and poly (diallyl dimethylammonium chloride) (PDDA, Mw 200,000–350,000 g/mol) were obtained from Sigma-Aldrich (St. Louis, MO, USA). Ultrapure water with a resistivity of 18.2 MΩ cm was obtained directly from a Milli-Q Plus water purification system (Millipore Corporation, Burlington, MA, USA). All other reagents used in the experiments were analytical grade and obtained from Guangfu Fine Chemical Research Institute (Tianjin, China).

2.2. Characterization and Instruments

The surface morphology of MnO_2@PSS/PDDA/PSS was observed using a transmission electron microscope (TEM, FEI Tecnai G2 20, San Diego, CA, USA) and a scanning electron microscope (SEM) (Merlin Compact, Tokyo, Japan). The FTIR spectra were measured using a Thermo Nicolet NEXUS FTIR spectrometer at room temperature to analyze the surface functional groups of samples. The oxidation state of elements in the samples was analyzed by XPS (ESCALAB 250Xi, Waltham, MA, USA). The concentration of the Pb^{2+}, Cu^{2+}, and Zn^{2+} solution was monitored with a UV-vis spectrometer (Shanghai Jinghua 756MC, Shanghai, China).

2.3. Preparation of Urchin-like MnO_2

Urchin-like MnO_2 were prepared based on the following protocol. Typically, 10.7817 g of $MnSO_4 \cdot H_2O$ and 14.6048 g of $(NH_4)_2S_2O_8$ were dissolved in 70.0 mL of deionized water and heated at 120 °C for 2 h. The dark precipitate was then centrifuged at 6000 rpm for 15 min, washed three times with DI water, and then dried at 70 °C for 12 h.

2.4. Layer-by-Layer Assembly of Polyelectrolytes on MnO_2

For the LBL deposition, the PSS and PDDA were coated in an adsorption-centrifugation cycle. In a typical procedure, 0.05 g of MnO_2 was added into 5.0 mL of the polyelectrolyte solution (5 mM). The particles were incubated at 25 °C for 30 min, placed in a centrifuge at 4500 rpm for 15 min, and then washed for three cycles. The final products were denoted as MnO_2@PSS/PDDA/PSS. The coating procedure was repeated 3 times, and finally, MnO_2@(PSS/PDDA)$_3$/PSS was obtained.

2.5. Adsorption Experimental Procedure

In a single system, the effect factors, including pH value, the initial concentration of Zn^{2+}, the dosage of adsorbent, and reaction time on adsorption were studied in a 50 mL conical flask with 20 mL of Zn^{2+} (Cu^{2+}/Pb^{2+}) solution. The mixture was stirred at a speed of 250 rpm/min for 24 h to reach adsorption equilibrium. The concentration of Zn^{2+} (Cu^{2+}/Pb^{2+}) in the solution was measured at a predetermined time. In the competition experiment with the presence of co-existing ions, adsorption performance was investigated in the solution of Zn^{2+}, Cu^{2+} and Pb^{2+} (the concentration of each metal ion was 50 mg/L). Each adsorption was replicated three times. For each set of data present, standard statistical methods were used to determine the mean values and standard deviations. Confidence intervals of 95% were calculated for each set of samples to determine the margin of error.

2.6. Modeling of Adsorption Kinetics

The adsorption kinetics were evaluated using pseudo-first-order (1) and pseudo-second-order (2) equations in this study [32,33]:

$$\ln(q_e - q_t) = \ln q_e - k_1 t \tag{1}$$

$$\frac{t}{q_t} = \frac{t}{q_e} + \frac{1}{h} \tag{2}$$

where q_e and q_t are the amount of adsorbed Zn^{2+} at equilibrium and time t (mg/g), k_1 (min^{-1}) and k_2 ($g \cdot mg^{-1} \cdot min^{-1}$) are rate constants for pseudo-first-order and second-order kinetics, respectively. The equation $h = k_2 q_e^2$ gives the initial adsorption rate when t approaches 0.

2.7. Modeling of Adsorption Isotherm

The Langmuir and Freundlich models are used in this case study. The non-linear form of the Langmuir and Freundlich equations are presented as [34,35]:

$$\frac{C_e}{q_e} = \frac{C_e}{q_{max}} + \frac{1}{K_L q_{max}} \tag{3}$$

$$\ln q_e = \frac{1}{n} \ln C_e + \ln K_F \tag{4}$$

where q_e is the equilibrium adsorption capacity (mg/g), C_e is the equilibrium concentration (mg/L), q_{max} is the maximum adsorption capacity (mg/g), which is the amount of adsorbate adsorbed per unit weight (mg/g of adsorbent), and K_L is the Langmuir constant related to the adsorption energy. K_F and n are the Freundlich constants.

3. Results and Discussion

3.1. Characterization of MnO_2, MnO_2/PSS/PDDA/PSS and MnO_2/(PSS/PDDA)$_3$/PSS

Figure 1 shows the TEM images of MnO_2@PSS/PDDA/PSS and MnO_2@(PSS/PDDA)$_3$/PSS. It can be seen that the thickness of the covering layer on the branches of urchin-like MnO_2 increases significantly from about 1 ± 0.3 nm (Figure 1A) to 5 ± 0.7 nm (Figure 1B) as the number of coating layers increases, which might be due to the deposition of different amounts of polyelectrolytes. It was also found that the shape of the urchin-like MnO_2 hardly changed after coating with one and three layers of polyelectrolytes.

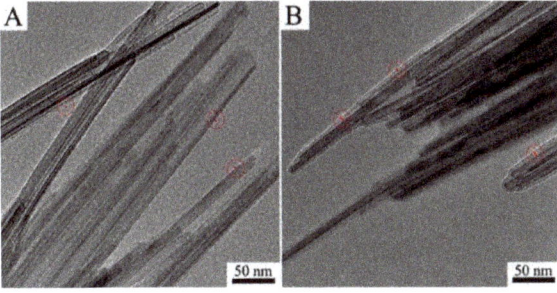

Figure 1. TEM images of the MnO_2@PSS/PDDA/PSS (**A**) and MnO_2@(PSS/PDDA)$_3$/PSS (**B**).

Figure 2A shows the successful preparation of urchin-like MnO$_2$. Figure 2B,C shows the conformal coating, with one layer and three layers of polyelectrolytes (MnO$_2$@PSS/PDDA/PSS and MnO$_2$@(PSS/PDDA)$_3$/PSS, respectively. With the three-layer coating, the thickness of the branches increases to 300 ± 58 nm (Figure 2F), which is much larger than that of pristine MnO$_2$ (20 ± 4 nm, Figure 2D) and MnO$_2$@PSS/PDDA/PSS (50 ± 11 nm, Figure 2E).

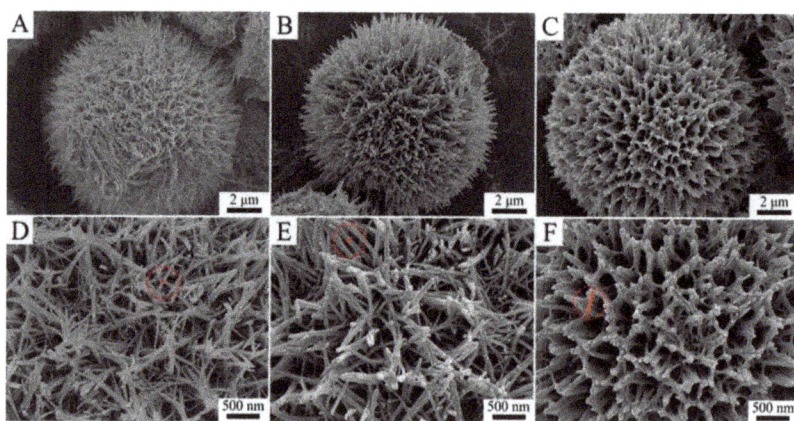

Figure 2. SEM images of MnO$_2$ (**A,D**), MnO$_2$@PSS/PDDA/PSS (**B,E**) and MnO$_2$@(PSS/PDDA)$_3$/PSS (**C,F**).

The FT-IR spectra of MnO$_2$ and MnO$_2$@PSS/PDDA/PSS are shown in Figure 3. The peaks at 515, 518.3 cm^{-1} can be attributed to Mn-O vibration. The band at nearly 1390 cm^{-1} can be assigned to the Mn=O stretching vibration, which decreased dramatically after the LBL deposition of polyelectrolytes. It might be due to the possible reaction between Mn=O and coated polyelectrolytes or the attraction between MnO$_2$ and oppositely charged polymers, which is not currently fully understood [36]. The peaks at 1178, 1129, 1034.6 cm^{-1} are related to the –S=O=S– and –SO$_3$ symmetric vibrations of PSS, indicating the coating of PSS/PDDA film on the surface of urchin-like MnO$_2$ [37,38]. The analysis of IR is consistent with the previous reports [30,36].

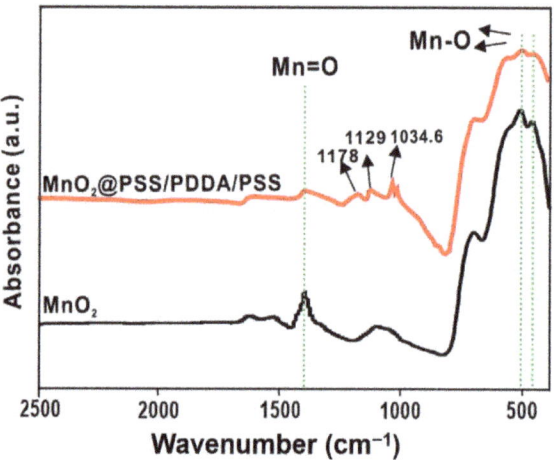

Figure 3. FT-IR spectra of MnO$_2$ and MnO$_2$@PSS/PDDA/PSS.

As shown in Figure 4A, the presence of Mn, N, and S elements in the sample of MnO$_2$@PSS/PDDA/PSS is due to the deposition of PSS/PDDA on the surface of manganese oxides. Moreover, the polymer coating might decrease the intensity of the Mn peak. Figure 4B shows the spectrum of Mn2p, in which the peaks of 652.4 eV and 641.6 eV are in agreement with an earlier report on MnO$_2$ [39]. The peak in Figure 4C appearing at 397.4 eV is assigned to N1s, which would come from the N-enriched polymer (PDDA). Figure 4D shows the spectrum of the S2p$_{1/2}$ peak (166.6 eV). These results indicate the coating of polymer layers on MnO$_2$.

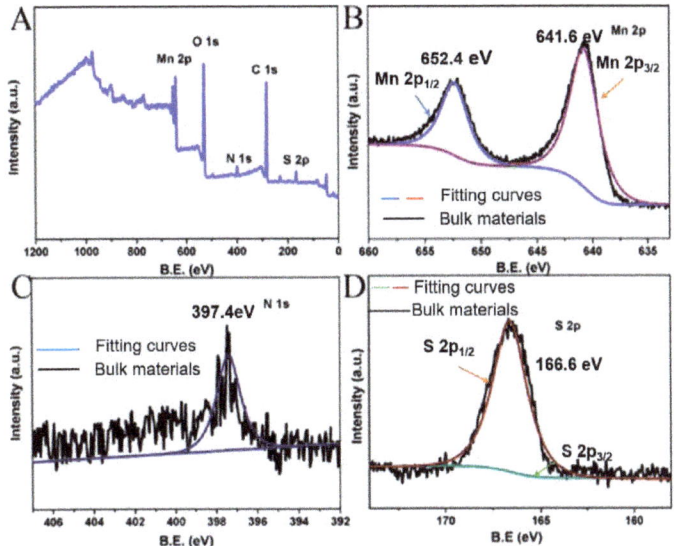

Figure 4. XPS spectra of MnO$_2$@PSS/PDDA/PSS: wide scan (**A**), Mn 2p spectra (**B**), N 1s spectra (**C**), S 2p spectra (**D**).

3.2. Adsorption of Zn^{2+} on MnO$_2$, MnO$_2$/PSS/PDDA/PSS, and MnO$_2$/(PSS/PDDA)$_3$/PSS

The adsorption performance on MnO$_2$, MnO$_2$/PSS/PDDA/PSS, and MnO$_2$/(PSS/PDDA)$_3$/PSS were tested at pH 13.0. The results are shown in Figure 5. For the control experiment from Zn^{2+}-bearing alkaline solutions (pH of 13.0), there is no precipitate observed over 24 h in the absence of prepared materials, indicating that the removal of Zn^{2+} from these alkaline solutions is solely due to the presence of the prepared materials as adsorbents. As shown in Figure 5, after coating with one layer of PSS/PDDA/PSS on urchin-like MnO$_2$, the highest adsorption capacity was achieved (177.74 mg/g), and as the number of coating layers continuously increased to three, the adsorption capacity of Zn^{2+} decreased. Based on the measured surface area of MnO$_2$, MnO$_2$@PSS/PDDA/PSS and MnO$_2$@(PSS/PDDA)$_3$/PSS (121, 108 and 54 m^2/g, respectively), it could be concluded that the decreased adsorption capacity of MnO$_2$@(PSS/PDDA)$_3$/PSS might be caused by the ultra-dense coating of polyelectrolytes on urchin-like MnO$_2$, which would significantly reduce the surface area of MnO$_2$ and also block the active sites of MnO$_2$. The number of coated polymers in MnO$_2$@PSS/PDDA/PSS and MnO$_2$@(PSS/PDDA)$_3$/PSS samples was calculated to be 2.173 g and 5.515 g/g of MnO$_2$, respectively, based on their FTIR spectra. As each Zn^{2+} would bind two unit-charge sites of PSS, theoretically, the amount of polymer coating on 0.05 g of MnO$_2$@PSS/PDDA/PSS would combine approximately 2.701 mmol of Zn^{2+}, which is close to the experimental adsorption capacity (177.74 mg/g). Thus, MnO$_2$/PSS/PDDA/PSS was chosen for the following experiments.

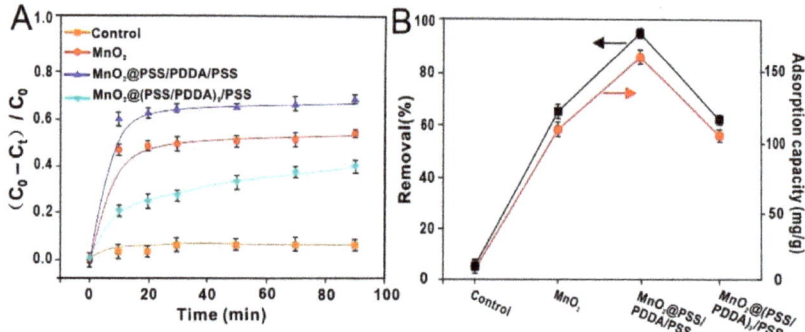

Figure 5. Zn^{2+} adsorption on MnO_2, MnO_2/PSS/PDDA/PSS, and MnO_2/(PSS/PDDA)$_3$/PSS with an initial concentration of 100 mg/L at 25 °C (**A**), and their corresponding adsorption capacity and removal efficiency (**B**).

3.3. Effect of Solution pH

The effect of pH on Zn^{2+} adsorption was investigated in the pH range from 5.0 to 13.0. As shown in Figure 6, the removal efficiency of Zn^{2+} continued to increase as the pH increased. At pH 13.0, the highest adsorption capacity of Zn^{2+} on MnO_2/PSS/PDDA/PSS was obtained. This is mainly because of the electrostatic attraction between Zn^{2+} and negatively charged PSS film. When the solution was acidic, there would be more H^+ in the solution, which would compete with Zn^{2+} to occupy the active sites.

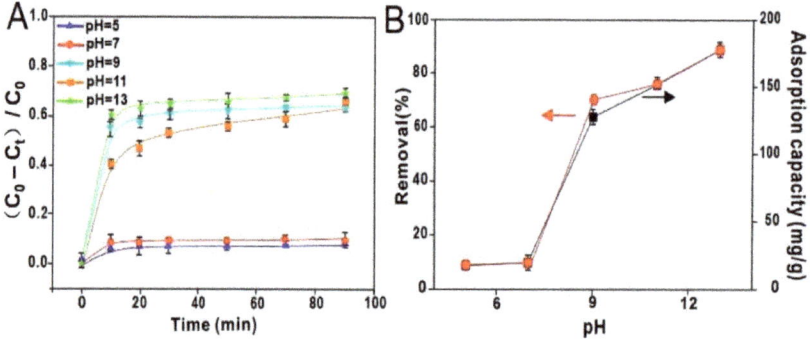

Figure 6. Effect of pH on Zn^{2+} adsorption with an initial concentration of 100 mg/L at 25 °C (**A**). Adsorption capacity and removal of Zn^{2+} on MnO_2@PSS/PDDA/PSS at different pH values (**B**).

3.4. Effect of Initial Concentration

The effect of the initial Zn^{2+} concentration was investigated at a pH of 13.0. The concentrations were studied at 20, 50, 100, 200, and 300 mg/L. The results of the initial concentration experiment are shown in Figure 7. It was found that the adsorption capacity was highest at an initial concentration of 100 mg/L. The removal rate increased as the initial Zn concentration increased from 20 mg/L to 100 mg/L, and then began to decrease sharply. This might be due to the fact that the adsorption site was occupied quickly; metal ion adsorption involves higher energy sites at low metal-ion concentrations. With an increase in the initial Zn^{2+} concentration (20 to 100 mg/L), the large concentration difference between the solution and the materials drives greater binding of Zn^{2+} and increases the removal rate [40]. Therefore, an optimal zinc concentration of 100 mg/L was selected for further experiments.

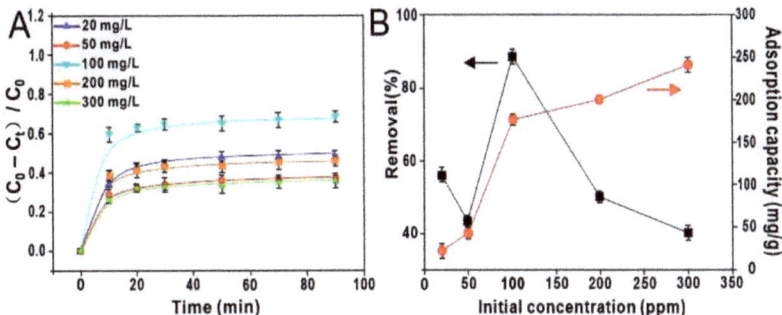

Figure 7. The effect of initial concentration on Zn^{2+} adsorption under a pH of 13.0 at 25 °C (**A**). Adsorption capacity and removal of Zn^{2+} on MnO_2@PSS/PDDA/PSS, with different initial Zn^{2+} concentrations at 25 °C (**B**).

3.5. Effect of Adsorbent Dosage

Figure 8 shows the effect of adsorbent dosage on Zn^{2+} removal by MnO_2/PSS/PDDA/PSS. The highest removal efficiency of Zn^{2+} was reached when the dose of MnO_2/PSS/PDDA/PSS was 0.5 g/L. When the adsorbent dosage is lower than 0.5 g/L, less surface area is available for adsorption due to there being fewer active sites present, leading to a decreased adsorption efficiency. With an increase in the adsorbent dose, the adsorption capacity, q_e, decreased. This is mainly because with the increase in the amount of adsorbent, more unoccupied adsorptive sites were left and their mass could still be used for the calculation of adsorption capacity [41,42]. Therefore, the amount of adsorbent used in the experiments was selected to be 0.5 g/L.

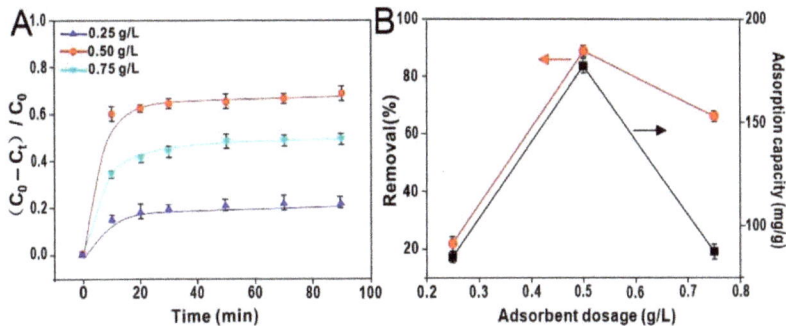

Figure 8. The effect of adsorbent dosage on Zn^{2+} adsorption with an initial concentration of 100 mg/L at 25 °C (**A**). The adsorption capacity and removal of Zn^{2+} on MnO_2@PSS/PDDA/PSS with different adsorbent dosages at 25 °C (**B**).

3.6. Adsorption Kinetics

The pseudo-first-order and pseudo-second-order kinetic models were applied to describe the experimental data. The relevant kinetic parameters for Zn^{2+} adsorption are displayed in Table 1. The results show that the correlation coefficient of the pseudo-second-order kinetic equation was 0.9989, higher than that of the first-order kinetic curve, indicating that the experimental data closely conformed to the second-order model.

Table 1. Parameters of the kinetics model for the adsorption of Zn^{2+} with MnO_2@PSS/PDDA/PSS, with an initial concentration of 100 mg/L, under a pH of 13.0 at 25 °C.

Initial conc. mg/L	q_{exp} mg/g	Pseudo-First-Order			Pseudo-Second-Order		
		$k_1 \times 10^{-2}$ min^{-1}	q_e mg/g	R^2	$K_2 \times 10^{-3}$ g/(g·min)	q_e mg/g	R^2
100	177.74 ± 0.21	0.173 ± 0.02	177.56 ± 0.44	0.9321	7.26 ± 0.01	94.97 ± 0.37	0.9989

3.7. Adsorption Isotherm Models

The fitted results of the Langmuir and Freundlich isotherm models in this study are presented in Table 2. The results showed that the Langmuir model with R^2 higher than 0.99 was a better fit than the Freundlich model, indicating that Zn^{2+} adsorption onto MnO_2@PSS/PDDA/PSS can be considered to be a monolayer adsorption process, mainly achieved via electrostatic attraction.

Table 2. Parameters of the isotherm model for the adsorption of Zn^{2+} onto MnO_2@PSS/PDDA/PSS.

Temperature K	Langmuir			Freundlich		
	q_{max} (mg/g)	b (L/mg)	R^2	K_f (L/mg)	1/n	R^2
298	246.91 ± 0.22	0.296 ± 0.025	0.9990	3.4261 ± 0.097	0.8339 ± 0.082	0.9469

3.8. Adsorption of Other Heavy Metals in Alkaline Solution

MnO_2@PSS/PDDA/PSS was used as an adsorbent to test the removal of Pb^{2+} and Cu^{2+} from alkaline water. The results are shown in Figure 9A. It can be seen that the maximum adsorption capacities of Pb^{2+} and Cu^{2+} were 177.63 mg/g and 150.93 mg/g, respectively, indicating the efficient removal of Zn^{2+}, Pb^{2+}, and Cu^{2+} from alkaline water when using MnO_2@PSS/PDDA/PSS as an adsorbent material. It may be concluded that the adsorption affinity of metals onto MnO_2@PSS/PDDA/PSS occurs in the following order: $Zn^{2+} \approx Pb^{2+} > Cu^{2+}$. Moreover, the competition experiments were conducted with the presence of Pb^{2+}, Cu^{2+}, and Zn^{2+} in the solution. The results show that the adsorption performance of Zn^{2+} slightly decreased in the presence of Pb^{2+} and Cu^{2+} (Figure 9B), which is likely due to the substitution of Zn^{2+} already adsorbed on the adsorption sites with Pb^{2+}. To simulate a real-life application, we collected tap water in the lab and Yellow River water in the city of Lanzhou, then prepared each solution of Zn^{2+}, Pb^{2+}, and Cu^{2+} with an initial concentration of 100 mg/L. The adsorption of Zn^{2+}, Pb^{2+}, and Cu^{2+} from tap water and Yellow River water was investigated. As shown in Table 3, the adsorption capacity of Zn^{2+}, Pb^{2+}, and Cu^{2+} in the tap water and Yellow River water was comparable to that in DI water, indicating the possible real application of this process in wastewater treatment.

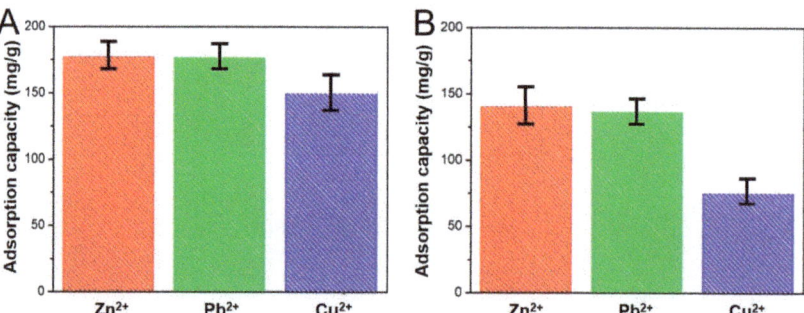

Figure 9. The adsorption of Pb^{2+} and Cu^{2+} on MnO_2@PSS/PDDA/PSS in a single system (**A**) and ternary system (**B**).

Table 3. The adsorption capacity of Zn^{2+}, Pb^{2+}, and Cu^{2+} in tap water and Yellow River water.

Metal Ions	Adsorption Capacity (mg/g)	
	Tap Water	Yellow River Water
Zn^{2+}	178.32 ± 0.89	171.66 ± 1.28
Pb^{2+}	174.85 ± 0.61	169.15 ± 1.72
Cu^{2+}	146.17 ± 0.50	138.22 ± 2.80

4. Conclusions

In this work, 3D urchin-like MnO_2@PSS/PDDA/PSS particles were prepared via the layer-by-layer (LBL) assembly of polyelectrolytes on MnO_2 for the extraction of Zn^{2+} from alkaline media. The characteristics of the pH effect, adsorbent dosage, the initial Zn^{2+} concentrations, and contact time for MnO_2@PSS/PDDA/PSS were tested. The results showed that MnO_2@PSS/PDDA/PSS was very effective in removing Zn^{2+} from an aqueous solution at pH 13. Adsorption kinetics and equilibrium studies were applied to investigate the adsorption behavior of MnO_2@PSS/PDDA/PSS. The results showed that the experimental data fitted well with the second-order equation, and the adsorption isotherm was closely related to the Langmuir model. It was found that both the urchin-like structure of MnO_2 and the surface coating of negatively charged PSS contributed to the efficient adsorption process. The competitive adsorption investigation suggests that Zn^{2+} adsorption could be interfered with by other cations present in wastewater. MnO_2@PSS/PDDA/PSS can be considered as a promising alternative for the adsorption of Zn^{2+}, Pb^{2+}, and Cu^{2+} from alkaline wastewater. We anticipate that more studies will take place on the efficient adsorption of Zn^{2+} using non-synthetic wastewater for real-life applications in future work.

Author Contributions: Z.N.: writing—review and editing; D.C.: concept and methodology, conducting experiments and writing—original draft. All authors have read and agreed to the published version of the manuscript.

Funding: This research received no external funding.

Institutional Review Board Statement: Not applicable.

Informed Consent Statement: Not applicable.

Data Availability Statement: Not applicable.

Conflicts of Interest: The authors declare no conflict of interest.

References

1. Huang, Q.; Zhang, Y.; Zhou, W.; Huang, X.; Chen, Y.; Tian, X.; Yu, T. Amorphous molybdenum sulfide mediated EDTA with multiple active sites to boost heavy metal ions removal. *Chin. Chem. Lett.* **2021**, *32*, 2797–2802. [CrossRef]
2. Pang, J.J.; Du, R.H.; Lian, X.; Yao, Z.Q.; Xu, J.; Bu, X.H. Selective sensing of CrVI and FeIII ions in aqueous solution by an exceptionally stable TbIII-organic framework with an AIE-active ligand. *Chin. Chem. Lett.* **2021**, *32*, 2443–2447.
3. Li, K.T.; Wu, G.H.; Wang, M.; Zhou, X.H.; Wang, Z.Q. Efficient Removal of Lead Ions from Water by a Low-Cost Alginate-Melamine Hybrid Sorbent. *Appl. Sci.* **2018**, *8*, 1518. [CrossRef]
4. Wang, Z.Q.; Wu, A.G.; Ciacchi, L.C.; Wei, G. Recent Advances in Nanoporous Membranes for Water Purification. *Nanomaterials* **2018**, *8*, 65. [CrossRef] [PubMed]
5. Zou, Y.D.; Wang, X.X.; Khan, A.; Wang, P.Y.; Liu, Y.H.; Alsaedi, A.; Hayat, T.; Wang, X.K. Environmental Remediation and Application of Nanoscale Zero-Valent Iron and Its Composites for the Removal of Heavy Metal Ions: A Review. *Environ. Sci. Technol.* **2016**, *50*, 7290–7304. [CrossRef] [PubMed]
6. Wang, Q.Q.; Hu, R.; Fang, Z.Q.; Shi, G.; Zhang, S.; Zhang, M. A multifunctional upconversion nanoparticles probe for Cu^{2+} sensing and pattern recognition of biothiols. *Chin. Chem. Lett.* **2021**; in press. [CrossRef]
7. Yan, X.; Ma, J.; Yu, K.; Li, J.; Yang, L.; Liu, J.; Wang, J.; Cai, L. Highly green fluorescent Nb2C MXene quantum dots for Cu^{2+} ion sensing and cell imaging. *Chin. Chem. Lett.* **2020**, *31*, 3173–3177. [CrossRef]
8. Chiririwa, H.; Naidoo, E.B. Removal Efficiency of Cu^{2+}, Zn^{2+}, Fe^{2+}, Al^{3+} and Mn^{2+} from Aqueous Solution in Presence of Bentonite Using Column Adsorption. *Asian J. Chem.* **2017**, *29*, 2536–2540.
9. Hossain, M.D.F.; Akther, N.; Zhou, Y. Recent advancements in graphene adsorbents for wastewater treatment: Current status and challenges. *Chin. Chem. Lett.* **2021**, *31*, 2525–2538. [CrossRef]

10. Zhao, B.; Jiang, L.; Jia, Q. Advances in cyclodextrin polymers adsorbents for separation and enrichment: Classification, mechanism and applications. *Chin. Chem. Lett.* **2022**, *33*, 11–21. [CrossRef]
11. Fernandez, Y.; Maranon, E.; Castrillon, L.; Vazquez, I. Removal of Cd and Zn from inorganic industrial waste leachate by ion exchange. *J. Hazard. Mater.* **2005**, *126*, 169–175. [CrossRef]
12. Alyuz, B.; Veli, S. Kinetics and equilibrium studies for the removal of nickel and zinc from aqueous solutions by ion exchange resins. *J. Hazard. Mater.* **2009**, *167*, 482–488. [CrossRef] [PubMed]
13. Kurniawan, T.A.; Chan, G.Y.S.; Lo, W.H.; Babel, S. Physico-chemical treatment techniques for wastewater laden with heavy metals. *Chem. Eng. J.* **2006**, *118*, 83–98. [CrossRef]
14. Charerntanyarak, L. Heavy metals removal by chemical coagulation and precipitation. *Water Sci. Technol.* **1999**, *39*, 135–138. [CrossRef]
15. Mansoorian, H.J.; Mahvi, A.H.; Jafari, A.J. Removal of lead and zinc from battery industry wastewater using electrocoagulation process: Influence of direct and alternating current by using iron and stainless steel rod electrodes. *Sep. Purif. Technol.* **2014**, *135*, 165–175. [CrossRef]
16. Liu, Y.; Liu, F.; Ding, N.; Hu, X.; Shen, C.; Li, F.; Huang, M.; Wang, Z.; Sand, W.; Wang, C.C. Recent advances on electroactive CNT based membranes for environmental applications: The perfect match of electrochemistry and membrane separation. *Chin. Chem. Lett.* **2020**, *31*, 2539–2548. [CrossRef]
17. Oh, H.; Song, J.; Jang, J. Fabrication of polyrhodanine nanotubes modified anodic aluminum oxide membrane and its application for heavy metal ions removal. *Abstr. Pap. Am. Chem. Soc.* **2011**, *242*, 1.
18. Fu, F.L.; Wang, Q. Removal of heavy metal ions from wastewaters: A review. *J. Environ. Manag.* **2011**, *92*, 407–418.
19. Liu, Y.; Sun, X.M.; Li, B.H. Adsorption of Hg^{2+} and Cd^{2+} by ethylenediamine modified peanut shells. *Carbohydr. Polym.* **2010**, *81*, 335–339. [CrossRef]
20. Zhang, Y.; Li, Y.F.; Yang, L.Q.; Ma, X.J.; Wang, L.Y.; Ye, Z.F. Characterization and adsorption mechanism of Zn^{2+} removal by PVA/EDTA resin in polluted water. *J. Hazard. Mater.* **2010**, *178*, 1046–1054. [CrossRef]
21. Iljina, A.; Eisinas, A.; Baltakys, K.; Bankauskaitė, A.; Šiaučiūnas, R. Adsorption capacity of clinoptilolite for Zn^{2+} ions in acidic solution. *Chem. Technol.* **2013**, *63*, 10–14. [CrossRef]
22. Iljina, A.; Eisinas, A.; Baltakys, K.; Bankauskaitė, A. Adsorption capacity of clinoptilolite for Zn^{2+} ions in alkaline solution. *Chem. Technol.* **2013**, *63*, 15–20. [CrossRef]
23. Iljina, A.; Baltakys, K.; Eisinas, A. Gyrolite adsorption of Zn^{2+} ions in acidic and alkaline solutions. *Mater. Sci.-Medzg.* **2015**, *21*, 123–128.
24. Zhang, Y.; Jing, L.; He, X.; Li, Y.; Ma, X. Sorption enhancement of TBBPA from water by fly ash-supported nanostructured γ-MnO_2. *J. Ind. Eng. Chem.* **2015**, *21*, 610–619. [CrossRef]
25. Tripathy, S.S.; Kanungo, S.B. Adsorption of Co^{2+}, Ni^{2+}, Cu^{2+} and Zn^{2+} from 0.5 M NaCl and major ion sea water on a mixture of delta-MnO_2 and amorphous FeOOH. *J. Colloid Interface Sci.* **2005**, *284*, 30–38. [CrossRef]
26. Ohta, A.; Kawabe, I. REE(III) adsorption onto Mn dioxide (delta-MnO_2) and Fe oxyhydroxide: Ce(III) oxidation by delta-MnO_2. *Geochim. Cosmochim. Acta* **2001**, *65*, 695–703. [CrossRef]
27. Montjoy, D.G.; Bahng, J.H.; Eskafi, A.; Hou, H.; Kotov, N.A. Omnidispersible Hedgehog Particles with Multilayer Coatings for Multiplexed Biosensing. *J. Am. Chem. Soc.* **2018**, *140*, 7835–7845. [CrossRef]
28. Zhang, Z.Q.; Mu, J. Hydrothermal synthesis of gamma-MnOOH nanowires and alpha-MnO_2 sea urchin-like clusters. *Solid State Commun.* **2007**, *141*, 427–430. [CrossRef]
29. Zeng, J.H.; Wang, Y.F.; Yang, Y.; Zhang, J. Synthesis of sea-urchin shaped γ-MnO_2 nanostructures and their application in lithium batteries. *J. Mater. Chem.* **2010**, *20*, 10915–10918. [CrossRef]
30. Wang, Y.; Wang, S.; Xiao, M.; Han, D.; Hickner, M.A.; Meng, Y. Layer-by-layer self-assembly of PDDA/PSS-SPFEK composite membrane with low vanadium permeability for vanadium redox flow battery. *RSC Adv.* **2013**, *3*, 15467–15474. [CrossRef]
31. Ge, A.; Matsusaki, M.; Qiao, L.; Akashi, M.; Ye, S. Salt Effects on Surface Structures of Polyelectrolyte Multilayers (PEMs) Investigated by Vibrational Sum Frequency Generation (SFG) Spectroscopy. *Langmuir* **2016**, *32*, 3803–3810. [CrossRef]
32. Nassar, N.N. Rapid removal and recovery of Pb(II) from wastewater by magnetic nanoadsorbents. *J. Hazard. Mater.* **2010**, *184*, 538–546. [CrossRef] [PubMed]
33. Li, X.; Qi, Y.; Li, Y.; Zhang, Y.; He, X.; Wang, Y. Novel magnetic beads based on sodium alginate gel crosslinked by zirconium(IV) and their effective removal for Pb^{2+} in aqueous solutions by using a batch and continuous systems. *Bioresour. Technol.* **2013**, *142*, 611–619. [CrossRef] [PubMed]
34. Repo, E.; Warchol, J.K.; Kurniawan, T.A.; Sillanpaa, M.E.T. Adsorption of Co(II) and Ni(II) by EDTA- and/or DTPA-modified chitosan: Kinetic and equilibrium modeling. *Chem. Eng. J.* **2010**, *161*, 73–82. [CrossRef]
35. Jing, L.; Li, X. Facile synthesis of PVA/CNTs for enhanced adsorption of Pb^{2+} and Cu^{2+} in single and binary system. *Desalin. Water Treat.* **2016**, *57*, 21391–21404. [CrossRef]
36. Li, H.; Jia, L.-P.; Ma, R.-N.; Jia, W.-L.; Wang, H.-S. Electrodeposition of PtNPs on the LBL assembled multilayer films of (PDDA-GS/PEDOT:PSS)n and their electrocatalytic activity toward methanol oxidation. *RSC Adv.* **2017**, *7*, 16371–16378. [CrossRef]
37. Yang, J.; Zou, L.; Song, H. Preparing MnO_2/PSS/CNTs composite electrodes by layer-by-layer deposition of MnO_2 in the membrane capacitive deionization. *Desalination* **2012**, *286*, 108–114. [CrossRef]
38. Liu, R.; Duay, J.; Lee, S.B. Redox Exchange induced MnO_2 nanoparticle enrichment in poly(3,4-ethylenedioxythiophene) nanowires for electrochemical energy storage. *ACS Nano* **2010**, *4*, 4299–4307. [CrossRef]

39. Dicastro, V.; Furlani, C.; Gargano, M.; Rossi, M. XPS Characterization of the CuO/MnO$_2$ catalyst. *Appl. Surf. Sci.* **1987**, *28*, 270–278. [CrossRef]
40. Kilpimaa, S.; Runtti, H.; Kangas, T.; Lassi, U.; Kuokkanen, T. Physical activation of carbon residue from biomass gasification: Novel sorbent for the removal of phosphates and nitrates from aqueous solution. *J. Ind. Eng. Chem.* **2015**, *21*, 1354–1364.
41. Mohan, D.; Singh, K.P. Single- and multi-component adsorption of cadmium and zinc using activated carbon derived from bagasse—An agricultural waste. *Water Res.* **2002**, *36*, 2304–2318. [CrossRef]
42. Wang, Y.; Qi, Y.; Li, Y.; Wu, J.; Ma, X.; Yu, C.; Ji, L. Preparation and characterization of a novel nano-absorbent based on multi-cyanoguanidine modified magnetic chitosan and its highly effective recovery for Hg(II) in aqueous phase. *J. Hazard. Mater.* **2013**, *260*, 9–15. [CrossRef] [PubMed]

Article

The Preparation and Electrochemical Pseudocapacitive Performance of Mutual Nickel Phosphide Heterostructures

Shao-Bo Guo, Wei-Bin Zhang *, Ze-Qin Yang, Xu Bao, Lun Zhang, Yao-Wen Guo, Xiong-Wei Han and Jianping Long *

College of Materials and Chemistry & Chemical Engineering, Chengdu University of Technology, Chengdu 610059, China; guoshoo@163.com (S.-B.G.); yangzeqin1@icloud.com (Z.-Q.Y.); 2019020453@stu.cdut.edu.cn (X.B.); 2019020460@stu.cdut.edu.cn (L.Z.); guoyaowen@stu.cdut.edu.cn (Y.-W.G.); hxw@stu.cdut.edu.cn (X.-W.H.)
* Correspondence: zhangweibin17@cdut.edu.cn (W.-B.Z.); longjianping@cdut.edu.cn (J.L.)

Abstract: Transition metal phosphide composite materials have become an excellent choice for use in supercapacitor electrodes due to their excellent conductivity and good catalytic activity. In our study, a series of nickel phosphide heterostructure composites was prepared using a temperature-programmed phosphating method, and their electrochemical performance was tested in 2 mol L^{-1} KOH electrolyte. Because the interface effect can increase the catalytic active sites and improve the ion transmission, the prepared $Ni_2P/Ni_3P/Ni$ (Ni/P = 7:3) had a specific capacity of 321 mAh g^{-1} under 1 A g^{-1} and the prepared Ni_2P/Ni_5P_4 (Ni/P = 5:4) had a specific capacity of 218 mAh g^{-1} under 1 A g^{-1}. After the current density was increased from 0.5 A g^{-1} to 5 A g^{-1}, 76% of the specific capacity was maintained. After 7000 cycles, the capacity retention rate was above 82%. Due to the phase recombination effect, the electrochemical performance of $Ni_2P/Ni_3P/Ni$ and Ni_2P/Ni_5P_4 was much better than that of single-phase N_2P. After assembling the prepared composite and activated carbon into a supercapacitor, the $Ni_2P/Ni_3P/Ni//AC$ had an energy density of 22 W h kg^{-1} and a power density of 800 W kg^{-1} and the $Ni_2P/Ni_5P_4//AC$ had an energy density of 27 W h kg^{-1} and a power density of 800 W kg^{-1}.

Keywords: nickel phosphide heterostructure; electrochemical pseudocapacitance; temperature programming; supercapacitor

Citation: Guo, S.-B.; Zhang, W.-B.; Yang, Z.-Q.; Bao, X.; Zhang, L.; Guo, Y.-W.; Han, X.-W.; Long, J. The Preparation and Electrochemical Pseudocapacitive Performance of Mutual Nickel Phosphide Heterostructures. *Crystals* **2022**, *12*, 469. https://doi.org/10.3390/cryst12040469

Academic Editor: Sergio Brutti

Received: 17 January 2022
Accepted: 11 February 2022
Published: 28 March 2022

Publisher's Note: MDPI stays neutral with regard to jurisdictional claims in published maps and institutional affiliations.

Copyright: © 2022 by the authors. Licensee MDPI, Basel, Switzerland. This article is an open access article distributed under the terms and conditions of the Creative Commons Attribution (CC BY) license (https://creativecommons.org/licenses/by/4.0/).

1. Introduction

With the rapid development of human society, the demand for energy has increased substantially. At the same time, the energy crisis is becoming increasingly more serious. In the future, exploring and developing clean and renewable energy such as wind and hydropower will become the mainstream [1–6]. However, due to the instability of power generation, this type of energy has not yet fully met the demand [7,8]. Through electrochemical energy storage devices, the electricity generated from these renewable energy sources can be stored for effective use [9,10]. Presently, there are two main types of commercial electrochemical energy storage devices: batteries, such as lithium-ion batteries [11,12], and supercapacitors, such as pseudocapacitance supercapacitors [13,14]. Supercapacitors have many advantages, such as higher power density, long cycle life, and fast charging and discharging, so they have attracted much attention [15–17]. However, there are still some challenges associated with supercapacitors, including how to increase their energy density while maintaining the above advantages [18,19].

Electrode material is the core factor of the supercapacitor system and affects the performance of the supercapacitor directly [20]. According to their characteristics, electrode materials can be divided into two categories [21]. The first is a variety of carbon materials including activated carbon, graphene, and carbon nanotubes; the energy storage mechanism is electrostatic adsorption and desorption [9]. These electrode materials provide good stability, porous characteristics, and good electrical conductivity [22], but the relatively low energy density associated with this group limits its wide application [23]. The second category is pseudocapacitive materials including transition metal oxides and transition metal phosphides; the energy storage mechanism is rapid reversible redox or intercalation reactions on or near the electrode surface [24,25]. Compared to transition metal oxides, transition metal phosphides have better electrical conductivity and a lower cost. They also have a specific capacity comparable to or even beyond that of transition metal oxides [26,27]. This is due to some of the characteristics of transition metal phosphides.

Phosphorus atoms can enter the transition metal crystals to form intermetallics [28]. The presence of phosphorus atoms pulls the electron delocalization of metal phosphide, which improves its catalytic activity. The higher the metal content, the more free electrons it contains and the better it is for conducting electricity [29]. Therefore, this compound with metallic properties exhibits high electrical conductivity and specific capacity [30,31]. Nickel phosphide has been a good choice for electrode materials for supercapacitors because of its high electrical conductivity, fast charge transfer ability, good reaction kinetics, and abundant earth reserves [32–34]. Single-phase nickel phosphide is difficult to prepare, and its electrochemical performance is unsatisfactory [35]. Therefore, many researchers have shown great interest in preparing nickel phosphide composites and applying them to supercapacitors [36,37]. Nickel phosphide compounds can be prepared using available nickel chemical plating methods on a nickel phosphide surface coated with a layer of amorphous nickel or by mechanically mixing graphene and nickel phosphide. It is also possible to compound nickel phosphide with other compounds through chemical precipitation to make composite materials [38–40]. Other types of composites used in the study of the electrochemical performance of supercapacitors include nickel-cobalt oxide modified with reduced graphene oxide, $ZnFe_2O_4$ nanorods on reduced graphene oxide, $NiCo_2O_4/Ni_2P$, nitrogen-doped $Ni_2P/Ni_{12}P_5/Ni_3S_2$, and MoS_2–ReS_2/rGO [41–45]. Due to the synergistic effect between different components and the interfacial effect, the electrochemical performance of such materials is satisfactory. The preparation of composite nickel phosphide shows good electrochemical performance, which provides inspiration for exploring such electrode materials.

In our study, nickel phosphide composite was prepared in one step by temperature programmed phosphating. Since no polymer binder was added during the preparation of the electrode, the electrode had maximum conductivity and catalytic activity. We used 2 mol L^{-1} KOH aqueous solution as the electrolyte for electrochemical performance testing. The best comprehensive electrochemical performance was observed when the stoichiometric ratio was 5:4. At this stoichiometric ratio, the nickel phosphide composite formed a Ni_2P/Ni_5P_4 heterostructure with a specific capacity of 218 mAh g^{-1} and a rate performance of 76%. After 7000 cycles, the capacity retention rate was above 82%. After combining it with activated carbon to form an asymmetric supercapacitor, the energy density was 27 W h kg^{-1} while the power density was 800 W kg^{-1}, demonstrating good electrochemical performance. At the stoichiometric ratio of 7:3, it formed a $Ni_2P/Ni_3P/Ni$ heterostructure with a specific capacity of 321 mAh g^{-1} and a rate performance of 59%. After combining it with activated carbon to form an asymmetric supercapacitor, the energy density was 22 W h kg^{-1} while the power density was 800 W kg^{-1}.

2. Experimental Section

2.1. Chemicals

All chemicals used were of analytical grade. C_2H_5OH, KOH, H_2SO_4, and acetone were purchased from Kelon Chemicals Co. Ltd., Chengdu, China. Nickel foam was purchased from Shanghai (China) Hesen Electric Co. Ltd. Nickel powder and red phosphorus were from Aladdin.

2.2. Electrode Material Synthesis

The nickel foam was cut into small pieces, each with a length and width of 1 cm, and was then pretreated with dilute hydrochloric acid solution and acetone solution (V_{acid}:$V_{acetone}$ = 6:1) to remove oxides and organic pollutants. The surface was then rinsed with anhydrous ethanol and deionized water. The metal nickel powder was pretreated with dilute sulfuric acid to remove oxides and organic pollutants, and then rinsed with anhydrous ethanol and deionized water several times. It was put into the oven at a constant temperature of 60 °C until dried. About 5 g of treated metal nickel powder for separated for use. Metal nickel powder and red phosphorus were mixed according to the stoichiometric amounts; a 1.5% excess of red phosphorous was required.

To make electrodes, slightly more than 4 mg of mixed powder was pressed on the treated nickel foam; the applied pressure was 10 MPa. The foam nickel electrode pads loaded with mixed powder were put into a porcelain boat along with the remaining mixed powder. The porcelain boat was then put into a tubular furnace, washed to vacuum with nitrogen, gradually heated to 700 °C at 4 °C min^{-1}, and held for 6 h. The intermediate temperatures were 350 °C, 450 °C, and 550 °C. Each intermediate temperature was held for 1 h. According to the stoichiometric amount, we recorded them as (3:1), (5:2), (12:5), (7:3), (2:1), (5:4), and (1:1). Single-phase Ni_2P can be made from stoichiometry (5:2).

2.3. Material Characterization

The crystalline structures were confirmed by an X-ray diffraction (XRD, Dandong DX-2700, Dandong, China) with Cu-Kα radiation (2θ = 5~80°) operating at 40 kV. The morphologies and microstructures of the samples were characterized by field-emission scanning electron microscope (SEM, JEOL JSM-6701 F, Tokyo, Japan) and transmission electron microscope (TEM, JEOL JEM2010, Tokyo, Japan). The pore properties and Brunauer-Emmett-Teller specific surface area were investigated via N_2 adsorption–desorption (NAD, ASAPR 2020, Atlanta, GA, USA) test at -196 °C. The surface bonding state was tested by X-ray photoelectron spectroscopy (XPS, Thermo Scientific K-Alpha, New York, NY, USA).

2.4. Electrochemical Evaluation

All electrochemical measurements were executed on the electrochemical station, and the electrolyte was 2 mol L^{-1} aqueous KOH. Before the test, we soaked the undertested materials in 2 mol L^{-1} of KOH electrolyte for 24 h to activate them. The mass of the active substance was obtained by weighing the nickel foam before and after the reaction. The mass of the active material on the nickel foam electrode was observed as 4 mg. All electrochemical measurements for the single electrode were executed in a three-electrode system with a platinum plate electrode as a counter electrode and a Hg/HgO electrode as a reference electrode.

The formula of mass specific capacitance is as follows:

$$C_s = \frac{I_d \Delta t}{3.6} \quad (1)$$

Calculated by integrating the slope of the discharge curve for the asymmetric device, the formula for energy density (W h kg^{-1}) and power density (W kg^{-1}) is as follows [46]:

$$E = \frac{I_d}{3.6} \int E \, dt \quad (2)$$

$$P = \frac{3600E}{\Delta t} \quad (3)$$

The supercapacitor was assembled with activated carbon as a cathode and the prepared electrode as an anode. The mass balancing of cathode and anode will follow the equation:

$$\frac{m^+}{m^-} = \frac{C_{s-} \Delta V_-}{C_{s+} \Delta V_+} \quad (4)$$

where C_s (mAh g^{-1}) is the specific capacitance, I_d is the specific current, and Δt is the discharge time.

3. Results and Discussion

3.1. Characterization of Materials

The crystal structure of each Ni/P ratio and single-phase Ni$_2$P are shown in Figure 1. In Figure 1d, some diffraction peaks appear and correspond to the crystal planes of standard PDF card Ni$_3$P, Ni$_2$P, and Ni. This XRD pattern shows the formation of Ni$_2$P/Ni$_3$P/Ni heterostructures with a Ni/P ratio of 7:3. Due to the interaction between the different phases, the positions of the diffraction peaks are slightly shifted, which indicates the successful synthesis of the heterostructure and has a positive effect on the electrochemical performance of the electrode. Due to the presence of metallic nickel, the catalytic activity of this electrode was greatly improved. In Figure 1f, some diffraction peaks appear and correspond to the crystal planes of standard PDF card Ni$_2$P, and Ni$_5$P$_4$. This XRD pattern shows that Ni$_2$P/Ni$_5$P$_4$ heterostructures are formed at a Ni/P ratio of 5:4. The combination of different phases produces a certain lattice distortion, so the position of the corresponding diffraction peak is slightly shifted compared to the standard card. Although only two nickel phosphide phases exist, the interfacial effect between the phases can still improve the electrochemical performance of the electrode, and at the same time will improve its catalytic stability. It can be observed from Figure 1h that diffraction peaks appear at 17.463°, 26.330°, 30.489°, 31.771°, 35.350°, 40.714°, 44.611°, 47.362°, 54.196°, 54.998°, 66.371°, 72.719°, and 74.790°, corresponding to (100), (001), (110), (101), (200), (111), (201), (210), (300), (211), (310), (311), and (400) crystal planes of standard PDF card Ni$_2$P, respectively. This XRD pattern shows the formation of single-phase Ni$_2$P. It can be observed from Figure 1a–c,e,g that Ni$_3$P/Ni$_2$P/Ni, Ni$_5$P$_4$/Ni$_2$P, Ni$_5$P$_4$/Ni$_2$P/Ni, Ni$_5$P$_4$/NiP$_2$/Ni, and Ni$_5$P$_4$/Ni$_2$P heterostructures are formed at a Ni/P ratio of 3:1, 5:2, 12:5, 2:1, and 1:1, respectively. Such heterostructures will contribute to electrochemical performance.

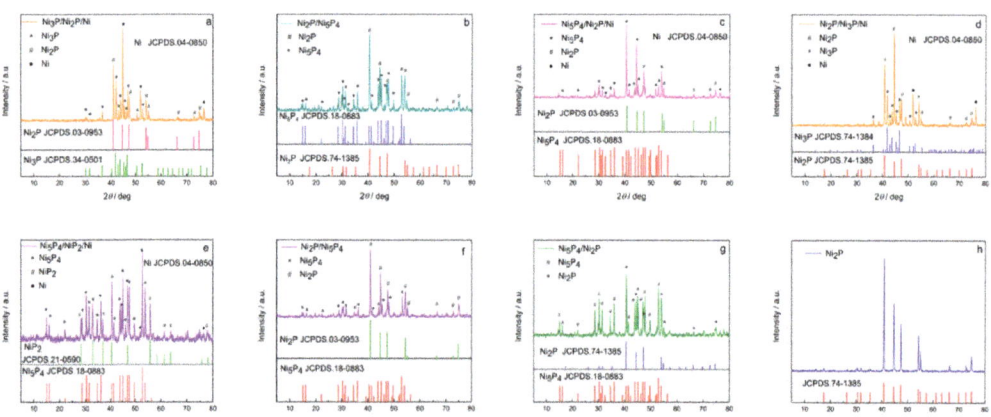

Figure 1. X-ray diffraction patterns of the samples in various Ni/P ratios. (**a–d**) XRD pattern of Ni/P ratio 3:1, 5:2, 12:5, and 7:3. (**e–h**) XRD pattern of Ni/P ratio 2:1, 5:4, 1:1 and single-phase Ni$_2$P.

The SEM images of each Ni/P ratio and single-phase Ni$_2$P are shown in Figure 2. In addition to the single-phase nickel phosphide, two distinct structures can be seen in each image, which confirms the successful preparation of the composite phase. Figure 2d is the SEM image of a Ni/P ratio of 7:3. It can be observed from this image that the linear structure is wrapped around the granular structure, increasing the contact area between the two. Figure 2f is the SEM image of a Ni/P ratio of 5:4. It can be observed from this image that the villous structure grows on the spherical particles, and there are small particles interspersed between them. Such a combination of different phases will produce lattice distortion at the interface, thereby improving the charge transfer efficiency and enhancing the catalytic activity.

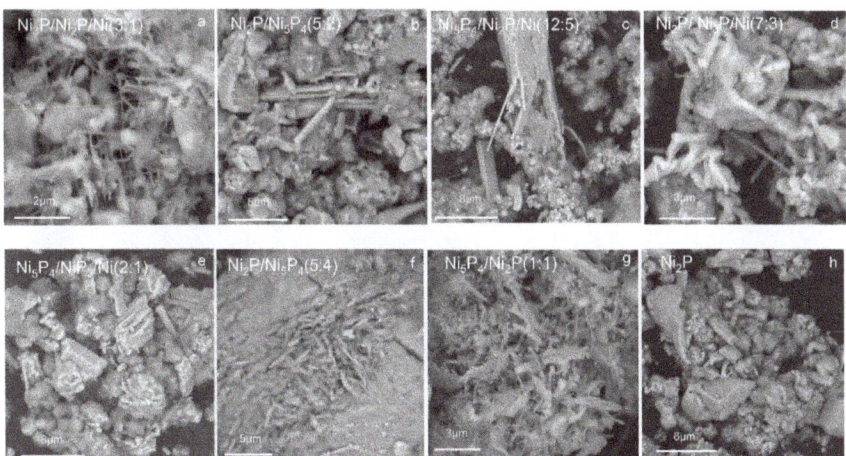

Figure 2. SEM images of various Ni/P ratios and single-phase Ni_2P. (**a**–**g**) SEM images of Ni/P ratio 3:1, 5:2, 12:5, 7:3, 2:1, 5:4, and 1:1. (**h**) SEM images of single-phase Ni_2P.

Figure 2a is the SEM image of a Ni/P ratio of 3:1. It can be observed from this image that braided linear structures and granular structures intersected with each other are generated under such condition. Figure 2b is the SEM image of a Ni/P ratio of 5:2. It can be observed from this image that under this condition, acicular and granular structures are formed, and most of the acicular structures are embedded between large and small particles. Figure 2c is the SEM image of a Ni/P ratio of 12:5. It can be observed from this image that prismatic and granular structures form under this condition, with granular structures surrounding the prism. It can also be observed that the prism is formed by a combination of small acicular structures. Figure 2e is the SEM image of a Ni/P ratio of 2:1. It can be observed from this image that the acicular and granular structures are bound together. Figure 2g is the SEM image of a Ni/P ratio of 1:1. It can be observed from this image that the bean-sprout-shaped structure is mixed with the lamellar structure. Figure 2h is the SEM image of single-phase Ni_2P. It can be observed from this image that large and small particles are not evenly distributed.

In order to explore the structure of the prepared nickel phosphide composite, we performed the TEM test on $Ni_2P/Ni_3P/Ni$ (Ni/P = 7:3) and Ni_2P/Ni_5P_4 (Ni/P = 5:4). Figure 3a,d are the TEM images of $Ni_2P/Ni_3P/Ni$ (Ni/P = 7:3) and Ni_2P/Ni_5P_4 (Ni/P = 5:4). Both images display different shapes of substances. This is the result of different phase recombinations of nickel phosphide, corresponding to the SEM image. Figure 3b is the HRTEM image of $Ni_2P/Ni_3P/Ni$ (Ni/P = 7:3). We can observe the (111) crystal plane of Ni_2P, (112) crystal planes of Ni_3P and (111) crystal planes of Ni, which prove the formation of the nickel phosphide composite phase. Figure 3e is the HRTEM image of Ni_2P/Ni_5P_4 (Ni/P = 5:4). We can observe the (111) crystal plane of Ni_2P and (103) crystal planes of Ni_5P_4, which further proves the formation of the nickel phosphide composite phase. The formation of the composite phase can effectively improve the catalytic activity of electrode materials. Different phases have different crystal lattices. The interface at the grain boundary when recombined is an inhomogeneous interface, and this causes lattice distortion due to the Jahn-Teller effect. The electronic state at the grain boundary is changed, and unpaired electrons may appear, which facilitates the transfer of charge. At the same time, the accumulated electrons can also generate a built-in electric field to promote the transmission of ions. Some randomly arranged atoms (marked by green dots) were also observed in image 3b,e; these disordered atoms rearrange in order to balance when the external environment changes. This generates more interfaces, greatly increases the efficiency of ion transport, and improves electrochemical performance [47]. We can observe the (311), (300) crystal plane of Ni_2P and (301) crystal planes of Ni_3P in Figure 3c, (102), (213) crystal plane of Ni_5P_4 and (311) crystal planes of Ni_2P in Figure 3f. These crystal planes correspond to XRD test results, which proves the formation of the nickel phosphide composite phase.

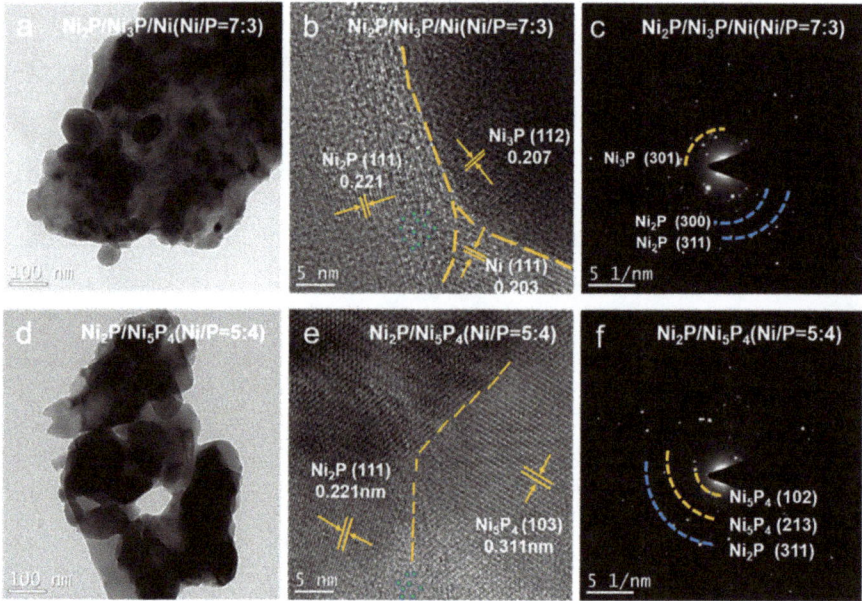

Figure 3. TEM images of the prepared sample. (a–c) The TEM, HRTEM, and SAED images of $Ni_2P/Ni_3P/Ni$ (Ni/P = 7:3). (d–f) The TEM, HRTEM, and SAED images of Ni_2P/Ni_5P_4 (Ni/P = 5:4).

Figure 4 displays the BET test and pore size distribution information. Figure 4a,c show the BET test information of $Ni_2P/Ni_3P/Ni$ (Ni/P = 7:3) and Ni_2P/Ni_5P_4 (Ni/P = 5:4), respectively. From that information we can conclude that the BET surface area of $Ni_2P/Ni_3P/Ni$ (Ni/P = 7:3) and Ni_2P/Ni_5P_4 (Ni/P = 5:4) are 1.32 $m^2\ g^{-1}$ and 0.54 $m^2\ g^{-1}$, respectively. Figure 4b,d show the pore size distribution information of $Ni_2P/Ni_3P/Ni$ (Ni/P = 7:3) and Ni_2P/Ni_5P_4 (Ni/P = 5:4), and their pore size distribution is between 2–50 nm, belonging to mesoporous materials. Such a porous structure can provide more ion channels for the catalytic reaction, thus having high-efficiency energy storage effects and good electrochemical performance.

X-ray photoelectron spectroscopy (XPS) was used to test and analyze the electronic structure of single-phase Ni_2P and Ni_2P/Ni_5P_4 (Ni/P = 5:4). Figure 5a is the XPS spectrum of single-phase Ni_2P and Ni_2P/Ni_5P_4 (Ni/P = 5:4) on the Ni 2p orbital. From the figure, we can observe that single-phase Ni_2P has peaks at the positions where the binding energy is 870.62 eV and 853.41 eV, which correspond to the Ni^{2+} $2p_{1/2}$ and Ni^{2+} $2p_{3/2}$ orbitals, respectively. Ni_2P/Ni_5P_4 (Ni/P = 5:4) has peaks at the positions where the binding energy is 870.22 eV and 853.09 eV, which correspond to the Ni^{2+} $2p_{1/2}$ and Ni^{2+} $2p_{3/2}$ orbitals, respectively. Compared with single-phase Ni_2P, the binding energy of Ni_2P/Ni_5P_4 (Ni/P = 5:4) has a negative offset of 0.4 eV and 0.32 eV in these two orbitals, respectively. This shows that the formation of the Ni_2P/Ni_5P_4 (Ni/P = 5:4) led to the occurrence of electronic remodeling, and the electrons flow from Ni_5P_4 to Ni_2P. This is due to the difference in Fermi energy levels. The electron density around the Ni_2P increases, and the electronic interaction between the Ni_2P and the Ni_5P_4 may increase the catalytic activity [48]. At the same time, it can be observed that single-phase Ni_2P has peaks at the positions where the binding energy is 875.4 eV and 857.46 eV and Ni_2P/Ni_5P_4 (Ni/P = 5:4) has peaks at the positions where the binding energy is 875 eV and 857.16 eV, corresponding to Ni^{3+} $2p_{1/2}$ and Ni^{3+} $2p_{3/2}$ orbitals, respectively. This proves that Ni has multiple valence states, so the electrons can transition and cause Debye relaxation [49], which facilitates the occurrence of redox reactions. In addition, studies have shown that the redistribution of interface electrons has a positive effect on conductivity [50,51]. The other peaks in this image are satellite peaks. Figure 5b is the XPS spectrum of single-phase Ni_2P and Ni_2P/Ni_5P_4 (Ni/P = 5:4) on the P 2p orbital. Single-phase Ni_2P and Ni_2P/Ni_5P_4 (Ni/P = 5:4) have peaks at binding energy 129.92 eV and 129.39 eV, respectively. Compared with single-phase Ni_2P, the binding energy of Ni_2P/Ni_5P_4 (Ni/P = 5:4) has a negative offset of 0.53 eV; this can once again prove the existence of charge transfer. As for the presence of P-O bonds, this is related to the surface reaction that occurs when the sample comes in contact with air during the transfer process [11,12].

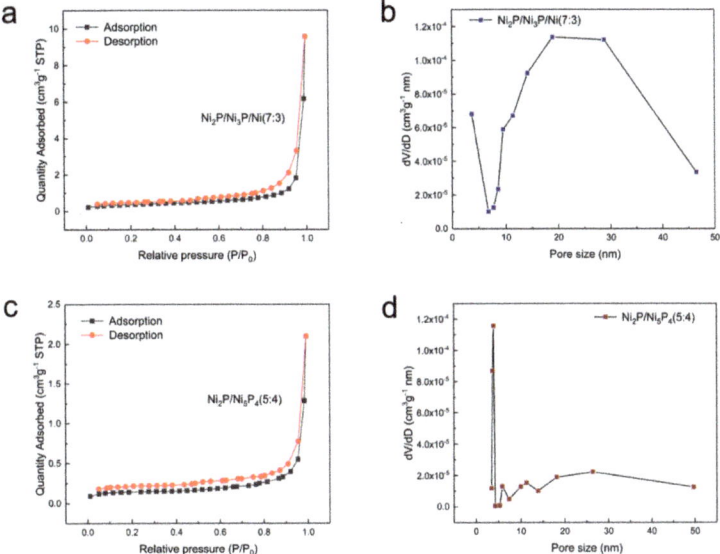

Figure 4. The BET test and pore size distribution information. (**a,c**) The BET test information of $Ni_2P/Ni_3P/Ni$ (Ni/P = 7:3) and Ni_2P/Ni_5P_4 (Ni/P = 5:4). (**b,d**) The pore size distribution information of $Ni_2P/Ni_3P/Ni$ (Ni/P = 7:3) and Ni_2P/Ni_5P_4 (Ni/P = 5:4).

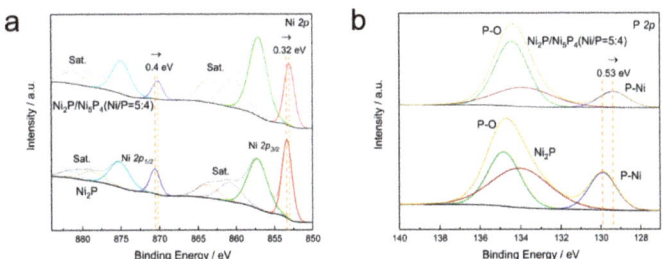

Figure 5. The XPS test and analysis of single-phase Ni_2P and Ni_2P/Ni_5P_4 (Ni/P = 5:4). (**a**) The XPS spectrum of single-phase Ni_2P and Ni_2P/Ni_5P_4 (Ni/P = 5:4) on the Ni 2p orbital. (**b**) The XPS spectra of single-phase Ni_2P and Ni_2P/Ni_5P_4 (Ni/P = 5:4) on the P 2p orbital.

3.2. Electrochemical Performance

The electrochemical test graphs of different nickel–phosphorus ratios are shown in Figure 6. Figure 6a shows the galvanostatic charge–discharge curves of electrodes prepared with different Ni/P ratios at 1 A g^{-1}. The curve of each electrode material has the appearance of a platform, which indicates that the redox reaction occurs during the charging and discharging process and corresponds to the redox peak in the cyclic voltammetry curve [30,31]. From this figure, it can be clearly seen that when the ratio of nickel to phosphorus is 7:3, it has the highest specific capacity of 321 mAh g^{-1}. Secondly, when the ratio of nickel to phosphorus is 5:4, the specific capacity is 218 mAh g^{-1}. The specific capacitance of each electrode material can be calculated according to Equation (1). The composites prepared with different raw material ratios and their electrochemical performances are shown in Table 1. The faradaic reaction corresponding to the redox peak is as follows [40]:

$$Ni^{2+} + 2OH^- \rightarrow Ni(OH)_2 \quad (5)$$

$$Ni(OH)_2 + OH^- \leftrightarrow NiOOH + H_2O + e^- \quad (6)$$

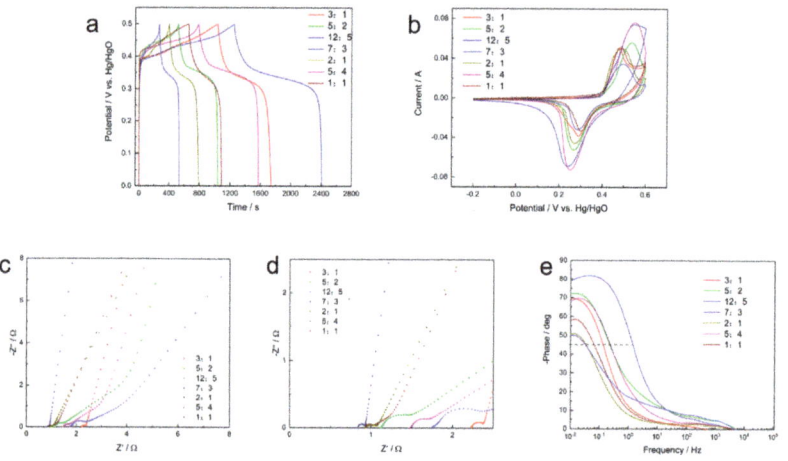

Figure 6. The electrochemical test graphs of various Ni/P ratios. (**a**) Galvanostatic charge-discharge curves of electrodes prepared with various Ni/P ratios at 1 A g^{-1}. (**b**) Cyclic voltammetry curves of electrode prepared with various Ni/P ratios at a sweep rate of 3 mV s^{-1}. (**c,d**) Nyquist plots of electrodes prepared by various Ni/P ratios. (**e**) Bode plots of electrodes prepared with various Ni/P ratios.

Table 1. The electrochemical performance of composites prepared with various raw material ratios.

Ratio of Raw (Ni:P)	Prepared Compounds	Specific Capacity (mAh g^{-1})	Rate Capability (%)
3:1	$Ni_3P/Ni_2P/Ni$	195	30
5:2	Ni_2P/Ni_5P_4	143	73
12:5	$Ni_5P_4/Ni_2P/Ni$	73	69
7:3	$Ni_2P/Ni_3P/Ni$	321	59
2:1	$Ni_5P_4/NiP_2/Ni$	108	69
5:4	Ni_2P/Ni_5P_4	218	76
1:1	Ni_5P_4/Ni_2P	119	53

Figure 6b is the cyclic voltammetry curve of various Ni/P ratios at 3 mV s^{-1}. The area is the largest when the ratio of nickel to phosphorus is 7:3, followed by a ratio of 5:4, which corresponds to its specific capacity. From the platform of the galvanostatic charge–discharge curve and the obvious redox peak of the cyclic voltammetry curve, it can be judged that the electrode material is a typical pseudocapacitive material. Figure 6c is the electrochemical impedance spectroscopy of different nickel–phosphorus ratios, and Figure 6d is an enlarged view of their high frequency region. Each curve has an intersection point in the high frequency area with the real axis, representing the equivalent series resistance (R_{es}). The resistance of this part is related to the interface resistance, internal resistance, and electrolyte ion resistance. It can be seen from Figure 6d that the high-frequency region has a shape similar to a semicircle, and the diameter of the semicircle can essentially represent the charge transfer resistance (R_{ct}). The linear slope in the low-frequency region is related to the diffusion resistance, which is related to the characteristics of the electrolyte and the material itself. With a larger slope comes a faster ion diffusion and stronger capacitance. It can be seen from the curve that when the ratio of nickel to phosphorus is 7:3, its equivalent series resistance and charge transfer resistance are the smallest, and its slope is the largest; therefore, it has good electrochemical performance. Figure 6e is the Bode graph of different ratios of nickel to phosphorus. At the phase angle of $-45°$, the nickel–phosphorus ratio of 7:3 corresponds to the largest frequency; therefore, its time constant is the smallest, followed by the time constant of the nickel–phosphorus ratio of 5:4. This proves that these two ratios need less time to reach the steady state. There are two main reasons for the relatively high electrochemical performance of the composites: electron transfer can occur between different components and the presence of heterojunctions improves the catalytic activity.

The electrochemical test graphs of $Ni_2P/Ni_3P/Ni$ (Ni/P = 7:3), Ni_2P/Ni_5P_4 (Ni/P = 5:4), and single-phase Ni_2P are shown in Figure 7. Figure 7a displays the galvanostatic charge–discharge curves of $Ni_2P/Ni_3P/Ni$ (Ni/P = 7:3), Ni_2P/Ni_5P_4 (Ni/P = 5:4), and single-phase Ni_2P at 1 A g^{-1}. According to Equation (1), their specific capacities are 321 mAh g^{-1}, 218 mAh g^{-1}, and 58 mAh g^{-1}, respectively. This shows that compared with single-phase Ni_2P, the preparation of the composite greatly improves its specific capacity. Figure 7b shows the cyclic voltammetry curves of $Ni_2P/Ni_3P/Ni$ (Ni/P = 7:3), Ni_2P/Ni_5P_4 (Ni/P = 5:4), and single-phase Ni_2P at sweep rate of 3 mV s^{-1}. The areas of $Ni_2P/Ni_3P/Ni$ (Ni/P = 7:3) and Ni_2P/Ni_5P_4 (Ni/P = 5:4) are much larger than that of the single-phase nickel phosphide, which corresponds to the specific capacity. Figure 7c shows the Nyquist plots of $Ni_2P/Ni_3P/Ni$ (Ni/P = 7:3), Ni_2P/Ni_5P_4 (Ni/P = 5:4), and single-phase Ni_2P. The equivalent series resistances (R_{es}) of $Ni_2P/Ni_3P/Ni$ (Ni/P = 7:3), Ni_2P/Ni_5P_4 (Ni/P = 5:4), and single-phase Ni_2P are 0.838 Ω, 1.473 Ω, and 0.862 Ω, respectively. The charge transfer resistances (R_{ct}) of $Ni_2P/Ni_3P/Ni$ (Ni/P = 7:3), Ni_2P/Ni_5P_4 (Ni/P = 5:4), and single-phase Ni_2P are 0.056 Ω, 0.111 Ω, and 0.178 Ω, respectively. Figure 7d is the bode plot of $Ni_2P/Ni_3P/Ni$ (Ni/P = 7:3), Ni_2P/Ni_5P_4 (Ni/P = 5:4), and single-phase Ni_2P. The time constants of $Ni_2P/Ni_3P/Ni$ (Ni/P = 7:3), Ni_2P/Ni_5P_4 (Ni/P = 5:4), and single-phase Ni_2P are 0.87 s, 4.01 s, and 43 s, respectively. Comprehensive electrochemical test results showed that the performance of nickel phosphide composite was better than that of single-phase nickel phosphide. The electrochemical performances of $Ni_2P/Ni_3P/Ni$ (Ni/P = 7:3), Ni_2P/Ni_5P_4 (Ni/P = 5:4), and single-phase nickel phosphide are listed in Table 2.

In order to further understand the electrochemical performance of nickel phosphide composites, further tests were done on materials with better performance. Figure 8 is the electrochemical test graph of $Ni_2P/Ni_3P/Ni$ (Ni/P = 7:3) and Ni_2P/Ni_5P_4 (Ni/P = 5:4). Figure 8a,b are the galvanostatic charge–discharge curves of $Ni_2P/Ni_3P/Ni$ (Ni/P = 7:3) under different current densities and cyclic voltammetry curves of $Ni_2P/Ni_3P/Ni$ (Ni/P = 7:3) at different scan rates. As the current density increases, the specific capacitance shows a downward trend, which is caused by the penetration of electrons and ions into the electrode surface [35]. The current density increased from 0.5 A g^{-1} to 5 A g^{-1} and its rate capability is 59%. The appearance of the redox peak is related to the surface Faraday reaction. As the sweep speed increases, the area of the cyclic voltammetry curve also increases, and the redox peaks move to higher and lower windows, respectively. This is related to the diffusion kinetics and internal resistance [30,31]. After 2000 cycles, the capacity retention rate was above 66% as shown in the Figure 8f. Figure 8d,e are the galvanostatic charge–discharge curves of Ni_2P/Ni_5P_4 (Ni/P = 5:4) under various current densities and the cyclic voltammetry curves of Ni_2P/Ni_5P_4 (Ni/P = 5:4) at various scan rates. The current density increased from 0.5 A g^{-1} to 5 A g^{-1}, and its rate capability was 76%. As the sweep speed increased, the area of the cyclic voltammetry curve also increased, and the oxidation peak gradually disappeared. This may be because the sweep speed was too fast, and the changes caused by the redox reaction were not recorded in time. After 7000 cycles, the capacity retention rate was above 82% as shown in the Figure 8f. For scientific analysis, we compared the electrochemical properties of our prepared electrodes with those of the electrode materials reported in the literature, as shown in Table 3. The electrochemical performance of our electrode material was relatively good.

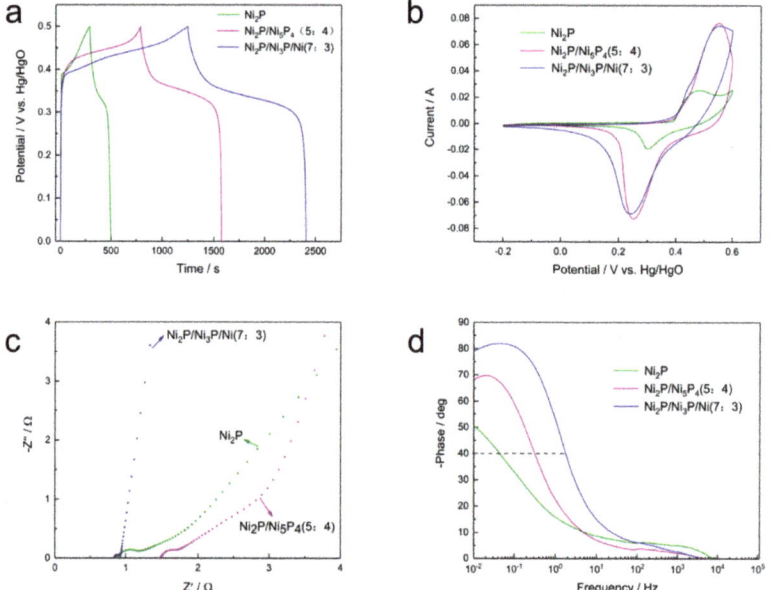

Figure 7. The electrochemical test graph of $Ni_2P/Ni_3P/Ni$ (Ni/P = 7:3), Ni_2P/Ni_5P_4 (Ni/P = 5:4), and single-phase Ni_2P. (**a**) Galvanostatic charge-discharge curves of $Ni_2P/Ni_3P/Ni$ (Ni/P = 7:3), Ni_2P/Ni_5P_4 (Ni/P = 5:4), and single-phase Ni_2P at 1 A g^{-1}. (**b**) Cyclic voltammetry curves of $Ni_2P/Ni_3P/Ni$ (Ni/P = 7:3), Ni_2P/Ni_5P_4 (Ni/P = 5:4), and single-phase Ni_2P at a sweep rate of 3 mV s^{-1}. (**c**) Nyquist plots of $Ni_2P/Ni_3P/Ni$ (Ni/P = 7:3), Ni_2P/Ni_5P_4 (Ni/P = 5:4), and single-phase Ni_2P. (**d**) Bode plots of $Ni_2P/Ni_3P/Ni$ (Ni/P = 7:3), Ni_2P/Ni_5P_4 (Ni/P = 5:4) and single-phase Ni_2P.

Table 2. The electrochemical performance of $Ni_2P/Ni_3P/Ni$ (Ni/P = 7:3), Ni_2P/Ni_5P_4 (Ni/P = 5:4), and single-phase nickel phosphide.

Electrode Materials	Specific Capacity (mAh g^{-1})	Rate Capability (%)	R_{es}	R_{ct}	Time Constant(s)
$Ni_2P/Ni_3P/Ni$ (Ni/P = 7:3)	321	59	0.838	0.056	0.87
Ni_2P/Ni_5P_4 (Ni/P = 5:4)	218	76	1.473	0.111	4.01
Ni_2P	58	33	0.862	0.178	43

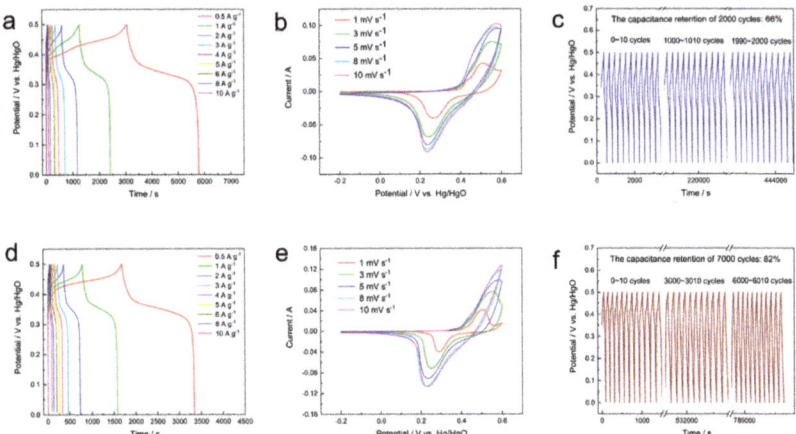

Figure 8. The electrochemical test graph of $Ni_2P/Ni_3P/Ni$ (Ni/P = 7:3) and Ni_2P/Ni_5P_4 (Ni/P = 5:4). (**a**,**b**) Galvanostatic charge-discharge curves of $Ni_2P/Ni_3P/Ni$ (Ni/P = 7:3) under various current densities and cyclic voltammetry curves of $Ni_2P/Ni_3P/Ni$ (Ni/P = 7:3) at various scan rates. (**d**,**e**) Galvanostatic charge-discharge curves Ni_2P/Ni_5P_4 (Ni/P = 5:4) under various current densities and cyclic voltammetry curves of Ni_2P/Ni_5P_4 (Ni/P = 5:4) at various scan rates. (**c**,**f**) Cyclic stability test graph of $Ni_2P/Ni_3P/Ni$ (Ni/P = 7:3) and Ni_2P/Ni_5P_4 (Ni/P = 5:4).

Table 3. Comparison of the electrochemical performance of our prepared electrodes with other reported electrode materials in the literature.

Electrode	Substrate	Electrolyte	Current Density	Specific Capacity	Refs.
$Ni_xCo_{1-x}O_y/Er$-Go	Stainless steel	1 M KOH	1 A g^{-1}	180 mAh g^{-1}	[42]
CoP/Ni_2P	NF	6 M KOH	1 A g^{-1}	1557 C g^{-1}	[30]
Ni-CoP	NF	6 M KOH	1 A g^{-1}	578 C g^{-1}	[30]
$NiCo_2O_4/Ni_2P$	NF	3 M KOH	8 mA cm^{-2}	2900 F g^{-1}	[41]
N-$Ni_2P/Ni_{12}P_5/Ni_3S_2$	NF	2 M KOH	20 mA cm^{-2}	12.71 F cm^{-2}	[44]
$Ni_2P@N$-C	NF	3 M KOH	10 A g^{-1}	1320.4 F g^{-1}	[27]
Co-Ni_2P	NF	3 M KOH	1 A g^{-1}	864 F g^{-1}	[35]
Fe-Ni_2P	NF	3 M KOH	1 A g^{-1}	856 F g^{-1}	[35]
NiCoP/CoP	NF	2 M KOH	1 A g^{-1}	152 mAh g^{-1}	[37]
$Ni_2P/Ni_3P/Ni$	NF	2 M KOH	1 A g^{-1}	321 mAh g^{-1}	This work
Ni_2P/Ni_5P_4	NF	2 M KOH	1 A g^{-1}	218 mAh g^{-1}	This work

In order to deeply evaluate the electrochemical performance of the prepared materials in practical applications, we assembled a supercapacitor with $Ni_2P/Ni_3P/Ni$ (Ni/P = 7:3) as the cathode and activated carbon as the anode. We also assembled a supercapacitor with Ni_2P/Ni_5P_4 (Ni/P = 5:4) as the cathode and activated carbon as the anode. Figure 9a shows the galvanostatic charge–discharge curves of the $Ni_2P/Ni_3P/Ni//AC$ and $Ni_2P/Ni_5P_4//AC$ supercapacitors at a current density of 1 A g^{-1}. According to Equation (1), the specific capacity of supercapacitor $Ni_2P/Ni_3P/Ni//AC$ at 1 A g^{-1} is 27.72 mAh g^{-1}, and according to Equations (2) and (3), it has an energy density of 22 W h kg^{-1} while its power density is 800 W kg^{-1}. The specific capacity of supercapacitor $Ni_2P/Ni_5P_4//AC$ at 1 A g^{-1} is 33.39 mAh g^{-1}, and it has an energy density of 27 W h kg^{-1} while its power density is 800 W kg^{-1}. Figure 9b displays the cyclic voltammetry curves of the $Ni_2P/Ni_3P/Ni//AC$ and $Ni_2P/Ni_5P_4//AC$ supercapacitors at a scan rate of 100 mV s^{-1}. The area of $Ni_2P/Ni_5P_4//AC$ is larger than that of $Ni_2P/Ni_3P/Ni//AC$, which corresponds to the specific capacity of the two supercapacitors. From the comprehensive test results, the electrochemical performance of $Ni_2P/Ni_5P_4//AC$ was better than that of Ni$_2$P/Ni$_3$P/Ni//AC; therefore, we further explored Ni$_2$P/Ni$_5$P$_4$//AC. Figure 9c is the galvanostatic charge–discharge curve of the Ni$_2$P/Ni$_5$P$_4$//AC supercapacitor at various current densities. We noticed that there are no obvious charging or discharging platforms in the galvanostatic charge–discharge curves, which indicates that the assembled device has good capacitance characteristics and a good electron transfer rate. Figure 9d shows the cyclic voltammetry curves of the Ni$_2$P/Ni$_5$P$_4$//AC supercapacitor at various scan rates. We noticed that the CV curve has no obvious redox peak, indicating that it is suitable for high-power output, and the shape of the curve does not change with the scanning speed, indicating that the assembled supercapacitor has a rate capability. The CV and GCD curves correspond to each other. With a larger specific capacitance comes a longer charge and discharge time in the GCD and a larger CV area. The relatively large specific capacitance of Ni$_2$P/Ni$_5$P$_4$//AC may be mainly due to its better stability.

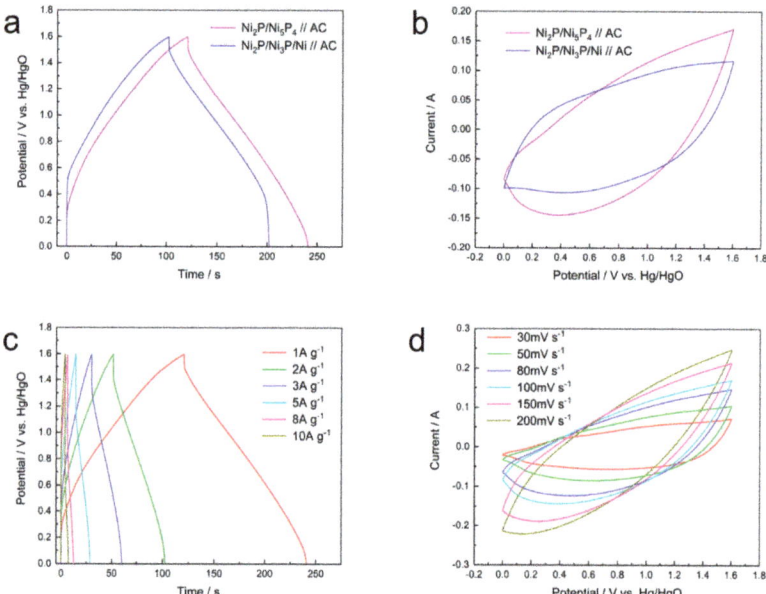

Figure 9. The electrochemical test graph of the assembled supercapacitors. (**a**) Galvanostatic charge-discharge curves of Ni$_2$P/Ni$_3$P/Ni//AC and Ni$_2$P/Ni$_5$P$_4$//AC supercapacitors at a current density of 1 A g^{-1}. (**b**) Cyclic voltammetry curves of Ni$_2$P/Ni$_3$P/Ni//AC and Ni$_2$P/Ni$_5$P$_4$//AC supercapacitors at a scan rate of 100 mV s^{-1}. (**c**,**d**) Galvanostatic charge-discharge curves of Ni$_2$P/Ni$_5$P$_4$//AC supercapacitor under various current densities and cyclic voltammetry curves of Ni$_2$P/Ni$_5$P$_4$//AC supercapacitor at various scan rates.

In order to further study the internal reasons for the difference in performance of the different electrode materials prepared, we performed the work shown in Figure 10 including the linear sweep volt–ampere curve, schematic diagram of heterogeneous interface dangling bonds (unpaired electronics) generation, and energy band diagrams of metal and semiconductors. Figure 10a is the current–voltage curve of Ni$_2$P/Ni$_3$P/Ni (Ni/P = 7:3), Ni$_2$P/Ni$_5$P$_4$ (Ni/P = 5:4), and single phase Ni$_2$P tested by linear sweep voltammetry. Clearly, the conductivity of electrode Ni$_2$P/Ni$_3$P/Ni (Ni/P = 7:3) is the largest, followed by electrode Ni$_2$P/Ni$_5$P$_4$ (Ni/P = 5:4), and the conductivity of electrode single-phase Ni$_2$P is the worst. We recognize that there is an interface between the two-phase composite. Due to the difference in the phase structure, the lattice mismatch at the interface is inevitable. At this time, a part of the unsaturated bond will appear in the semiconductor material with a smaller crystal lattice at the interface as shown in the Figure 10b. These unsaturated bonds are dangling bonds, which form unpaired electrons. The bond density is affected by different semiconductor lattice constants and the crystal plane as the interface, which is determined by the following formula:

$$\Delta N_s = N_{s1} - N_{s2} \tag{7}$$

N_{s1}, N_{s2} are the bond density of the two semiconductor materials at the interface, respectively.

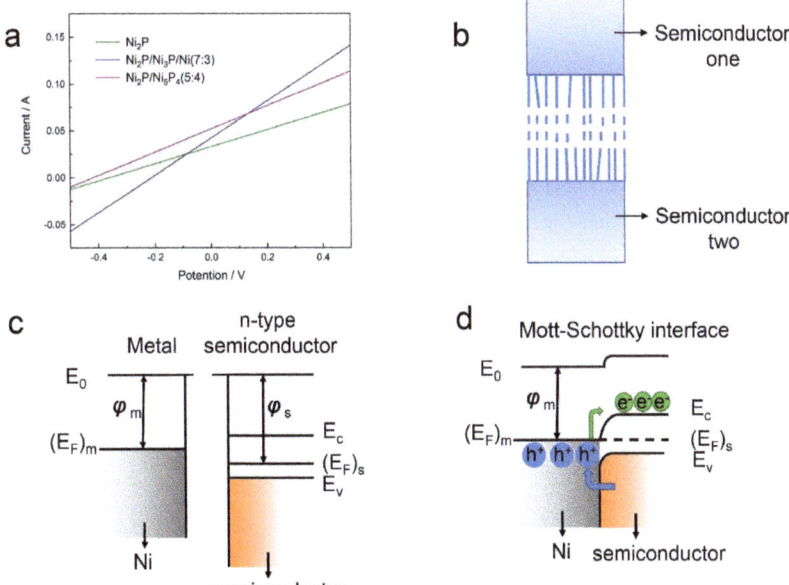

Figure 10. Linear sweep volt–ampere curve, schematic diagram of heterogeneous interface dangling bonds (unpaired electronics) generation, energy band diagrams of metal and semiconductors. (**a**) The Linear sweep volt–ampere curve of $Ni_2P/Ni_3P/Ni$ (Ni/P = 7:3), Ni_2P/Ni_5P_4 (Ni/P = 5:4), and single phase Ni_2P. (**b**) The schematic diagram of dangling bonds of heterogeneous interface. (**c**,**d**) The energy band diagrams of metal and semiconductors.

These dangling bonds, i.e., the unpaired electrons, are in a relatively free state and are easily excited to become free electrons, which provides excellent conditions for high-efficiency ion transmission and more active sites, thereby improving the electrochemical performance of the electrode. The prepared working electrodes with heterostructures all have such an interface as Ni_2P/Ni_5P_4 (Ni/P = 5:4). As shown in Figure 10c, the work function of the metal is greater than the work function of the semiconductor. The Mott-Schottky interface shown in Figure 10d is formed after the metal is in contact with the semiconductor. Due to the difference in work function, electrons flow from the metal to the semiconductor, carriers flow from the semiconductor to the metal, and the accumulation of electrons in the semiconductor forms an internal potential field, which provides a high-speed channel for the continuous transmission of free electrons [52]. This improves the utilization of active sites, promotes the occurrence of redox reactions, and makes the working electrode have better capacitance characteristics. The prepared working electrodes with metallic nickel all have such an effect as $Ni_2P/Ni_3P/Ni$ (Ni/P = 7:3).

4. Conclusions

In summary, the nickel phosphide composites were prepared in one step using the temperature-programmed phosphating method, and the prepared nickel phosphide composite showed good electrochemical performance. The prepared compound $Ni_2P/Ni_3P/Ni$ (Ni/P = 7:3) had a specific capacity of 321 mAh g^{-1} under 1 A g^{-1}, and the prepared compound Ni_2P/Ni_5P_4 (Ni/P = 5:4) had a specific capacity of 218 mAh g^{-1} under 1 A g^{-1} and a 76% rate performance. After 7000 cycles, the capacity retention rate was above 82%. After assembling the prepared composite and activated carbon into a supercapacitor, $Ni_2P/Ni_3P/Ni//AC$ was found to have an energy density of 22 W h kg^{-1} and a power density of 800 W kg^{-1}, while $Ni_2P/Ni_5P_4//AC$ had an energy density of 27 W h kg^{-1} and a power density of 800 W kg^{-1}. These results provide ideas for further research.

Author Contributions: Conceptualization, Methodology, Software, Investigation, Writing-original draft. S.-B.G.; Supervision, Resources, Writing—review & editing. W.-B.Z. and J.L.; Visualization, Formal analysis, Validation. X.B., L.Z. and Y.-W.G.; Software, Validation. Z.-Q.Y. and X.-W.H. All authors have read and agreed to the published version of the manuscript.

Funding: This research received no external funding.

Institutional Review Board Statement: Not applicable.

Informed Consent Statement: Not applicable.

Data Availability Statement: Not applicable.

Acknowledgments: This work was financially supported by the Sichuan Science and Technology Project (No.2020YJ0163) and the Research Foundation for Teacher Development of Chengdu University of Technology (No.10912-2019KYQD-06847).

Conflicts of Interest: The authors declare no conflict of interest.

Highlights:

- A relatively simple method was used to prepare energy storage electrode materials with good electrochemical properties.
- The Ni_2P/Ni_5P_4 electrode possessed satisfactory specific capacitance and excellent electrochemical kinetics.
- The assembled $Ni_2P/Ni_5P_4//AC$ based on nickel phosphide heterostructure showed remarkable performance.

References

1. Winter, M.; Brodd, R. What Are Batteries, Fuel Cells, and Supercapacitors? *Chem. Rev.* **2004**, *104*, 4245–4269. [CrossRef] [PubMed]
2. Zhang, L.; Zhao, X.S. Carbon-based materials as supercapacitor electrodes. *Chem. Soc. Rev.* **2009**, *38*, 2520–2531. [CrossRef] [PubMed]
3. Li, S.; Yu, C.; Yang, J.; Zhao, C.; Zhang, M.; Huang, H.; Liu, Z.; Guo, W.; Qiu, J. A superhydrophilic "nanoglue" for stabilizing metal hydroxides onto carbon materials for high-energy and ultralong-life asymmetric supercapacitors. *Energy Environ. Sci.* **2017**, *10*, 1958–1965. [CrossRef]
4. Bao, X.; Zhang, W.B.; Zhang, Q.; Zhang, L.; Ma, X.J.; Long, J. Interlayer material technology of manganese phosphate toward and beyond electrochemical pseudocapacitance over energy storage application. *J. Mater. Sci. Technol.* **2021**, *71*, 109–128. [CrossRef]
5. Han, X.-W.; Zhang, W.-B.; Ma, X.-J.; Zhou, X.; Zhang, Q.; Bao, X.; Guo, Y.-W.; Zhang, L.; Long, J. Review—Technologies and Materials for Water Salinity Gradient Energy Harvesting. *J. Electrochem. Soc.* **2021**, *168*, 90505. [CrossRef]
6. Yu, X.; Ren, X.; Zhang, Y.; Yuan, Z.; Sui, Z.; Wang, M. Self-supporting hierarchically micro/nano-porous Ni_3P-Co_2P-based film with high hydrophilicity for efficient hydrogen production. *J. Mater. Sci. Technol.* **2021**, *65*, 118–125. [CrossRef]
7. Simon, P.; Gogotsi, Y. Materials for electrochemical capacitors. *Nat. Mater.* **2008**, *7*, 845–854. [CrossRef]
8. Shao, Y.; El-Kady, M.F.; Sun, J.; Li, Y.; Zhang, Q.; Zhu, M.; Wang, H.; Dunn, B.; Kaner, R.B. Design and Mechanisms of Asymmetric Supercapacitors. *Chem. Rev.* **2018**, *118*, 9233–9280. [CrossRef]
9. Dou, Q.; Park, H.S. Perspective on High-Energy Carbon-Based Supercapacitors. *Energy Environ. Mater.* **2020**, *3*, 286–305. [CrossRef]
10. Zhang, L.; Zhang, W.B.; Chai, S.S.; Han, X.W.; Zhang, Q.; Bao, X.; Guo, Y.W.; Zhang, X.L.; Zhou, X.; Guo, S.B.; et al. Review—Clay Mineral Materials for Electrochemical Capacitance Application. *J. Electrochem. Soc.* **2021**, *168*, 70558. [CrossRef]
11. Ran, Z.; Shu, C.; Hou, Z.; Cao, L.; Liang, R.; Li, J.; Hei, P.; Yang, T.; Long, J. $Ni_3Se_2/NiSe_2$ heterostructure nanoforests as an efficient bifunctional electrocatalyst for high-capacity and long-life Li–O_2 batteries. *J. Power Sources* **2020**, *468*, 228308. [CrossRef]
12. Ran, Z.; Shu, C.; Hou, Z.; Zhang, W.; Yan, Y.; He, M.; Long, J. Modulating electronic structure of honeycomb-like $Ni_2P/Ni_{12}P_5$ heterostructure with phosphorus vacancies for highly efficient lithium-oxygen batteries. *Chem. Eng. J.* **2021**, *413*, 127404. [CrossRef]

13. Abdollahifar, M.; Liu, H.W.; Lin, C.H.; Weng, Y.T.; Sheu, H.S.; Lee, J.F.; Lu, M.L.; Liao, Y.F.; Wu, N.L. Enabling Extraordinary Rate Performance for Poorly Conductive Oxide Pseudocapacitors. *Energy Environ. Mater.* **2020**, *3*, 405–413. [CrossRef]
14. Brezesinski, T.; Wang, J.; Tolbert, S.H.; Dunn, B. Ordered mesoporous alpha-MoO$_3$ with iso-oriented nanocrystalline walls for thin-film pseudocapacitors. *Nat. Mater.* **2010**, *9*, 146–151. [CrossRef] [PubMed]
15. He, X.; Li, Z.; Hu, Y.; Li, F.; Huang, P.; Wang, Z.; Jiang, J.; Wang, C. Three-dimensional coral-like Ni$_2$P-ACC nanostructure as binder-free electrode for greatly improved supercapacitor. *Electrochim. Acta* **2020**, *349*, 136259. [CrossRef]
16. Wang, D.; Kong, L.B.; Liu, M.C.; Zhang, W.B.; Luo, Y.C.; Kang, L. Amorphous Ni–P materials for high performance pseudocapacitors. *J. Power Sources* **2015**, *274*, 1107–1113. [CrossRef]
17. Gou, J. Ni2P/NiS2 Composite with Phase Boundaries as High-Performance Electrode Material for Supercapacitor. *J. Electrochem. Soc.* **2017**, *164*, A2956–A2961. [CrossRef]
18. Zhang, L.; Zhang, W.B.; Zhang, Q.; Bao, X.; Ma, X.J. Electrochemical capacitance of intermetallic vanadium carbide. *Intermetallics* **2020**, *127*, 106976. [CrossRef]
19. Zong, Q.; Yang, H.; Wang, Q.; Zhang, Q.; Xu, J.; Zhu, Y.; Wang, H.; Wang, H.; Zhang, F.; Shen, Q. NiCo$_2$O$_4$/NiCoP nanoflake-nanowire arrays: A homogeneous hetero-structure for high performance asymmetric hybrid supercapacitors. *Dalton Trans.* **2018**, *47*, 16320–16328. [CrossRef]
20. Zhang, Q.; Zhang, W.B.; Hei, P.; Hou, Z.; Yang, T.; Long, J. CoP nanoprism arrays: Pseudocapacitive behavior on the electrode-electrolyte interface and electrochemical application as an anode material for supercapacitors. *Appl. Surf. Sci.* **2020**, *527*, 146682. [CrossRef]
21. Elshahawy, A.M.; Guan, C.; Li, X.; Zhang, H.; Hu, Y.; Wu, H.; Pennycook, S.J.; Wang, J. Sulfur-doped cobalt phosphide nanotube arrays for highly stable hybrid supercapacitor. *Nano Energy* **2017**, *39*, 162–171. [CrossRef]
22. Shang, M.; Zhang, J.; Liu, X.; Liu, Y.; Guo, S.; Yu, S.; Filatov, S.; Yi, X. N, S self-doped hollow-sphere porous carbon derived from puffball spores for high performance supercapacitors. *Appl. Surf. Sci.* **2021**, *542*, 148697. [CrossRef]
23. Shao, Y.; Zhao, Y.; Li, H.; Xu, C. Three-Dimensional Hierarchical NixCo1-xO/NiyCo2-yP@C Hybrids on Nickel Foam for Excellent Supercapacitors. *ACS Appl. Mater. Interfaces* **2016**, *8*, 35368–35376. [CrossRef] [PubMed]
24. Fleischmann, S.; Mitchell, J.B.; Wang, R.; Zhan, C.; Jiang, D.E.; Presser, V.; Augustyn, V. Pseudocapacitance: From Fundamental Understanding to High Power Energy Storage Materials. *Chem Rev.* **2020**, *120*, 6738–6782. [CrossRef] [PubMed]
25. Wang, S.; Huang, Z.; Li, R.; Zheng, X.; Lu, F.; He, T. Template-assisted synthesis of NiP@CoAl-LDH nanotube arrays with superior electrochemical performance for supercapacitors. *Electrochim. Acta* **2016**, *204*, 160–168. [CrossRef]
26. Ruan, Y.; Wang, C.; Jiang, J. Nanostructured Ni compounds as electrode materials towards high-performance electrochemical capacitors. *J. Mater. Chem. A* **2016**, *4*, 14509–14538. [CrossRef]
27. Liu, Y.; Zhang, X.; Matras-Postolek, K.; Yang, P. Ni2P nanosheets modified N-doped hollow carbon spheres towards enhanced supercapacitor performance. *J. Alloys Compd.* **2021**, *854*, 157111. [CrossRef]
28. Li, X.; Elshahawy, A.M.; Guan, C.; Wang, J. Metal Phosphides and Phosphates-based Electrodes for Electrochemical Supercapacitors. *Small* **2017**, *13*, 1701530. [CrossRef]
29. Aziz, S.T.; Kumar, S.; Riyajuddin, S.; Ghosh, K.; Nessim, G.D.; Dubal, D.P. Bimetallic Phosphides for Hybrid Supercapacitors. *J. Phys. Chem. Lett.* **2021**, *12*, 5138–5149. [CrossRef]
30. Zhang, W.B.; Zhang, Q.; Bao, X.; Ma, X.J.; Han, X.W.; Zhang, L.; Guo, Y.; Long, J.; Nian, C. Enhancing pseudocapacitive performance of CoP coating on nickel foam via surface Ni$_2$P modification and Ni (II) doping for supercapacitor energy storage application. *Surf. Coat. Technol.* **2021**, *421*, 127469. [CrossRef]
31. Zhang, Q.; Zhang, W.B.; Ma, X.J.; Zhang, L.; Bao, X.; Guo, Y.W.; Long, J. Boosting pseudocapacitive energy storage performance via both phosphorus vacancy defect and charge injection technique over the CoP electrode. *J. Alloys Compd.* **2021**, *864*, 158106. [CrossRef]
32. Zhao, H.; Yuan, Z.Y. Transition metal–phosphorus-based materials for electrocatalytic energy conversion reactions. *Catal. Sci. Technol.* **2017**, *7*, 330–347. [CrossRef]
33. Liu, M.C.; Hu, Y.M.; An, W.Y.; Hu, Y.X.; Niu, L.Y.; Kong, L.B.; Kang, L. Construction of high electrical conductive nickel phosphide alloys with controllable crystalline phase for advanced energy storage. *Electrochim. Acta* **2017**, *232*, 387–395. [CrossRef]
34. Zhou, K.; Zhou, W.; Yang, L.; Lu, J.; Cheng, S.; Mai, W.; Tang, Z.; Li, L.; Chen, S. Ultrahigh-Performance Pseudocapacitor Electrodes Based on Transition Metal Phosphide Nanosheets Array via Phosphorization: A General and Effective Approach. *Adv. Funct. Mater.* **2015**, *25*, 7530–7538. [CrossRef]
35. Ayom, G.E.; Khan, M.D.; Choi, J.; Gupta, R.K.; van Zyl, W.E.; Revaprasadu, N. Synergistically enhanced performance of transition-metal doped Ni$_2$P for supercapacitance and overall water splitting. *Dalton Trans* **2021**, *50*, 11821–11833. [CrossRef]
36. An, C.; Wang, Y.; Wang, Y.; Liu, G.; Li, L.; Qiu, F.; Xu, Y.; Jiao, L.; Yuan, H. Facile synthesis and superior supercapacitor performances of Ni$_2$P/rGO nanoparticles. *RSC Adv.* **2013**, *3*, 4628–4633. [CrossRef]
37. Dang, T.; Wei, D.; Zhang, G.; Wang, L.; Li, Q.; Liu, H.; Cao, Z.; Zhang, G.; Duan, H. Homologous NiCoP/CoP hetero-nanosheets supported on N-doped carbon nanotubes for high-rate hybrid supercapacitors. *Electrochim. Acta* **2020**, *341*, 135988. [CrossRef]
38. Du, W.; Wei, S.; Zhou, K.; Guo, J.; Pang, H.; Qian, X. One-step synthesis and graphene-modification to achieve nickel phosphide nanoparticles with electrochemical properties suitable for supercapacitors. *Mater. Res. Bull.* **2015**, *61*, 333–339. [CrossRef]
39. Lu, Y.; Liu, J.K.; Liu, X.Y.; Huang, S.; Wang, T.Q.; Wang, X.L.; Gu, C.D.; Tu, J.P.; Mao, S.X. Facile synthesis of Ni-coated Ni$_2$P for supercapacitor applications. *Cryst. Eng. Comm.* **2013**, *15*, 7071–7079. [CrossRef]

40. Hu, Y.M.; Liu, M.C.; Hu, Y.X.; Yang, Q.Q.; Kong, L.B.; Han, W.; Li, J.J.; Kang, L. Design and synthesis of $Ni_2P/Co_3V_2O_8$ nanocomposite with enhanced electrochemical capacitive properties. *Electrochim. Acta* **2016**, *190*, 1041–1049. [CrossRef]
41. Jia, H.; Li, Q.; Li, C.; Song, Y.; Zheng, H.; Zhao, J.; Zhang, W.; Liu, X.; Liu, Z.; Liu, Y. A novel three-dimensional hierarchical $NiCo_2O_4/Ni_2P$ electrode for high energy asymmetric supercapacitor. *Chem. Eng. J.* **2018**, *354*, 254–260. [CrossRef]
42. Adán-Más, A.; Silva, T.M.; Guerlou-Demourgues, L.; Bourgeois, L.; Labrugere-Sarroste, C.; Montemor, M.F. Nickel-cobalt oxide modified with reduced graphene oxide: Performance and degradation for energy storage applications. *J. Power Sources* **2019**, *419*, 12–26. [CrossRef]
43. Askari, M.B.; Salarizadeh, P.; Seifi, M.; Ramezan zadeh, M.H.; Di Bartolomeo, A. $ZnFe_2O_4$ nanorods on reduced graphene oxide as advanced supercapacitor electrodes. *J. Alloys Compd.* **2021**, *860*, 158497. [CrossRef]
44. Qin, Y.; Lyu, Y.; Chen, M.; Lu, Y.; Qi, P.; Wu, H.; Sheng, Z.; Gan, X.; Chen, Z.; Tang, Y. Nitrogen-doped $Ni_2P/Ni_{12}P_5/Ni_3S_2$ three-phase heterostructure arrays with ultrahigh areal capacitance for high-performance asymmetric supercapacitor. *Electrochim. Acta* **2021**, *393*, 139059. [CrossRef]
45. Salarizadeh, P.; Askari, M.B. MoS_2–ReS_2/rGO: A novel ternary hybrid nanostructure as a pseudocapacitive energy storage material. *J. Alloys Compd.* **2021**, *874*, 159886. [CrossRef]
46. Oyedotun, K.O.; Mirghni, A.A.; Fasakin, O.; Tarimo, D.J.; Kitenge, V.N.; Manyala, N. High-Energy Asymmetric Supercapacitor Based on the Nickel Cobalt Oxide ($NiCo_2O_4$) Nanostructure Material and Activated Carbon Derived from Cocoa Pods. *Energy Fuels* **2021**, *35*, 20309–20319. [CrossRef]
47. Liang, R.; Shu, C.; Hu, A.; Li, M.; Ran, Z.; Zheng, R.; Long, J. Interface engineering induced selenide lattice distortion boosting catalytic activity of heterogeneous $CoSe_2$@$NiSe_2$ for lithium-oxygen battery. *Chem. Eng. J.* **2020**, *393*, 124592. [CrossRef]
48. Hu, A.; Lv, W.; Lei, T.; Chen, W.; Hu, Y.; Shu, C.; Wang, X.; Xue, L.; Huang, J.; Du, X.; et al. Heterostructured $NiS_2/ZnIn_2S_4$ Realizing Toroid-like Li_2O_2 Deposition in Lithium–Oxygen Batteries with Low-Donor-Number Solvents. *ACS Nano* **2020**, *14*, 3490–3499. [CrossRef]
49. Liang, L.L.; Song, G.; Liu, Z.; Chen, J.P.; Xie, L.J.; Jia, H.; Kong, Q.Q.; Sun, G.H.; Chen, C.M. Constructing $Ni_{12}P_5/Ni_2P$ Heterostructures to Boost Interfacial Polarization for Enhanced Microwave Absorption Performance. *ACS Appl. Mater. Interfaces* **2020**, *12*, 52208–52220. [CrossRef]
50. An, L.; Li, Y.; Luo, M.; Yin, J.; Zhao, Y.Q.; Xu, C.; Cheng, F.; Yang, Y.; Xi, P.; Guo, S. Atomic-Level Coupled Interfaces and Lattice Distortion on CuS/NiS_2 Nanocrystals Boost Oxygen Catalysis for Flexible Zn-Air Batteries. *Adv. Funct. Mater.* **2017**, *27*, 1703779. [CrossRef]
51. Liu, Y.; Hua, X.; Xiao, C.; Zhou, T.; Huang, P.; Guo, Z.; Pan, B.; Xie, Y. Heterogeneous Spin States in Ultrathin Nanosheets Induce Subtle Lattice Distortion to Trigger Efficient Hydrogen Evolution. *J. Am. Chem. Soc.* **2016**, *138*, 5087–5092. [CrossRef] [PubMed]
52. Sun, Z.; Wang, Y.; Zhang, L.; Wu, H.; Jin, Y.; Li, Y.; Shi, Y.; Zhu, T.; Mao, H.; Liu, J.; et al. Simultaneously Realizing Rapid Electron Transfer and Mass Transport in Jellyfish-Like Mott–Schottky Nanoreactors for Oxygen Reduction Reaction. *Adv. Funct. Mater.* **2020**, *30*, 1910482. [CrossRef]

Article

Zeolitic Imidazolate Framework 67-Derived Ce-Doped CoP@N-Doped Carbon Hollow Polyhedron as High-Performance Anodes for Lithium-Ion Batteries

Yanjun Zhai [1,*], Shuli Zhou [1], Linlin Guo [1], Xiaole Xin [1], Suyuan Zeng [1], Konggang Qu [1], Nana Wang [2] and Xianxi Zhang [1,*]

[1] School of Chemistry and Chemical Engineering, Shandong Provincial Key Laboratory/Collaborative Innovation Center of Chemical Energy Storage and Novel Cell Technology, Liaocheng University, Liaocheng 252059, China; zhouli1996@126.com (S.Z.); 15506436037@163.com (L.G.); xxl15865517198@163.com (X.X.); drzengsy@163.com (S.Z.); qukonggang@lcu.edu.cn (K.Q.)

[2] Institute for Superconducting and Electronic Materials, University of Wollongong Innovation Campus, Wollongong, NSW 2500, Australia; nanaw@uow.edu.au

* Correspondence: zhaiyanjun@lcu.edu.cn (Y.Z.); xxzhang3@126.com (X.Z.)

Abstract: Zeolitic Imidazolate Framework 67 (ZIF-67) and its derivates have attracted extensive interest for lithium-ion batteries (LIBs). Here, Cerium-doped cobalt phosphide@nitrogen-doped carbon (Ce-doped CoP@NC) with hollow polyhedron structure materials were successfully synthesized via ionic-exchange with Co and Ce ions using the ZIF-67 as a template followed with a facile low-temperature phosphorization treatment. Benefitting from the well-designed hollow polyhedron, steady carbon network, and Ce-doping structural merits, the as-synthesized Ce-doped CoP@NC electrode demonstrated superior performance as the anode in LIBs: a superior cyclability (400 mA h g^{-1} after 500 cycles) and outstanding rate-capability (590 mA h g^{-1}, reverted to 100 mA g^{-1}). These features not only produced more lithium-active sites for LIBs anode and a shorter Li-ion diffusion pathway to expedite the charge transfer, but also the better tolerance against volume variation of CoP during the repeated lithiation/delithiation process and greater electronic conductivity properties. These results provide a methodology for the design of well-organized ZIFs and rare earth element-doped transition metal phosphate with a hollow polyhedron structure.

Keywords: hollow polyhedron; Ce-doped CoP; N-doped carbon; lithium-ion batteries; anode

Citation: Zhai, Y.; Zhou, S.; Guo, L.; Xin, X.; Zeng, S.; Qu, K.; Wang, N.; Zhang, X. Zeolitic Imidazolate Framework 67-Derived Ce-Doped CoP@N-Doped Carbon Hollow Polyhedron as High-Performance Anodes for Lithium-Ion Batteries. *Crystals* **2022**, *12*, 533. https://doi.org/10.3390/cryst12040533

Academic Editor: Konrad Świerczek

Received: 10 March 2022
Accepted: 6 April 2022
Published: 11 April 2022

Publisher's Note: MDPI stays neutral with regard to jurisdictional claims in published maps and institutional affiliations.

Copyright: © 2022 by the authors. Licensee MDPI, Basel, Switzerland. This article is an open access article distributed under the terms and conditions of the Creative Commons Attribution (CC BY) license (https:// creativecommons.org/licenses/by/ 4.0/).

1. Introduction

The growing demands on energy storage technology for commercial energy storage markets drive the exploration of LIBs with outstanding rate-capability as well as longer cycling stability [1–5]. At present, extensive efforts have been devoted to developing new-type high-capacity anode materials in LIBs. Transition metal phosphides (TMPs) have been widely investigated as an alternative anode for lithium storage due to their high theoretical capacities, low costs, and the lithiation product of Li$_3$P from TMPs with higher conductivity (Li$_3$P: ~1 × 10^{-4} S cm^{-1}) (Li$_2$O from similar metal oxides: ~5 × 10^{-8} S cm^{-1}; Li$_2$S from similar metal sulfides: ~1 × 10^{-13}) [6–10]. Among the various TMPs for LIBs, cobalt phosphide (CoP) with a high theoretical capacity (~894 mA h g^{-1}) and relatively low redox potential (~0.6 V) attracts great interests [11–14]. However, the dramatic volume variation of CoP during cycling results in the destruction of the electrode structure and rapid capacity fading. Simultaneously, its intrinsically inferior electrical conductivity usually causes poor rate capability. To address these issues, various improvement measures have been reported to tune the structure at nanoscale-to-microscale and improve the conductivity of CoP [15–20].

ZIF-67 materials have been widely used as precursors to synthesize diverse functional materials via pyrolysis reactions at different temperatures or a series of chemical reactions with relevant chemical reagents. ZIF-67-derived nitrogen-doped carbon-based carbides, oxides, phosphides, and chalcogenides have the characteristics of high chemical/mechanical stabilities, controllable structures/compositions, and large surface area, utilizing the modification of physical/chemical properties and the improvement of energy storage performance. The construction of hollow-structured carbon frame materials is beneficial for the rapid diffusion of electrolytes, which effectively alleviate the volume changes during the electrochemical reaction process. Simultaneously, it can bring more active storage sites in the composites. [10,21–24]. For instance, Liu's group employed ZIF-67 as a template for the first time to design CoP nanoparticles embedded in nitrogen-doped carbon (NC). It delivered a high reversible discharge special capacity of 522.6 mA h g^{-1} at 200 mA g^{-1} after 750 cycles and outstanding cycling stability up to 2000 cycles at 500 mA g^{-1} [10]. Li and his co-workers investigated the electrochemical properties of cage-structured CoP@N, P-doped double carbon from the confined phosphorization of ZIFs@CNCs, demonstrating a high initial columbic efficiency (ICE) of 96% and a superior cycle performance (1215.2 mA h g^{-1} after 1000 cycles at 200 mA g^{-1}) [19]. Although great achievements have been made in the synthesis of hollow-structured carbon-based CoP composite materials, more intensive studies are required to identify the relationship between the complex structures, controllable compositions, and the improvement of lithium storage.

As recently as the last century, it has been generally accepted that doping an appropriate amount of rare earth elements on the micro-nanostructured materials is a facile strategy to tune their morphology and electronic structures [18,25–27]. The addition of some dopant in the CoP lattices would lead to some defects, which are suitable for higher ionic conductance than that of undoped CoP materials. Rare earth elements, with their large radius, high charge, and 4f electron orbit, are expected to bring further improvements in LIBs. For example, Cerium (Ce) or Rubidium (Ru) doped into CoP used as hydrogen evolution reaction (HER) materials could promise its Pt-like HER catalytic performance by remarkably reducing the hydrogen binding energy and modulating their electronic structures and electronic conductivity [28,29]. Ce, benefiting from the strengths of the most abundant rare earth element and unique properties of a half-full 4f electron orbit, has been widely used to improve the mechanical stability and electronic conductivity in electrocatalysis and LIBs [25,30–33]. The hierarchy on hollow, polyhedron structural, doping of rare earth elements and the complementary effects between different ions determine the functionalities and performances of the materials. Therefore, the control of the hierarchy at each level is crucial [34–37].

In this work, the hollow polyhedron structured Ce-doped CoP@NC composites were rationally designed via chemical etching of ZIF-67 polyhedron with Ce(NO$_3$)$_3$ following a phosphorization procedure via gas-solid reactions. The as-prepared Ce-doped CoP@NC hollow polyhedron materials exhibited good cycle stability, which is comparable to that of CoP@NC.

2. Materials and Methods

2.1. The synthesis of Hollow Polyhedron Structured Ce-CoP@NC Composite

ZIF-67 polyhedrons were synthesized using a previously reported [10]. In the typical procedure, 5 mmol of Co(NO$_3$)$_2$·6H$_2$O and 20 mmol of 2-methylimidazole (C$_4$H$_6$N$_2$) were dissolved in 30 mL of methanol, respectively. Then, the clear solution of C$_4$H$_6$N$_2$ was quickly poured into the clear solution of Co(NO$_3$)$_2$ while stirring vigorously for 1 h. Then, it was aged in atmosphere for 24 h. In the end, the product was collected by centrifugation, washed with methanol three times, and dried at 60 °C for 4 h.

First, 0.2 g ZIF-67 powder was dispersed in 25 mL methanol under ultrasonication to obtain a homogeneous dispersion. Then 0.2 g (0.5 mmol) Ce(NO$_3$)$_3$·6H$_2$O was added to the above dispersion. After stirring at room temperature for 12 h, the Ce-doped ZIF-67 precursor was formed and dried. The obtained product was put into a tube furnace and calcined at 600 °C for 2 h in a nitrogen atmosphere with a heating rate of 2 °C min^{-1}. Then, the Ce-Co/NC product was obtained.

Ce-CoP@NC was synthesized via a phosphidation process with NaH_2PO_2 as the phosphorus source. The Ce-Co@NC product and NaH_2PO_2 (a mass ratio of 1: 5) were put into two separate quartz boats. The quartz boat for NaH_2PO_2 was placed in the position of direction with gas entry in a tube furnace. Afterward, the material was calcined in N_2 atmosphere at 350 °C for 2 h with 2 °C min^{-1}. Finally, Ce-CoP@NC was obtained.

2.2. Materials Characterization

The phase structures of the sample were determined using a Rigaku SmartLab (9) with Cu Ka radiation. The morphology and microstructures were characterized by scanning electron microscopy (SEM, JEOL-JSM-6700F) and transmission electron microscopy (TEM, FEI-Talos F200X) equipped with an aberration-corrector for the imaging lenses for high-resolution TEM (HRTEM), a high-angle annular dark-field (HAADF) detector for STEM, and an energy dispersive X-ray (EDX) detector for composition mapping. X-ray photoelectron spectroscopy (XPS) was performed by the Thermo Fisher Scientific ESCALAB Xi+ to determine the composition of samples. Thermogravimetric analysis (TGA) was conducted on a Mettler Toledo TGA/SDTA 851 thermal analyzer with a heating rate of 10 °C min^{-1} under N2 atmosphere. The specific surface areas of the samples were evaluated by using the Brunauere Emmette Teller test (BET, micromeritics-ASAP2460).

2.3. Electrochemical Measurements

The electrochemical tests were measured on a Land battery test system (CT2001A, Wuhan, China) at room temperature. The working electrodes were fabricated by pasting the mixed slurry that consisted of active materials (70 wt%), acetylene black (20 wt%), and sodium carboxy methyl cellulose (CMC, 10 wt%) onto a copper foil. The electrodes were dried at 80 °C for 6 h in a vacuum oven. The final loading of the active materials for each electrode was about 0.8~1.0 mg cm^{-2}. The separator was a Celgard 2500 microporous polypropylene membrane. The residual space of the battery was filled by the nickel foam (7 mm in radius, ~2 mm in thickness). The electrolyte used in the cells was 1 M $LiPF_6$ in an ethylene carbonate/diethyl carbonate containing fluoroethylene carbonate (EC/DEC = 1:1 v/v + 5% FEC). Lithium foil was used as a counter electrode. The coin cells of type CR2032 assembly were carried out in an argon-filled glove box with both the moisture and the oxygen content below 2 ppm. Galvanostatic discharge/charge cycled in the voltage range of 0.01~3.0 V (vs. Li^+/Li). The cyclic voltammetry (CV) profiles and electrochemical impedance spectroscopy (EIS) were carried out by an electrochemical work station (Gamry reference 600+).

3. Results

3.1. Composition and Microstructures of the Composite Materials

The formation process of Ce-doped CoP@NC is schematically shown in Figure 1. Firstly, uniform polyhedral nanocrystals of ZIF-67 were synthesized with a typical method [10]. Then, through an ionic exchange experiment, the ZIF-67 particles were reacted with $Ce(NO_3)_3 \cdot 6H_2O$ to form the Ce-doped ZIF-67 precursor, without changing the uniform polyhedral shape of ZIF-67 particles. Subsequently, the as-prepared precursor composites were heated in a nitrogen atmosphere at 600 °C for 2 h to transform Ce-doped ZIF-67 into Ce-doped Co@NC. In the final process, Ce-doped-CoP@NC hollow polyhedron composites were achieved through a phosphorization procedure via calcination treatment at 350 °C for 2 h under a nitrogen atmosphere. Figure S1a,b (See Supplementary Materials) illustrate the X-ray diffraction (XRD) patterns of samples. After incorporating Ce ion into the ZIF-67 matrix, the obtained Ce-doped ZIF-67 exhibited a similar XRD diffraction pattern with ZIF-67 except the peaks shifted to larger angles. This phenomenon indicates that the Ce ion successfully inserted into the lattice structure of ZIF-67 and unchanged the crystal structure. The EDX elemental mapping image of the Ce-doped ZIF-67 shows the homogeneous distribution of Ce, Co, and C throughout the whole polyhedron (Figure S1c). The scanning electron microscopy (SEM) image of the Ce-doped ZIF-67 (Figure S2) illustrates it has a smooth surface with rounded edges and corners (similar to the corresponding transmission electron microscopy (TEM) image in Figure 1) and without visible nanoparticles. The thermal behavior of Ce-doped ZIF-67 was investigated through the thermogravimetric analysis (TGA) method (Figure S3), which indicated that Ce-doped ZIF-67 underwent a weight loss lower than 600 °C and remained at 67% of its original weight when subjected to heat treatment under a flow of N_2. Therefore, we adopted a calcination process to pyrolyze Ce-doped ZIF-67 from room temperature to 600 °C at a ramping rate of 2 °C min^{-1}, and then stabilized at 600 °C for 2 h. The TEM bright-field (BF) image, high-angle annular dark-field (HAADF)-STEM image, and corresponding mapping images of Ce-doped Co/NC (Figure S4) indicated that Ce, Co, N, and C elements were uniformly distributed in the Ce-doped Co/NC hollow shell.

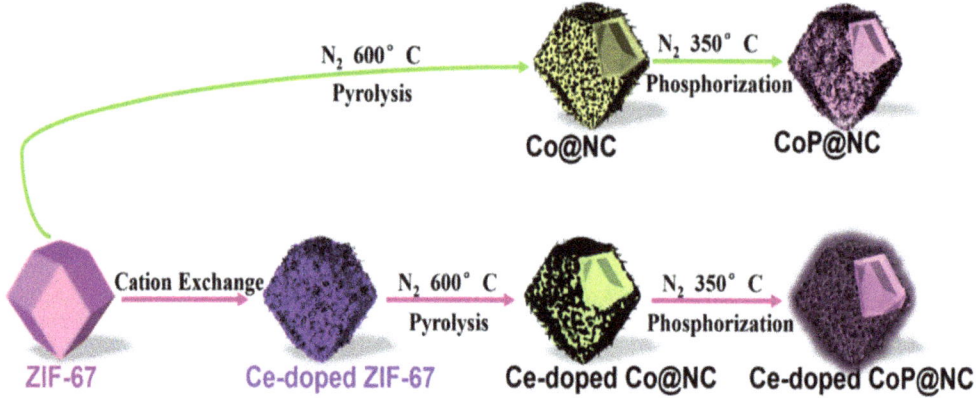

Figure 1. Schematic illustration showing the growth process of CoP@NC and Ce-doped CoP@NC hollow polyhedron.

Figure 2a shows the XRD patterns of the CoP@NC and Ce-doped CoP@NC samples. The diffraction peaks of the two samples were basically the same, which can be indexed as the orthorhombic CoP (JCPDS No. 29-0497). The broad diffraction peaks located at ~26° are well matched to amorphous carbon. It is worth mentioning that no XRD patterns related to CeO_2 species were observed. The enlarged views of XRD patterns are displayed in Figure 2b. It can be observed that the peaks at (011), (111), (112), and (211) of Ce-doped CoP@NC shifted to relatively lower angles compared with those of pristine CoP@NC. The phenomenon is due to the expansion of the lattice constant when the Co^{3+} (~0.58 Å) and Co^{2+} (0.65 Å) were partially substituted by Ce^{3+} with a larger radius for ~1.02 Å [29].

Figure 3 shows the X-ray photoelectron spectroscopy (XPS) of Ce-doped CoP@NC. From the full survey scan spectrum (Figure 3a), the Ce-doped CoP@NC sample consists of Co, Ce, P, C, and N elements, very compatible with the results of the XRD and elemental mapping. The high-resolution spectrum of Co 2p is presented in Figure 3b, which can be deconvoluted into two spin-orbit doublets. The first doublet centered at 782.3 and 784.0 eV are indexed to Co $2p_{3/2}$. The second doublet located at 798.5 and 800.2 eV are identified to Co $2p_{1/2}$. These peaks can be assigned to the Co element in Co-P and the surface cobalt oxide species. The peaks at 786.6 and 803.8 eV are the satellite peaks (denoted as "Sat"), which correspond to the Co^{2+} state. Besides, the peaks at 779.2 and 797.3 eV may be attributed to the presence of metallic Co in the composites [10,38–40]. In the Ce 3d spectrum (Figure 3c), two sets of spin-orbital multiples could be attributed to the Ce $3d_{5/2}$ and Ce $3d_{3/2}$. The weak signals at 885.5 and 904.0 eV are the characteristic peaks of Ce-ion with the chemical valence of +3 [41–43]. Besides, there is no XPS pattern of CeO_2, which is in good agreement with the XRD results in Figure 2a. In the high-resolution curve of P 2p (Figure 3d), four peaks at 129.3, 130.2, 133.9, and 134.8 eV can be identified and assigned to the P $2P_{1/2}$, P $2P_{3/2}$, C-P bond, and O-P bond, respectively. The O-P bond might result from the surface oxidation of the Ce-doped CoP@NC due to exposure to air. [10,44]. Similar to other carbonized ZIF structures, the N element originated from the high-temperature pyrolysis of nitrogen-rich 2-mIM. The N 1s spectra (Figure 3e) can be decomposed into three major peaks at binding energies of 401.6, 400.5, and 398.7 eV, which are attributed to the graphitic-N (N-C=C), pyrrolic-N (N=C), and pyridinic-N (N-H) in the composites, respectively. According to the previous reports, the incorporation of N atoms into carbon rings plays a key role in the charge delocalization of C atoms to improve the electrical conductivity and provides more lithium-active sites. Thus, the N-doped C will greatly improve the immobility of C, surface wettability, and stable electrode integrity. Thus, this reasonable design can effectively improve the stable cycle life of carbon-based materials for LIBs [45–47].

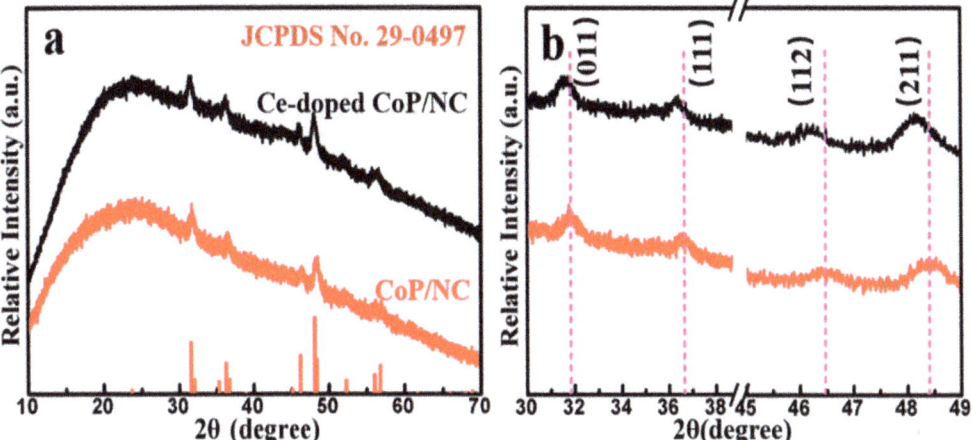

Figure 2. (**a**) XRD patterns, (**b**) the magnified diffraction peaks of the (011), (111), (112), and (211) of Ce-doped CoP@NC and CoP@NC.

Figure 4a,b show the typical SEM images of Ce-doped CoP@NC. After carbonization and phosphidation of Ce-doped ZIF-67, the irregular polyhedral structure was well-reserved with the size of about 600~1000 nm, with a slight surface shrinkage due to the decomposition process. Ce-doped CoP nanoparticles are embedded in an overlayer of the NC matrix. The TEM image (Figure 4c) of the Ce-doped CoP@NC displays it was an obvious hollow polyhedral structure. The SEM and TEM images of CoP@NC (Figure S5) show the hollow polyhedral structure with approximately 400–1000 nm. The curves exhibit the specific surface areas (BET) of CoP@NC and Ce-doped CoP@NC samples (Figure S6) are 55.467 and 96.848 m^2 g^{-1}, respectively. The hollow polyhedron structure and lager BET of Ce-doped CoP@NC facilitate the diffusion of electrolyte ions to the active sites and smart charge transport for reversible redox reactions, following an enhancement in lithium storage properties [46–48]. Based on the Barrett-Joyner-Halenda (BJH) plots, the corresponding pore size distributions are described in the inset of the Figure S6. They clearly illustrate that the Ce-doping resulted in a smaller pore size. The meso-sized pores in the Ce-doped CoP@NC provide optimal accessibility for transportation of electrolyte during LIB reaction. Figure 3d reveals that the Ce-doped CoP nanoparticles were surrounded by conductive carbon materials. The high-magnifcation HRTEM images (Figure 4e,f) display interlayer distances of around 0.247 and 0.350 nm corresponding to the (111) planes of CoP and (200) planes of amorphous carbon, respectively. The selected area electron diffraction (SAED) pattern (Figure 4g) consists of four concentric diffraction rings that belong to (011), (111), (211), and (301) diffraction planes of the CoP phase. The elemental mapping images (Figure 4h) display the uniform distribution of five elements (Ce, Co, C, N, P). The successful phosphidation and presence of bimetallic Ce and Co in the Ce-doped CoP@NC composites are worth noting. From the HAADF-STEM image (Figure S7a,b), the elemental mapping result for Ce element indicated its uniform distribution within the shells. Energy dispersive X-ray spectroscopy analysis (EDX) from Figure S6c revealed the content of Ce in Ce-doped CoP@NC at approximately 2.23%.

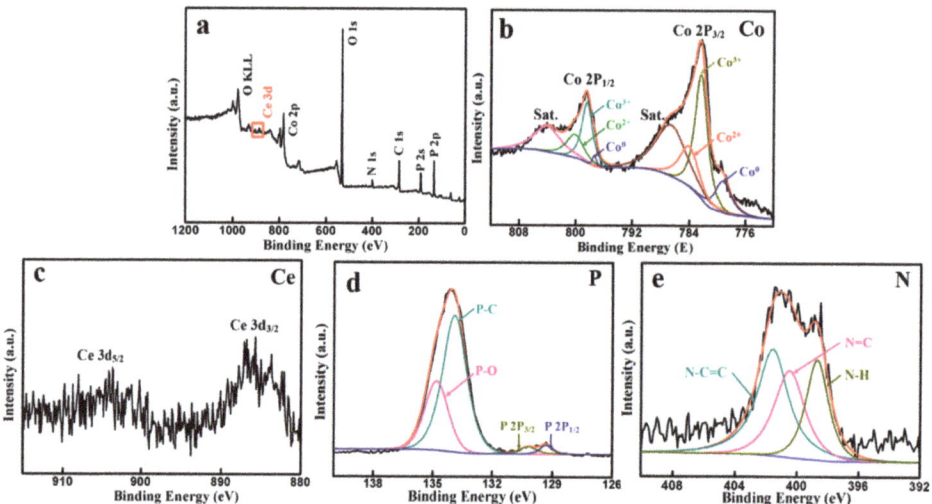

Figure 3. (**a**) XPS full spectrum, high-resolution XPS spectrum of (**b**) Co 2p, (**c**) Ce 3d, (**d**) P 2p, and (**e**) N 1s of Ce-doped CoP@NC.

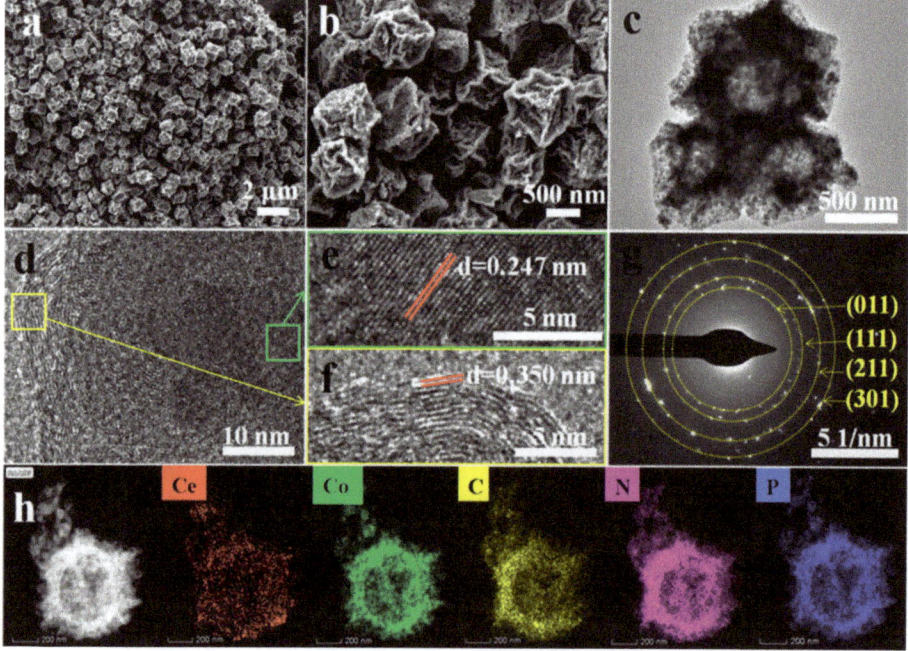

Figure 4. (**a**,**b**) SEM images, (**c**) TEM image, (**d**–**f**) HRTEM images, (**g**) selected-area diffraction pattern, and (**h**) EDX elemental mapping images of Ce-doped CoP@NC.

3.2. Electrochemical Property in Half-Cells

Cyclic voltammogram (CV) at a scan rate of 0.1 mV s^{-1} was used to investigate the electrochemical behavior of the Ce-doped CoP@NC anodes. As shown in Figure 5a, in the first cathodic scan, a major reduction peak at around 1.10 V corresponded to the conversion reaction of CoP + 3Li$^+$ + 3e$^-$ → Co + Li$_3$P [18,21,47]. The small reduction peak near 0.47 V was assigned to some irreversible reaction together with the formation of the solid electrolyte interphase (SEI) film on the surface of electrode materials, which disappeared during the subsequent curves. This may result in capacity loss after the first cycle [10,47,49]. In the faction peaks, approximately 1.05–1.16 V is attributed to the decomposition of Li$_3$P (Li$_3$P → LiP + 2Li$^+$ + 2e$^-$). For the subsequent scans, the reduction peak at 1.10 V shifted to 0.60 V, which may be related to the activation and redistribution of CoP in the initial cathodic scan, as reported in the literature [50]. These peaks after the first scan became steady, revealing the highly electrochemical reaction reversibility of Ce-doped CoP@NC during the repeated Li$^+$ intercalation-deintercalation process. Figure 5b demonstrates the galvanostatic discharge and charge curves of Ce-doped CoP@NC for different cycles. Two discharge platforms and one discharge platform were visualized for the first cycle, which is consistent with the first CV curve. In the subsequent cycles, the curves almost overlapped even for the 50th cycle. It indicated the Ce-doped CoP@NC electrode possesses superior cyclability and reversibility. Figure 5c exhibits the galvanostatic cycling measurements of the Ce-doped CoP@NC electrode at 100 mA g^{-1}. It exhibits an initial discharge and charge specific capacity of 1229 and 741 mA h g^{-1} with an ICE of 60.3%. In the initial 20 cycles, the specific capacity slightly declined, which may be attributed to the gradual activation process, including the phase and structure transformations and the surface SEI film [18,31]. Subsequently, the specific capacity stabilized around 675 mA h g^{-1}. Additionally, the electrode performed more outstanding rate capabilities.

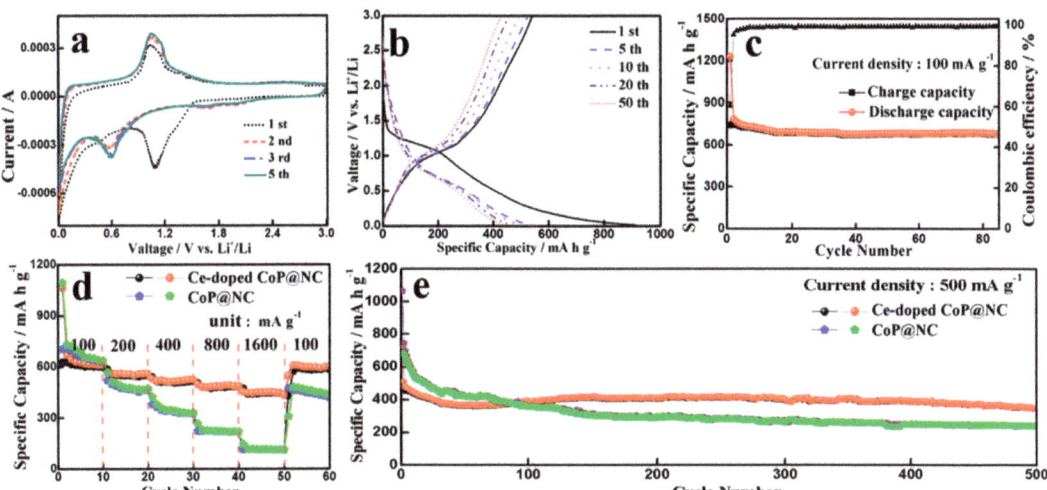

Figure 5. Electrochemical characterization for lithium storage electrodes. (**a**) CV of Ce-doped CoP@NC at a scan rate of 0.1 mV s^{-1} in the voltage range of 0.01~3.0 V. (**b**) Galvanostatic discharge and charge curves of Ce-doped CoP@NC for different cycles. (**c**) Cyclic behavior of Ce-doped CoP@NC. (**d**) Rate capabilities of Ce-doped CoP@NC and CoP@NC at different current densities. (**e**) Long-term cycling performance of Ce-doped CoP@NC and CoP@NC at 500 mA g^{-1}.

Figure 5d compares the rate capability of Ce-doped CoP@NC and CoP@NC electrode at varied current densities from 100 to 1600 mA g^{-1}. The average reversible discharge specific capacities of Ce-doped CoP@NC were 620, 560, 510, 480, and 440 mA h g^{-1} at the increasing current of 100, 200, 400, 800, and 1600 mA g^{-1}, respectively. Furthermore, when the current density was reverted to 100 mA g^{-1}, the specific capacity could return to 590 mA h g^{-1}. However, the pure CoP@NC electrode exhibited inferior rate performance. At rates of 100, 200, 400, 800, and 1600 mA g^{-1}, it presented reversible capacities of 660, 470, 330, 230, and 125 mA h g^{-1}, respectively. When the rate returned to 100 mA g^{-1}, the specific capacity decayed rapidly to 450 mA h g^{-1}. To further evaluate the high-rate and long-term cycling stability, the Ce-doped CoP@NC and pure CoP@NC electrodes were cycled at a high current density of 500 mA g^{-1} (Figure 5e). It can be intuitively observed that the discharge specific capacity of Ce-doped CoP@NC gradually tended to be stable and maintained a discharge specific capacity of ~400 mA h g^{-1} after 500 cycles. The cycling stability of pure CoP@NC was rather poor, only retaining capacity of ~250 mA h g^{-1} after 500 cycles. Therefore, ion doping might strengthen the structural integrity and stability of the electrode under high current densities [18,31].

To further investigate the charge transfer kinetics behavior of the electrode, electrochemical impedance spectroscopy (EIS) was studied. Figure S8 displays the Nyquist plots and the inset presents the corresponding equivalent circuit. In the Nyquist plots, the intercept on the X-axis in the high-frequency region corresponds to the electrolyte resistance (Rs); the diameter of the semicircle in the high/medium-frequency region is indexed to the charge-transfer resistance (R_{ct}) and the double-layer capacitance of the constant phase element (CPE), which occurs at the electrode/electrolyte interfaces; the slope of the inclined line in the low-frequency region corresponds to the Warburg impedance (Z_w) induced by the diffusion impedance of Li$^+$ intercalation/deintercalation into the active anode [18,51]. As expected, the R_s and R_{ct} value of Ce-doped CoP@NC electrode was about 6.3 and 81.5 Ω, which is smaller than that of pure CoP@NC. All these explicitly designated that the introduction of Ce-ion could induce fast diffusion of Li$^+$ and active sites for rapid ion diffusion.

4. Conclusions

In conclusion, the Ce-doped CoP@NC hollow polyhedron composites with a good lifespan and rate capability were successfully fabricated by the ionic-exchange, carbonization, and phosphorization methods. Ce-doped CoP nanoparticles were embedded in the NC hollow polyhedron skeleton. This unique nanoarchitecture and compositional merits can effectively hamper aggregation and buffer the volumetric expansion of CoP nanoparticles, fully immerse the structure in the electrolyte, and provide ample active sites for Li$^+$ insertion and fast transport paths for ion/electron during cycling. As a result, the Ce-doped CoP@NC electrode exhibited high reversible capacity (690 mA h g^{-1} at 100 mA g^{-1}) and excellent cyclability (the capacity of ~400 mA h g^{-1} is maintained at 500 mA g^{-1} after 500 cycles) for LIBs. We believe that the rare earth metal ion doping for M_xP_y is a promising and capable LIBs anode material for good rate capability and excellent cycling stability.

Supplementary Materials: The following supporting information can be downloaded at: https://www.mdpi.com/article/10.3390/cryst12040533/s1, Figure S1: (a) XRD patterns of ZIF-67 and Ce-doped ZIF-67. (b) the magnified region of 2θ = 16.7–18.0° showing a peak shift. (c) EDX elemental mapping images of Ce, Co and C for Ce-doped ZIF-67; Figure S2: SEM images of Ce-doped ZIF-67; Figure S3: The TGA curves of Ce-doped ZIF-67; Figure S4: (a) TEM bright field (BF) image, (b) HAADF-STEM image of Ce-doped Co/CN; Figure S5: (a,b) SEM images and (c,d) TEM images of CoP; Figure S6: N$_2$ adsorption/desorption isotherm of (a) CoP@NC and (b) Ce-doped CoP@NC; Figure S7: (a,b) HAADF-STEM images and (b) EDX analysis of Ce-doped CoP@NC; Figure S8: Nyquist plot and equivalent circuit model of Ce-doped CoP@NC and pure CoP@NC electrodes.

Author Contributions: Conceptualization, Y.Z. and S.Z. (Suyuan Zeng); methodology, Y.Z., S.Z. (Shuli Zhou), L.G. and X.X.; formal analysis, Y.Z. and S.Z. (Shuli Zhou); investigation, Y.Z., L.G. and X.X.; writing—original draft preparation, Y.Z.; writing—review and editing, S.Z. (Suyuan Zeng), K.Q., N.W. and X.Z.; supervision, Y.Z. and X.Z. All authors have read and agreed to the published version of the manuscript.

Funding: This research was funded by the Research Fund for the Doctoral Program of Liaocheng University, grant number 318051638; the Development Project of Youth Innovation Team in Shandong Colleges and Universities, grant number 2019KJC031; the Science and Technology Innovation Foundation for the University or College Students, grant number 201710447035, S202110447055, CXCY2021023.

Institutional Review Board Statement: Not applicable.

Informed Consent Statement: Not applicable.

Data Availability Statement: The data presented in this study are available on request from the corresponding author.

Conflicts of Interest: The authors declare no conflict of interest.

References

1. Dunn, B.; Kamath, H.; Tarascon, J.M. Electrical energy storage for the grid: A battery of choices. *Science* **2011**, *334*, 928–935. [CrossRef] [PubMed]
2. Li, M.; Lu, J.; Chen, Z.W.; Amine, K. 30 Years of Lithium-Ion Batteries. *Adv. Mater.* **2018**, *30*, 1800561. [CrossRef] [PubMed]
3. Zong, H.; Hu, L.; Wang, Z.G.; Qi, R.J.; Yu, K.; Zhu, Z.Q. Metal-organic frameworks-derived CoP anchored on MXene toward an efficient bifunctional electrode with enhanced lithium storage. *Chem. Eng. J.* **2021**, *416*, 129102. [CrossRef]
4. Zhan, R.M.; Wang, X.; Chen, Z.H.; She, Z.W.; Wang, L.; Sun, Y.M. Promises and challenges of the practical implementation of prelithiation in lithium-ion batteries. *Adv. Energy Mater.* **2021**, *11*, 2101565. [CrossRef]
5. Lu, Y.; Yu, L.; Lou, X.W.D. Nanostructured Conversion-type Anode Materials for Advanced Lithium-Ion Batteries. *Chem* **2018**, *4*, 972–996. [CrossRef]
6. Muhammad, I.; Jabeen, M.; Wang, P.R.; He, Y.S.; Liao, X.Z.; Ma, Z.F. Spray-dried assembly of 3D N, P-Co-doped graphene microspheres embedded with core–shell CoP/MoP@C nanoparticles for enhanced lithium-ion storage. *Dalton Trans.* **2021**, *50*, 4555–4566. [CrossRef]
7. Wang, N.N.; Bai, Z.C.; Fang, Z.W.; Zhang, X.; Xu, X.; Du, Y.; Liu, L.; Dou, S.; Yu, G. General synthetic strategy for pomegranate-like transition-metal phosphides@N-doped carbon nanostructures with high lithium storage capacity. *ACS Mater. Lett.* **2019**, *1*, 265–271. [CrossRef]
8. Chen, K.; Guo, H.N.; Li, W.Q.; Wang, Y.J. MOF-derived core-shell CoP@NC@TiO$_2$ composite as a high-performance anode material for Li-ion batteries. *Chem. Asian J.* **2021**, *16*, 322–328. [CrossRef]
9. Shi, Y.M.; Li, M.Y.; Yu, Y.F.; Zhang, B. Recent advances in nanostructured transition metal phosphides: Synthesis and energy-related applications. *Energy Environ. Sci.* **2020**, *13*, 4564–4582. [CrossRef]
10. Zhu, K.J.; Liu, J.; Li, S.T.; Liu, L.L.; Yang, L.Y.; Liu, S.L.; Wang, H.; Xie, T. Ultrafine cobalt phosphide nanoparticles embedded in nitrogen-doped carbon matrix as a superior anode material for lithium ion batteries. *Adv. Mater. Interfaces* **2017**, *4*, 1700377. [CrossRef]
11. Zhang, Z.; Zhu, P.P.; Li, C.; Yu, J.; Cai, J.X.; Yang, Z.Y. Needle-like cobalt phosphide arrays grown on carbon fiber cloth as a binder-free electrode with enhanced lithium storage performance. *Chin. Chem. Lett.* **2021**, *32*, 154–157. [CrossRef]
12. Wang, Z.J.; Wang, F.; Liu, K.Y.; Zhu, J.F.; Chen, T.G.; Gu, Z.Y.; Yin, S. Cobalt phosphide nanoparticles grown on Ti$_3$C$_2$ nanosheet for enhanced lithium ions storage performances. *J. Alloys Compd.* **2021**, *853*, 157136. [CrossRef]
13. Shang, F.F.; Yu, W.; Shi, R.T.; Wan, S.H.; Zhang, H.; Wang, B.; Cao, R. Enhanced lithium storage performance guided by intricate-cavity hollow cobalt phosphide. *Appl. Surf. Sci.* **2021**, *563*, 150395. [CrossRef]
14. Xu, X.J.; Liu, J.; Hu, R.Z.; Liu, J.W.; Ouyang, L.Z.; Zhu, M. Self-Supported CoP nanorod arrays grafted on stainless steel as an advanced integrated anode for stable and long-life lithium ion batteries. *Chem. Eur. J.* **2017**, *23*, 5198–5204. [CrossRef] [PubMed]
15. Wang, B.B.; Chen, K.; Wang, G.; Liu, X.J.; Wang, H.; Bai, J.T. A multidimensional and hierarchical carbon confined cobalt phosphide nanocomposite as an advanced anode for lithium and sodium storage. *Nanoscale* **2019**, *11*, 968–985. [CrossRef]
16. Wang, X.X.; Na, Z.L.; Yin, D.M.; Wang, C.L.; Wu, Y.M.; Huang, G.; Wang, L. Phytic acid-assisted formation of hierarchical porous CoP/C nanoboxes for enhanced lithium storage and hydrogen generation. *ACS Nano* **2018**, *12*, 12238–12246. [CrossRef]
17. Liu, Z.L.; Yang, S.J.; Sun, B.X.; Chang, X.H.; Zheng, J.; Li, X.G. A Peapod-like CoP@C nanostructure from phosphorization in a low temperature molten salt for high-performance lithium-ion batteries. *Angew. Chem. Int. Ed.* **2018**, *130*, 10344–10348. [CrossRef]
18. Ni, L.S.; Chen, G.; Liu, X.H.; Han, J.; Xiao, X.; Zhang, N.; Liang, S.; Qiu, G.; Ma, R. Self-supported Fe-doped CoP nanowire arrays grown on carbon cloth with enhanced properties in lithium-ion batteries. *ACS Appl. Energy Mater.* **2019**, *2*, 406–412. [CrossRef]
19. Li, W.L.; Zhao, R.F.; Zhou, K.H.; Shen, C.; Zhang, X.E.; Wu, H.Y.; Ni, L.; Yan, H.; Diao, G.; Chen, M. Cage-structured MxPy@CNCs (M = Co and Zn) from MOF confined growth in carbon nanocages for superior lithium storage and hydrogen evolution performance. *J. Mater. Chem. A* **2019**, *7*, 8443–8450. [CrossRef]
20. Guo, K.K.; Xi, B.J.; Wei, R.C.; Li, H.B.; Feng, J.K.; Xiong, S.L. Hierarchical microcables constructed by CoP@C⊂Carbon framework intertwined with carbon nanotubes for efficient lithium storage. *Adv. Energy Mater.* **2020**, *10*, 1902913. [CrossRef]
21. Zhu, P.P.; Zhang, Z.; Zhao, P.F.; Zhang, B.W.; Cao, X.; Yu, J.; Cai, J.; Huang, Y.; Yang, Z. Rational design of intertwined carbon nanotubes threaded porous CoP@carbon nanocubes as anode with superior lithium storage. *Carbon* **2019**, *142*, 269. [CrossRef]

22. Liu, X.J.; Zhou, L.; Huang, L.; Chen, L.L.; Long, L.; Wang, S.Y.; Xu, X.; Liu, M.; Yang, W.; Jia, J. ZIF-67 derived hierarchical hollow sphere-like CoNiFe phosphide for enhanced performances in oxygen evolution reaction and energy storage. *Electrochim. Acta* **2019**, *318*, 883–891. [CrossRef]
23. Li, X.R.; Yang, X.C.; Xue, H.G.; Pang, H.; Xu, Q. Metal–organic frameworks as a platform for clean energy applications. *EnergyChem* **2020**, *2*, 100027. [CrossRef]
24. Zhong, M.; Kong, L.; Li, N.; Liu, Y.Y.; Zhu, J.; Bu, X.H. Synthesis of MOF-derived nanostructures and their applications as anodes in lithium and sodium ion batteries. *Coord. Chem. Rev.* **2019**, *388*, 172–201. [CrossRef]
25. Zhai, Y.J.; Xu, L.Q.; Qian, Y.T. Ce-doped α-FeOOH nanorods as high-performance anode material for energy storage. *J. Power Sources* **2016**, *327*, 423–431. [CrossRef]
26. Ren, H.X.; Bai, Y.; Wang, X.R.; Ni, Q.; Wang, Z.H.; Li, Y.; Chen, G.; Wu, F.; Xu, H.; Wu, C. High-capacity interstitial Mn-incorporated $Mn_xFe_{3-x}O_4$/graphene nanocomposite for sodium-ion battery anodes. *ACS Appl. Mater. Interfaces* **2019**, *11*, 37812–37821. [CrossRef]
27. Luo, P.; Zhuge, F.W.; Zhang, Q.F.; Chen, Y.Q.; Lv, L.; Huang, Y.; Li, H.; Zhai, T. Doping engineering and functionalization of two-dimensional metal chalcogenides. *Nanoscale Horiz.* **2019**, *4*, 26–51. [CrossRef]
28. Gao, W.; Yan, M.; Cheung, H.Y.; Xia, Z.M.; Zhou, X.M.; Qin, Y.B.; Wong, C.Y.; Ho, J.C.; Chang, C.R.; Qu, Y. Modulating electronic structure of CoP electrocatalysts towards enhanced hydrogen evolution by Ce chemical doping in both acidic and basic media. *Nano Energy* **2017**, *38*, 290–296. [CrossRef]
29. Yan, Y.Z.; Huang, J.Z.; Wang, X.J.; Gao, T.L.; Zhang, Y.M.; Yao, T.; Song, B. Ruthenium incorporated cobalt phosphide nanocubes derived from a prussian blue analog for enhanced hydrogen evolution. *Front. Chem.* **2018**, *6*, 521. [CrossRef]
30. Han, X.Y.; Cui, Y.P.; Liu, H.W. Ce-doped Mn_3O_4 as high-performance anode material for lithium ion batteries. *J. Alloys Compd.* **2020**, *814*, 152348–152356. [CrossRef]
31. Chen, P.Y.; Zheng, G.T.; Guo, G.Z.; Wang, Z.C.; Tang, J.; Li, S.; Wen, Z.; Ji, S.; Sun, J. Ce-doped V_2O_5 microspheres with improved electrochemical performance for high-power rechargeable lithium ion batteries. *J. Alloys Compd.* **2019**, *784*, 574–583. [CrossRef]
32. Song, Y.Y.; Li, J.M.; Qiao, R.; Dai, X.; Jing, W.T.; Song, J.X.; Chen, Y.; Guo, S.; Sun, J.; Tan, Q.; et al. Binder-free flexible zinc-ion batteries: One-step potentiostatic electrodeposition strategy derived Ce doped-MnO_2 cathode. *Chem. Eng. J.* **2022**, *431*, 133387. [CrossRef]
33. Song, X.Y.; Zhang, Y.H.; Sun, P.P.; Gao, J.; Shi, F.N. Lithium-lanthanide bimetallic metal-organic frameworks towards negative electrode materials for lithium-ion batteries. *Chem. Eur. J.* **2020**, *26*, 5654–5661. [CrossRef] [PubMed]
34. Ge, P.; Yuan, S.H.; Zhao, W.Q.; Zhao, L.M.; Yang, Y.; Xie, L.L.; Zhu, L.; Cao, X. Rare earth metal La-doped induced electrochemical evolution of LiV_3O_8 with an oxygen vacancy toward a high energy-storage capacity. *J. Mater. Chem. A* **2021**, *9*, 1845–1858. [CrossRef]
35. Yang, D.D.; Xu, M.; Liang, X.; Wang, J.Y.; Fang, W.Y.; Zhu, C.G.; Wang, F. Facile synthesis of Pr-doped Co_3O_4 nanoflakes on the nickel-foam for high performance supercapacitors. *Electrochim. Acta* **2022**, *406*, 139815. [CrossRef]
36. Dong, C.F.; Guo, L.J.; Li, H.B.; Zhang, B.; Gao, X.; Tian, F.; Qian, Y.; Wang, D.; Xu, L. Rational fabrication of CoS_2/Co_4S_3@N-doped carbon microspheres as excellent cycling performance anode for half/full sodium ion batteries. *Energy Storage Mater.* **2020**, *25*, 679–686. [CrossRef]
37. Du, F.H.; Li, S.Q.; Yan, Y.; Lu, X.M.; Guo, C.F.; Ji, Z.Y.; Hu, P.; Shen, X. Facile fabrication of $Fe_{0.8}Mn_{1.2}O_3$ with various nanostructures for high-performance lithium-ion batteries. *Chem. Eng. J.* **2022**, *427*, 131697. [CrossRef]
38. Kang, L.B.; Ren, H.P.; Xing, Z.; Zhao, Y.L.; Ju, Z.C. Hierarchical porous $Co_xFe_{3-x}O_4$ nanocubes obtained by calcining prussian blue analogous as anode for lithium-ion batteries. *New J. Chem.* **2020**, *44*, 12546–12555. [CrossRef]
39. Zhou, D.; Yi, J.G.; Zhao, X.D.; Yang, J.Q.; Lu, H.R.; Fan, L.Z. Confining ultrasmall CoP nanoparticles into nitrogen-doped porous carbon via synchronous pyrolysis and phosphorization for enhanced potassium-ion storage. *Chem. Eng. J.* **2021**, *413*, 127508. [CrossRef]
40. Tao, S.; Xu, J.Q.; Xie, T.H.; Chu, S.Q.; Wu, D.J.; Qian, B. Regulating the electronic structure of CoP nanoflowers by molybdenum incorporation for enhanced lithium and sodium storage. *J. Power Sources* **2021**, *500*, 229975. [CrossRef]
41. Chen, T.Y.; Fu, Y.Y.; Liao, W.H.; Zhang, Y.Q.; Qian, M.; Dai, H.J.; Tong, X.; Yang, Q. Fabrication of Cerium-Doped CoMoP/MoP@C Heterogeneous Nanorods with High Performance for Overall Water Splitting. *Energy Fuels* **2021**, *35*, 14169–14176. [CrossRef]
42. Wei, C.Z.; Liu, K.F.; Tao, J.; Kang, X.T.; Hou, H.Y.; Cheng, C.; Zhang, D. Self-template synthesis of hybrid porous Co_3O_4-CeO_2 hollow polyhedrons for high performance supercapacitors. *Chem. Asian J.* **2018**, *4*, 111–117. [CrossRef] [PubMed]
43. Xing, H.N.; Long, G.K.; Zheng, J.M.; Zhao, H.Y.; Zong, Y.; Li, X.H.; Wang, Y.; Zhu, X.; Zhang, M.; Zheng, X. Interface engineering boosts electrochemical performance by fabricating CeO_2@CoP schottky conjunction for hybrid supercapacitors. *Electrochim. Acta* **2020**, *337*, 135817. [CrossRef]
44. Vijayakumar, E.; Ramakrishnan, S.; Sathiskumar, C.; Dong, J.Y.; Balamurugan, J.; Noh, H.S.; Kwon, D.; Kim, Y.H.; Lee, H. MOF-derived CoP-nitrogen-doped carbon@NiFeP nanoflakes as an efficient and durable electrocatalyst with multiple catalytically active sites for OER, HER, ORR and rechargeable zinc-air batteries. *Chem. Eng. J.* **2022**, *428*, 131115. [CrossRef]
45. Zhang, H.Y.; Xin, S.S.; Li, J.; Cui, H.T.; Liu, Y.Y.; Yang, Y.Z.; Wang, M. Synergistic regulation of polysulfides immobilization and conversion by MOF-derived CoP-HNC nanocages for high-performance lithium-sulfur batteries. *Nano Energy* **2021**, *85*, 106011. [CrossRef]

46. Zhang, Y.F.; Pan, A.Q.; Ding, L.; Zhou, Z.L.; Wang, Y.P.; Niu, S.Y.; Liang, S.; Cao, G. Nitrogen-doped yolk−shell-structured CoSe/C dodecahedra for high-performance sodium ion batteries. *ACS Appl. Mater. Interfaces* **2017**, *9*, 3624–3633. [CrossRef]
47. Li, H.H.; Zhu, Y.Q.; Zhao, K.J.; Fu, Q.; Wang, K.; Wang, Y.P.; Wang, N.; Lv, X.; Jiang, H.; Chen, L. Surface modification of coordination polymers to enable the construction of CoP/N, P-codoped carbon nanowires towards high-performance lithium storage. *J. Colloid Interface Sci.* **2020**, *565*, 503–512. [CrossRef]
48. Liu, Q.N.; Hu, Z.; Liang, Y.R.; Li, L.; Zou, C.; Jin, H.L.; Wang, S.; Lu, H.; Gu, Q.; Chou, S.L.; et al. Facile synthesis of hierarchical hollow CoP@C composites with superior performance for sodium and potassium storage. *Angew. Chem. Int. Ed.* **2020**, *59*, 5159–5164. [CrossRef]
49. Chang, Q.Q.; Jin, Y.H.; Jia, M.; Yuan, Q.; Zhao, C.C.; Ji, M.Q. Sulfur-doped CoP@ nitrogen-doped porous carbon hollow tube as an advanced anode with excellent cycling stability for sodium-ion batteries. *J. Colloid Interface Sci.* **2020**, *575*, 61–68. [CrossRef]
50. Tabassum, H.; Zou, R.; Mahmood, A.; Liang, Z.; Wang, Q.; Zhang, H.; Gao, S.; Qu, C.; Guo, W.; Guo, S. A universal strategy for hollow metal oxide nanoparticles encapsulated into B/N co-doped graphitic nanotubes as high-performance lithium-ion battery anodes. *Adv. Mater.* **2018**, *30*, 1705441. [CrossRef]
51. Jiang, Y.L.; Zou, G.Q.; Hou, H.S.; Li, J.Y.; Liu, C.; Qiu, X.Q.; Ji, X. Composition engineering boosts voltage windows for advanced sodium-ion batteries. *ACS Nano* **2019**, *13*, 10787–10797. [CrossRef] [PubMed]

Article

In Situ Electrochemical Derivation of Sodium-Tin Alloy as Sodium-Ion Energy Storage Devices Anode with Overall Electrochemical Characteristics

Liangfeng Niu, Shoujie Guo, Wei Liang, Limin Song, Burong Song, Qianlong Zhang and Lijun Wu *

College of Chemistry and Chemical Engineering, Xuchang University, Xuchang 461000, China; hnbfnlf@126.com (L.N.); gshgjinchuan@126.com (S.G.); 22011003@xcu.edu.cn (W.L.); songlimin110@126.com (L.S.); s18739910420@163.com (B.S.); zql17550031007@163.com (Q.Z.)
* Correspondence: 12017017@xcu.edu.cn

Abstract: Inspired by the fermentation of multiple small bread embryos to form large bread embryos, in this study, the expansion of tin foil inlaid with sodium rings in the process of repeated sodium inlaid and removal was utilized to maximum extent to realize the formation of sodium-tin alloy anode and the improvement of sodium storage characteristics. The special design of Sn foil inlaid with Na ring realized the in-situ electrochemical formation of fluffy porous sodium-tin alloy, effectively alleviated the volume expansion and shrinkage of non-electrochemical active Sn metal, and inhibited the generation of sodium dendrites. The abundance of sodium ions provided by the Na metal ring compensated for the active sodium components consumed during the repeated formation of SEI. When sodium-tin alloy in situ derived by Sn foil inlaid with Na ring was used as negative electrodes matched with SCDC and $Na_{0.91}MnO_2$ hexagonal tablets (NMO HTs) positive electrodes, the as-assembled sodium-ion energy storage devices present high specific capacity and excellent cycle stability.

Keywords: in situ electrochemical derivatization; sodium-tin alloy; anode; sodium storage

1. Introduction

With the aggravation of the greenhouse effect and the increase of people's demand for clean energy, lithium-ion batteries (LIBs), which are scarce in resources, will not be able to meet the demand for large-scale energy storage due to the price rise and market competition [1–3]. Among many alternatives, sodium ion batteries/capacitors (SIBs/SICs) with similar working principle to LIBs/LICs, suitable Stoke radius (solvation radius, that of Na^+ is 4.6 Å), rich sodium resources, aluminum as a collector, low cost and other advantages have attracted extensive attention of researchers [4]. However, the larger radius of Na^+ (1.02 Å) than that of Li^+ (0.76 Å) makes it difficult to insert Na^+ into the interlayer of commercial graphite, which results the low theoretical capacity (35 mA h g^{-1}) and energy density of SIBs [5,6]. Although transition metal oxides/sulfides/selenides based on conversion-reaction mechanism can deliver fascinating sodium ion storage capacity, their realistic application is still limited because of the slow dynamics and inadequate cyclic stability [7–9]. Therefore, the exploration of anodes for rapid storage of Na^+ ions is still an important research area.

Sn, Sb, Te, Bi, Si and Se metal and their alloys with good electrical conductivity and high specific capacity were considered as rapid sodium ion storage anodes [10–19]. However, the large volume expansion of them often leads to capacity attenuation and insufficient cycle stability of SIBs during repeated charge and discharge [20–25]. Aiming at the volume expansion mentioned, many strategies such as heterogeneous element doping, carbon coating were taken. Although these strategies can effectively alleviate the problem of large-rate discharge, the electrochemical activity Na ions consumed by the repeated

formation of SEI film still cannot be instantaneously replenish during the process of charge and discharge, and their specific capacity and cyclic stability are still far from practical application [26–30]. Therefore, based on rational utilization of expansion of metal Sn during charge and discharge, this study purposefully designed a Sn foil inlaid Na ring structure to in situ achieve electrochemical derivatization of sodium-tin alloy anode. Subsequently, the sodium-tin alloy anodes derived by tin foil inlaid sodium ring were matched with the positive electrode of sodium citrate derived carbon (SCDC) and NMO HTs respectively to assemble button cells. The result demonstrated that these cells exhibited the high mass specific capacity and excellent cyclic longevity, which is about 100 mAh g^{-1} for SCDC after 10,000 cycles and 60 mAh g^{-1} for NMO HTs after 8400 cycles.

2. Materials and Methods

2.1. Materials and Chemicals

All the reagents used were manually ground for several minutes without further purification.

2.2. Preparation of $Na_{0.91}MnO_2$ Hexagonal Tablets

$Na_{0.91}MnO_2$ hexagonal tablets were prepared as follows: MnO_2 and anhydrous Na_2CO_3 with different molar ratio of 1:0.53 were separately added to a certain amount of deionized water, and subsequently were stirred and evaporated to dryness on the heated magnetic stirrer, then the homogenous mixture of MnO_2 and anhydrous Na_2CO_3 were obtained. Finally, the temperature of the mixture was procedurally raised to 350 °C at a rate of 3 °C per minute for 6 h in air, then to 850 °C at the same rate for 12 h in air. A kind of smoke grey powder was obtained.

2.3. Instrumentation and Sample Analysis

The morphologies and microstructure of the NMO HTs and sodium-tin alloy prepared by electrochemical in-situ method was detected by Field Emission Scanning Electronic Microscopy (FESEM, JSM-6701F). Transmission electron microscopy (TEM) and high-resolution TEM (HRTEM) were performed with HITACHI-H7650 and JEM-2100F (JEOL) instruments, and the kinetic energy of electrons was 200 kV. For SEM detection of sodium-tin alloy, it was particularly worth mentioning that the button battery after circulation was disassembled in the glove box with water and oxygen content less than 0.1 ppm, next the sodium-tin alloy was taken out. After washing with DMC solution, the sodium-tin alloy was placed in the glove box overnight to drying, and cut off the parts with scissors. Subsequently, one of the cut parts was paste on the conductive adhesive of the sample stand, and transferred into the test box filled with high purity argon gas. The entire transfer process was carried out in the high purity argon gas.

2.4. Electrochemical Measurements

The electrochemical tests were performed at constant temperature of 25 °C using double-electrode 2032-type coin cells with sodium disc or tin foil disc and sodium rings as the counter electrodes. For fabrication of SCDC and NMO HTs cathodes, 70 wt% of active material, 20 wt% of carbon black (Super-P) and 10 wt% of polyvinylidenefluoride (PVDF) in methyl-2-pyrrolidone (NMP) were well mixed and then coated on the Al foil which served as a current collector. After heated at 90 °C for 12 h under vacuum, the sheet was pressed and punched into 14 mm diameter electrodes with a mass loading of 1.0–2.0 mg. For electrochemical in situ fabrication of sodium-tin alloy anode: after inserting 2 mm sodium ring around the 14 mm tin foil, sodium-tin alloy is formed during repeated charge and discharge. The electrolyte used in our work was 1 M $NaClO_4$ in a 1:1:1 $(v/v/v)$ mixture of ethylene carbonate (EC), diethyl carbonate (DMC) and ethyl methyl carbonate (EMC), which was added 2% (mass ratio) fluoroethylene carbonate (FEC). The cell assembly was carried out in an argon-filled glove box with both the moisture and the oxygen content below 0.1 ppm. The CV tests were conducted with scan rates from 0.1 to 5 mV s^{-1}. Electrical impedance spectroscopy (EIS) studies were carried out using PGSTAT302N (Metrohm Co.,

Herisau, Switzerland) in the frequency range from 0.1 MHz to 0.05 Hz with an amplitude of 5 mV. Galvanostatic charge/discharge and cyclic stability characteristic were collected between 2.0 V and 4.0 V on a Neware BTS-5V10 mA (New well electronic Technology Co., LTD, Guangzhou, China).

3. Results and Discussions

As Figure 1 shown, inspired by the crowded arrangement and fermentation of multiple small embryos into a large embryo, this study aims to design the tin foil inlaid sodium ring structure to realize the in-situ electrochemical formation of sodium-tin alloy. The unique structure can alleviate the volume expansion and shrinkage of non-electrochemical active metal Sn and inhibit the generation of sodium dendrites. The embedded sodium ring can provide sufficient electrochemical activity Na$^+$ for the system, which can compensate for sodium ions consumed by the repeated formation of SEI during the charge and discharge process of anode and cathode materials. To avoid the influence of positive electrode reaction on the evolution process of sodium-tin alloy, the mature SCDC was selected as the positive electrode to assemble coin cells.

Figure 1. The schematic illustration for in situ electrochemical derivation of sodium-tin alloy.

Firstly, the morphology of SCDC was tested by SEM and TEM. As Figure 2a and inset 2a shown, SCDC presents a comb structure. Raman result present inset Figure 2a demonstrated that the ratio of D/G was 2.57, which indicated that SCDC was rich in defects. The existence of these defects is beneficial to improve the storage of sodium ions. To explore the optimum operating voltage range of coin cell with sodium tin alloy as anode and SCDC as cathode. Cyclic voltammetry of these cells was determined in different voltage range. As shown in Figure 2a, the polarization of cyclic voltammetry curves for SCDC positive is significant at 1.0 V–4.0 V compared with at 1.5–4.0 V and 2.0–4.0 V. Therefore, considering the influence of the operating voltage range on mass specific capacity of SCDC, 1.5 V–4.0 V is selected as the charge-discharge voltage range. As a comparison, the electrochemical properties of pure sodium and tin foil anode were also tested, as exhibited in Figure 2c. The results show that when sodium-tin alloy prepared by in situ electrochemical epitaxy is used as anode, SCDC exhibits the highest gradient mass specific capacity compared with that of pure Sn foil and pure Na metal disc as anode, just as shown in Figure 2d. It should be noted that the specific capacity of the pure sodium anode for the first 20 laps was lower than that of the sodium tin alloy anode because the voltage range for the first 10 laps is 2–4 V.

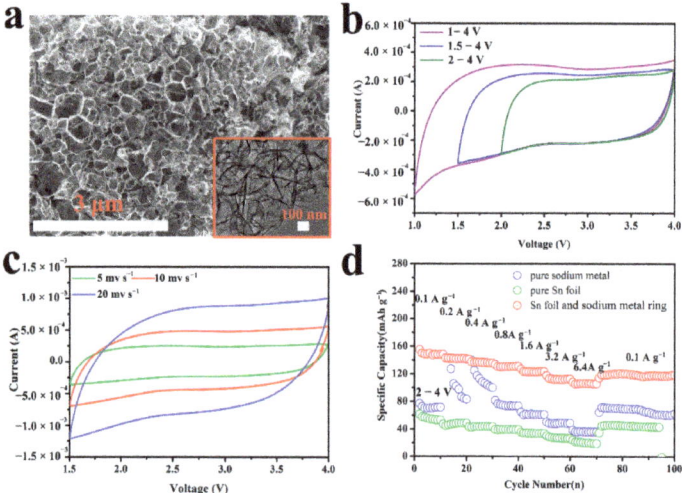

Figure 2. (**a**) The morphology of SCDC tested by FESEM, Raman (in the upper right of Figure 2a) and TEM (in the right bottom of Figure 2a); (**b**) Cyclic voltammetry curves of SCDC vs. Na/Na$^+$ in different voltage ranges of 1.0–4.0 V, 1.5–4.0 V and 2.0–4.0 V; (**c**) Cyclic voltammetry curves of SCDC vs. Na/Na$^+$ at different scan rate of 5 mV s^{-1}, 10 mV s^{-1} and 20 mV s^{-1}; (**d**) Comparison of gradient properties for pure Sn foil, pure Na metal disc and sodium-tin alloy anode in situ electrochemical derived from Sn foil inlaid with sodium ring.

The gradient characteristic (Figure 3a–c) and cyclic stability (Figure 3b–d) of SCDC vs. sodium-tin alloy anode were also determined. As presented in Figure 3a,b, SCDC exhibits the first mass specific capapcity of 280 mAh g^{-1}, and considerable mid-value voltage of 2.66 V. Figure 3c indicates that SCDC exhibited high gradient capacities of 250 mAh g^{-1}, 170 mAh g^{-1}, 155 mAh g^{-1}, 148 mAh g^{-1}, 145 mAh g^{-1}, 125 mAh g^{-1} and 120 mAh g^{-1} at the current density of 0.1 A g^{-1}, 0.2 A g^{-1}, 0.4 A g^{-1}, 0.8 A g^{-1}, 1.6 A g^{-1}, 3.2 A g^{-1} and 6.4 A g^{-1}, respectively. Along with high capacity of 100 mAh g^{-1} and ultra-long cyclic life of 10,000 cycles at 1.0 A g^{-1}, sodium-tin alloy in situ electrochemically derived will possibly be a candidate anode of devices with high mass specific capacity and excellent cyclic longevity. It basically attributes to the multi-hole and irregular surface of sodium-tin alloy, which can afford more storage space for electrochemically active Na$^+$ ions. While these electrochemical activity Na ions can instantaneously complement those consumed during the repeated formation of SEI film.

Considering the universal application of the sodium-tin alloy in situ electrochemically derived, the coin cells were also assembled, in which sodium-tin alloy in situ electrochemically derived by Sn foil (13 mm) inlaid with Na ring (1 mm) was negative, and Na$_{0.91}$MnO$_2$ hexagonal tablets (just as XRD demonstrated inset Figure 4b) was the counter electrode. The electrochemical performance of as-assembled coin cells presents that NMO HTs vs. sodium-tin alloy anode displays a high first-lap specific capacity of 140 mAh g^{-1} at 0.1 A g^{-1} current density, and it remains nearly 90 mAh g^{-1} after 570 cycles, as Figure 4a shown. The gradient capacity and the cyclic stability for NMO HTs cathode vs. sodium-tin alloy anode was also collected, as shown in Figure 4b,c. In the gradient range from 0.1 to 1.6 A g^{-1} for 10 laps at each stage, the discharge capacity of the coin cell with NMO HTs as cathode and sodium-tin alloy as anode keeps 96 mAh g^{-1} at current density of 0.1 A g^{-1} after 10 laps. When the current density stepwise increased to 0.2, 0.5, 1, and 2 A g^{-1}, the corresponding specific capacities present 85, 78, 67, and 55 mAh g^{-1}, respectively. When the current density turns back to 0.1 A g^{-1}, the specific capacity still maintains 140 mAh g^{-1} and ultimately holds 89 mAh g^{-1}. However, it cannot be neglected that the first coulombic efficiencies of these batteries are all lower than 70% and gradually increase to 100% as the charge-discharge cycle continues, which dominantly comes down to the concentration polarization caused by the difference of sodium ion concentration on the surface of metal tin anode at the initial stage of charge and discharge. But with the progress of the charging and discharging cycle, this polarization gradually weakens, so that NMO HTs still maintains nearly 60 mAh g^{-1} specific capacity after 8400 cycles at 1 A g^{-1} directly achieved after gradient cycles, as Figure 4c,d demonstrated. The excellent cyclic stability principally attributes to the improvement of Na$^+$ ion transport rate caused by thin lamellar of NMO HTs cathode and the exceptional conductivity and stability of sodium-tin alloy counter electrode in situ electrochemically derived by Sn foil (13 mm) inlaid with Na ring (1 mm).

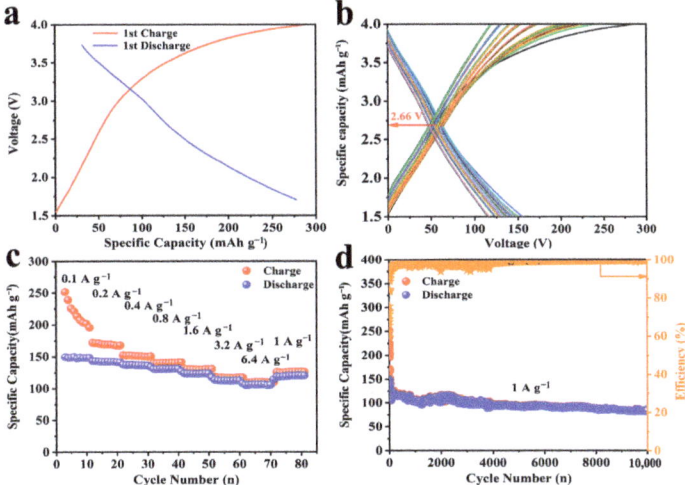

Figure 3. The electrochemical characteristic for sodium ion capacitors with SCDC as cathode and sodium-tin as anode; (**a**) curves of voltage vs. specific capacity for the first charge-discharge; (**b**) the gradient performance obtained from 0.1 to 6.4 A g^{-1} increasing by 2 times and circulates 10 laps for each current density; (**c**) the cyclic performance at 1 A g^{-1}; (**d**) curves of voltage vs. specific capacity for the first 100 cycles of (**c**).

To further explore the potential reasons for the excellent electrochemical performance of NMO HTs, we performed CV detection and fitting. Figure 5a illustrates the original CV curves at different scan rates of NMO HTs positive vs. sodium-tin alloy negative in situ electrochemically derived, it is found that there are one distinct reduction peak at 2.13 V and two oxidation peaks at 2.45 V and 3.6 V, respectively. CV curves scanned at different rates shown in Figure 5b are overlapped and almost the same shape, which further confirms the excellent reversibility of the above coin cell. Furthermore, the capacitive/battery contribution to the total charge storage of NMO HTs positive vs. sodium-tin alloy negative electrochemically derived was also quantified through the method of charging storage distinction according to the accurate and efficient analysis method of charge storage mechanism and pseudo contribution in cyclic voltammetry curve developed by Dunn et al. Accordingly, the total current at a fixed potential can be expressed as the combination of two parts of pseudocapacitive ($k_1 v$) and diffusion-controlled capacity ($k_2 v^{0.5}$) [31,32].

$$i(V) = k_1 v + k_2 v^{0.5} \quad (1)$$

For further mathematical treatment, Equation (1) slightly transfers to

$$i(V)/v^{0.5} = k_1 v^{0.5} + k_2 \quad (2)$$

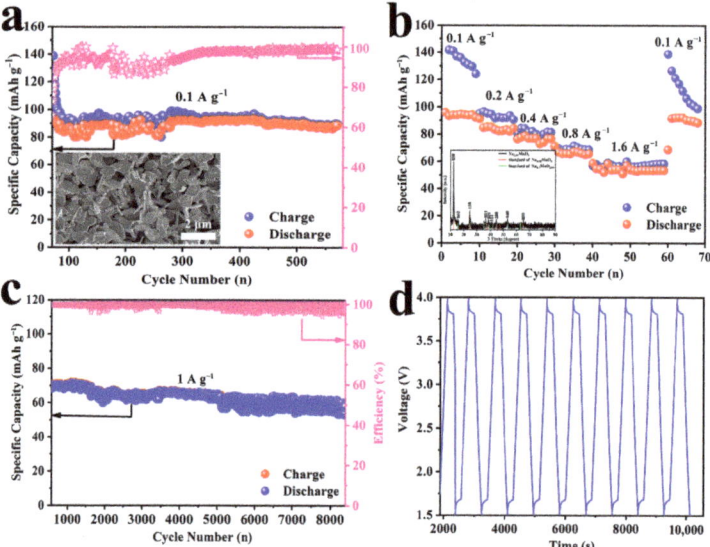

Figure 4. (**a**) Cycle performance of $Na_{0.91}MnO_2$ hexagonal tablets (NMO HTs) vs. sodium-tin alloy in situ electrochemically derived directly collected for 570 cycles at 0.1 A g^{-1}; (**b**) Capacity evolution of NMO HTs vs. sodium-tin alloy in situ electrochemically derived by evolving current densities ranging from 0.1, 0.2, 0.4, 0.8, 1.6 A g^{-1} and transfer to 0.1 A g^{-1} (inset, XRD of NMO HTs); (**c**) Cycle performance for 8400 cycles at 1 A g^{-1} directly collected after the various current densities for NMO HTs vs. sodium-tin alloy in situ electrochemically derived were tested; (**d**) The profiles of voltage vs. specific capacity for NMO HTs vs. sodium-tin alloy at 1.5–4.0 V between 4290th and 4300th cycles.

On account of the linear fitting (Figure 5c) for logarithm of voltage to logarithm of currents at each potential, the coefficients k_1 and k_2 can be determined, and the $k_1\nu$ and $k_2\nu^{0.5}$ were calculated. Then the pseudo capacitive ($k_1\nu$) and diffusion control contribution ($k_2\nu^{0.5}$) are distinguished [33–35]. As demonstrated in Figure 5d, the shaded regions stand for the contribution from charge controlled by the capacitive, while the blank areas represent the diffusion control contribution. The results demonstrate that the capacitive controlled charge occupied about 25% of the entire charge storage capacity of 1.3 mV s^{-1}. However, it suddenly increases to 30% at 1.5 mV s^{-1}, which is due to the coin cell holds for a whole night between 1.3 mV s^{-1} and 1.5 mV s^{-1} test. In particular, the capacitance charge ratio has turned slowly upward from 1.7–2.1 mV s^{-1} (Figure 5e). The above results indicate although the proportion of NMO HTs pseudocapacitive charge is less than 30% of the total charge, the dynamic defects of solid-state diffusion can still be effectively alleviated. This is also demonstrated by electrochemical impedance spectroscopy (EIS, Figure 5f). As Figure 5f indicated, the Nyquist plot in high frequency region is respectively contributed by physical impedance of device and interface impedance resistances between NMO HTs electrode and electrolyte, while the low-frequency impedance derived from the diffusion of Na ions, which is consistent with the equivalent current diagram inset Figure 5f.

In order to verify the in-situ electrochemical formation of sodium-tin alloy and compare the differences between sodium-tin alloy derived by (a) pure Sn foil and (b) Sn foil inlaid with Na metal ring. the according image were observed after repeated charge-discharge. As Figure 6a–c presented, the surface of sodium-tin alloy derived by Sn foil inlaid with Na metal ring become multi-hole and irregular compared with that of sodium-tin alloy derived by pure Sn foil. This unique structure is contributed to affording more storage space for electrochemically active Na^+ ions. Simultaneously, Na ions consumed by the repeated formation of SEI film can be instantaneously complement during the process of charge and discharge. Together with the uniform distribution of Na and Sn elements (Figure 6d,e), sodium-tin alloy in situ derived by Sn foil inlaid with Na ring was used as negative, and matched with SCDC and NMO HTs as positive present high specific capacity and remarkable cyclic stability. This provides a new avenue for the assembly of sodium storage energy devices with excellent electrochemical performance. The EDS (Figure 6f) of sodium-tin alloy further verified that the molar ratio of Na and Sn elements is 19.4/1, which is far higher than maximum theoretical mole ratio 15/4 for sodium embedding. It is indicated that in addition to contributing to the formation of $Na_{15}Sn_4$ alloy ($Na_{15}Sn_4 \Leftrightarrow 15Na^+ + 4Sn + 15e^-$), Na metal ring also compensated for sodium ions consumed by the repeated formation of SEI in reaction system during charge and discharge.

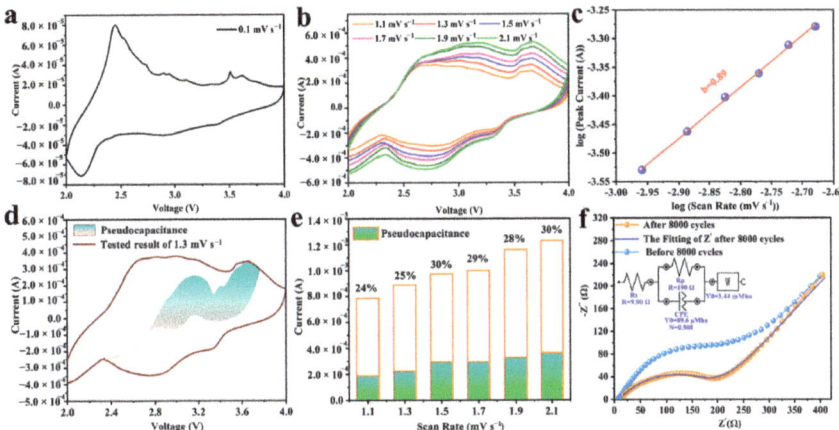

Figure 5. (**a**) The first cyclic voltammetry of NMO HTs vs. Na/Na$^+$ at 0.1 mV s^{-1}; (**b**) Cyclic voltammetry curves of NMO HTs vs. Na/Na$^+$ at different scan rate between 1.1 mV s^{-1} and 2.1 mV s^{-1} at intervals of 0.2 mV s^{-1}; (**c**) The b-value determination of the peak currents of cathode shows that charge storage of SCDC vs. sodium-tin alloy; (**d**) Cyclic voltammetry curves of NMO HTs vs. Na/Na$^+$ at 1.3 mV s^{-1} and the shadowed areas represent the capacitive contribution; (**e**) Separation of diffusion-controlled and capacitive charge at different sweep rates; (**f**) Nyquist dots of the NMO HTs vs. sodium-tin alloy in situ electrochemically derived by Sn foil (13 mm) inlaid with Na ring (1 mm) before and after 8400 cycles at 1 A g^{-1} (inset, the corresponding equivalent circuit diagram after 8400 cycles).

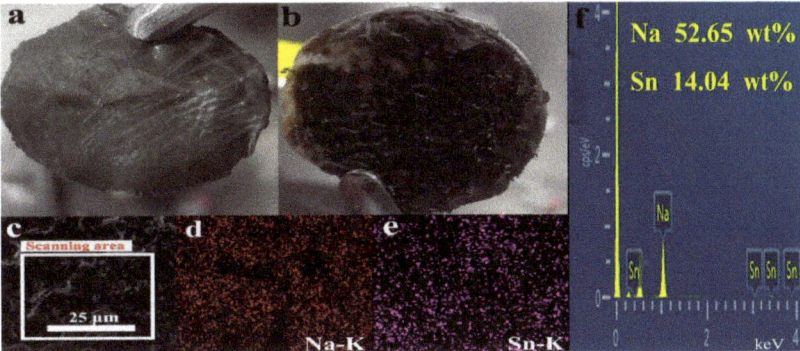

Figure 6. Low magnification FESEM images of sodium-tin alloy derived by (**a**) pure Sn foil and (**b**) Sn foil inlaid with Na metal ring; (**c**)The representative district scanning and corresponding element mappings: (**d**) Na and (**e**) Sn; (**f**) the electron diffraction spectra (EDS) of sodium-tin alloy.

4. Conclusions

To summery, the sodium-tin alloy was successfully in situ electrochemically derived through a simple and practicable design of Sn foil (13 mm) inlaid with Na ring (1 mm). In view of the distinctive design, the expansion of tin foil during the repeated sodium ions inlaid and removal was utilized to maximum extent, the formation of sodium-tin alloy anode and the improvement of sodium storage characteristics were realized, Na ions consumed by the repeated formation of SEI film were instantaneously complement. When the in-situ as-derived sodium-tin alloy was used as anode, SCDC as cathode, the as-assembled sodium ion capacitors exhibited about 100 mAh g^{-1} high specific capacity after 10,000 cycles; while when NMO HTs as cathode, the as-assembled sodium ion batteries exhibited favorable pseudocapacitive behavior and about 80 mAh g^{-1} specific capacity after 8400 cycles. The above results demonstrated that the strategy of in-situ electrochemical derivation can provide a reference for sodium ion energy storage devices with overall electrochemical characteristics.

Author Contributions: Formal analysis, writing—original draft preparation, writing—review and editing, L.W.; resources, funding acquisition, L.N.; data curation, visualization, W.L.; investigation, B.S., L.S. and Q.Z.; project administration, S.G. All authors have read and agreed to the published version of the manuscript.

Funding: This work was supported by the Nature Science Foundations of Henan Province (202300410349), High Education Teaching Reform Research and Practice Project of Henan Province (2021SJGLX227Y, Postgraduate Education), Scientific Research Project of Xuchang University (2021YB001).

Institutional Review Board Statement: Not applicable.

Informed Consent Statement: Not applicable.

Data Availability Statement: The data presented in this study are available on request from the corresponding author.

Conflicts of Interest: The authors declare no conflict of interest.

References

1. Sun, M.; Wang, Z.; Ni, J.; Li, L. Dual-doped hematite nanorod arrays on carbon cloth as a robust and flexible sodium anode. *Adv. Funct. Mater.* **2020**, *30*, 1910043. [CrossRef]
2. Yabuuchi, N.; Kubota, K.; Dahbi, M.; Komaba, S. Research development on sodium-ion batteries. *Chem. Rev.* **2014**, *114*, 11636–11682. [CrossRef] [PubMed]
3. Wang, G.; Yu, M.; Feng, X. Carbon materials for ion-intercalation involved rechargeable battery technologies. *Chem. Soc. Rev.* **2021**, *50*, 2388–2443. [CrossRef] [PubMed]
4. Usiskin, R.; Lu, Y.; Popovic, J.; Law, M.; Balaya, P.; Hu, Y.S.; Maier, J. Fundamentals, status and promise of sodium-based batteries. *Na. Rev. Mater.* **2021**, *6*, 1020–1035. [CrossRef]
5. Cao, Y.; Xiao, L.; Sushko, M.L.; Wang, W.; Schwenzer, B.; Xiao, J.; Nie, Z.; Saraf, L.V.; Yang, Z.; Liu, J. Sodium ion insertion in hollow carbon nanowires for battery applications. *Nano Lett.* **2012**, *12*, 3783–3787. [CrossRef]
6. Chen, Y.; Li, X.; Park, K.; Lu, W.; Wang, C.; Xue, W.; Yang, F.; Zhou, J.; Suo, L.; Lin, T. Nitrogen-doped carbon for sodium-ion battery anode by self-etching and graphitization of bimetallic MOF-based composite. *Chem* **2017**, *3*, 152–163. [CrossRef]
7. Zhang, B.; Kang, F.; Tarascon, J.M.; Kim, J.K. Recent advances in electrospun carbon nanofibers and their application in electrochemical energy storage. *Prog. Mater. Sci.* **2016**, *76*, 319–380. [CrossRef]
8. Wenzel, S.; Hara, T.; Janek, J.; Adelhelm, P. Room-temperature sodium-ion batteries: Improving the rate capability of carbon anode materials by templating strategies. *Energy Env. Sci.* **2011**, *4*, 3342–3345. [CrossRef]
9. Li, Z.; Ding, J.; Mitlin, D. Tin and tin compounds for sodium ion battery anodes: Phase transformations and performance. *Acc. Chem. Res.* **2015**, *48*, 1657–1665. [CrossRef]
10. Jing, W.T.; Yang, C.C.; Jiang, Q. Recent progress on metallic Sn- and Sb-based anodes for sodium-ion batteries. *J. Mater. Chem. A* **2020**, *8*, 2913–2933. [CrossRef]
11. Kim, C.; Kim, H.; Sadan, M.K.; Jeon, M.; Cho, G.; Ahn, J.; Kim, K.; Cho, K.; Ahn, H. Development and evaluation of Sn foil anode for sodium-ion batteries. *Small* **2021**, *17*, 2102618. [CrossRef]
12. Song, M.; Wang, C.; Du, D.; Li, F.; Chen, J. A high-energy-density sodium-ion full battery based on tin anode. *Sci. China Chem.* **2019**, *62*, 616–621. [CrossRef]
13. Chen, L.; He, X.; Chen, H.; Huang, S.; Wei, M. N-Doped carbon encapsulating Bi nanoparticles derived from metal-organic frameworks for high-performance sodium-ion batteries. *J. Mater. Chem. A* **2021**, *9*, 22048–22055. [CrossRef]

14. Cheng, D.; Yang, L.; Hu, R.; Liu, J.; Che, R.; Cui, J.; Wu, Y.; Chen, W.; Huang, J.; Zhu, M. Sn-C and Se-C Co-bonding SnSe/few-layered graphene micro-nano structure: Route to a densely compacted and durable anode for lithium/sodium-Ion Batteries. *ACS Appl. Mater. Interfaces* **2019**, *11*, 36685–36696. [CrossRef]
15. Cheng, Y.; Huang, J.; Li, R.; Xu, Z.; Cao, L.; Ouyang, H.; Li, J.; Qi, H.; Wang, C. Enhanced cycling performances of hollow Sn compared to solid Sn in Na-ion battery. *Electrochim. Acta* **2015**, *180*, 227–233. [CrossRef]
16. Lin, Y.M.; Abel, P.R.; Gupta, A.; Goodenough, J.B.; Heller, A.; Mullins, C.B. Sn-Cu nanocomposite anodes for rechargeable sodium-ion batteries. *ACS Appl. Mater. Interfaces* **2013**, *5*, 8273–8277. [CrossRef]
17. Liu, J.; Yu, L.; Wu, C.; Wen, Y.; Yin, K.; Chiang, F.K.; Hu, R.; Liu, J.; Sun, L.; Gu, L. New nanoconfined galvanic replacement synthesis of hollow Sb@C yolk-shell spheres constituting a stable anode for high-rate Li/Na-ion batteries. *Nano Lett.* **2017**, *17*, 2034–2042. [CrossRef]
18. Liu, S.; Luo, Z.; Guo, J.; Pan, A.; Cai, Z.; Liang, S. Bismuth nanosheets grown on carbon fiber cloth as advanced binder-free anode for sodium-ion batteries. *Electrochem. Commun.* **2017**, *81*, 10–13. [CrossRef]
19. Lu, Y.; Zhou, P.; Lei, K.; Zhao, Q.; Tao, Z.; Chen, J. Selenium phosphide (Se_4P_4) as a new and promising anode material for sodium-ion batteries. *Adv. Energy Mater.* **2017**, *7*, 1601973. [CrossRef]
20. Luo, B.; Qiu, T.; Ye, D.; Wang, L.; Zhi, L. Tin nanoparticles encapsulated in graphene backboned carbonaceous foams as high-performance anodes for lithium-ion and sodium-ion storage. *Nano Energy* **2016**, *22*, 232–240. [CrossRef]
21. Mao, J.; Fan, X.; Luo, C.; Wang, C. Building self-healing alloy architecture for stable sodium-ion battery anodes: A case study of tin anode materials. *ACS Appl. Mater. Interfaces* **2016**, *8*, 7. [CrossRef]
22. Nagulapati, V.M.; Lee, J.H.; Kim, H.S.; Oh, J.; Kim, I.T.; Hur, J.; Lee, S.G. Novel hybrid binder mixture tailored to enhance the electrochemical performance of SbTe bi-metallic anode for sodium ion batteries. *J. Electroanal. Chem.* **2020**, *865*, 114160. [CrossRef]
23. Oh, J.A.S.; Sun, J.; Goh, M.; Chua, B.; Zeng, K.; Lu, L. A robust solid-solid interface using sodium-tin alloy modified metallic sodium anode paving way for all-solid-state battery. *Adv. Energy Mater.* **2021**, *11*, 2101228. [CrossRef]
24. Qian, J.; Chen, Y.; Wu, L.; Cao, Y.; Ai, X.; Yang, H. High capacity Na-storage and superior cyclability of nanocomposite Sb/C anode for Na-ion batteries. *Chem. Commun. Camb* **2012**, *48*, 7070–7072. [CrossRef]
25. Sheng, M.; Zhang, F.; Ji, B.; Tong, X.; Tang, Y. A novel tin-graphite dual-ion battery based on sodium-ion electrolyte with high energy density. *Adv. Energy Mater.* **2017**, *7*, 1601963. [CrossRef]
26. Liu, Y.; Xu, Y.; Zhu, Y.; Culver, J.N.; Wang, C. Tin-coated viral nanoforests as sodium-ion battery anodes. *ACS Nano* **2013**, *7*, 3627–3634. [CrossRef]
27. Yu, H.; Seomoon, K.; Kim, J.; Kim, J.-K. Low-cost and highly safe solid-phase sodium ion battery with a Sn-C nanocomposite anode. *J. Ind. Eng. Chem.* **2021**, *100*, 112–118. [CrossRef]
28. Zhang, J.; Yin, Y.X.; Guo, Y.G. High-Capacity Te Anode Confined in Microporous Carbon for Long-Life Na-Ion Batteries. *ACS Appl. Mater. Interfaces* **2015**, *7*, 27838–27844. [CrossRef] [PubMed]
29. Zhang, Y.; Pan, A.; Ding, L.; Zhou, Z.; Wang, Y.; Niu, S.; Liang, S.; Cao, G. Nitrogen-Doped Yolk-Shell-Structured CoSe/C Dodecahedra for High-Performance Sodium Ion Batteries. *ACS Appl. Mater. Interfaces* **2017**, *9*, 3624–3633. [CrossRef]
30. Zhu, H.; Jia, Z.; Chen, Y.; Weadock, N.; Wan, J.; Vaaland, O.; Han, X.; Li, T.; Hu, L. Tin anode for sodium-ion batteries using natural wood fiber as a mechanical buffer and electrolyte reservoir. *Nano Lett.* **2013**, *13*, 3093–3100. [CrossRef] [PubMed]
31. Brezesinski, T.; Wang, J.; Tolbert, S.H.; Dunn, B. Ordered mesoporous alpha-MoO_3 with iso-oriented nanocrystalline walls for thin-film pseudocapacitors. *Nat. Mater.* **2010**, *9*, 146–151. [CrossRef]
32. Brezesinski, K.; Wang, J.; Haetge, J.; Reitz, C.; Steinmler, S.O.; Tolbert, S.H.; Smarsly, B.M.; Dunn, B.; Brezesinski, T. Pseudocapacitive Contributions to Charge Storage in Highly Ordered Mesoporous Group V Transition Metal Oxides with Iso-Oriented Layered Nanocrystalline Domains. *J. Am. Chem. Soc.* **2010**, *132*, 6982–6990. [CrossRef]
33. Pu, X.; Zhao, D.; Fu, C.; Chen, Z.; Cao, S.; Wang, C.; Cao, Y. Understanding and Calibration of Charge Storage Mechanism in Cyclic Voltammetry Curves. *Angew. Chem. Inter. Ed.* **2021**, *133*, 21480–21488. [CrossRef]
34. Chen, M.; Chen, L.; Hu, Z.; Liu, Q.; Zhang, B.; Hu, Y.; Gu, Q.; Wang, J.L.; Wang, L.Z.; Guo, X. Carbon-coated $Na_{3.32}Fe_{2.34}(P_2O_7)_2$ cathode material for high-rate and long-life sodium-ion batteries. *Adv. Mater.* **2017**, *29*, 1605535. [CrossRef]
35. Wu, L.; Ou Yang, J.; Guo, S.; Yao, L.; Li, H.; Zhang, S.; Yue, H.; Cai, K.; Zhang, C.; Yang, C. Pseudocapacitive trimetal $Fe_{0.8}CoMnO_4$ nanoparticles@carbon nanofibers as high-performance sodium storage anode with self-supported mechanism. *Adv. Funct. Mater.* **2020**, *30*, 2001718. [CrossRef]

Article

Bias-Voltage Dependence of Tunneling Decay Coefficient and Barrier Height in Arylalkane Molecular Junctions with Graphene Contacts as a Protecting Interlayer

Kyungjin Im, Dong-Hyoup Seo and Hyunwook Song *

Department of Applied Physics, Kyung Hee University, Yongin 17104, Korea; limkj0512@khu.ac.kr (K.I.); ehdguq1309@khu.ac.kr (D.-H.S.)
* Correspondence: hsong@khu.ac.kr

Abstract: We studied a molecular junction with arylalkane self-assembled monolayers sandwiched between two graphene contacts. The arrangement of graphene-based molecular junctions provides a stable device structure with a high yield and allows for extensive transport measurements at 78 K. We observed a temperature-independent current density–voltage (J–V) characteristic and the exponential dependency of the current density on the molecular length, proving that the charge transport occurs by non-resonant tunneling through the molecular barrier. Based on the Simmons model, the bias-voltage dependence of the decay coefficient and barrier height was extracted from variable-length transport characterizations. The J–V data measured were simulated by the Simmons model, which was modified with the barrier lowering induced by the bias voltage. Indeed, there is no need for adjustable fitting parameters. The resulting simulation was in remarkable consistency with experimental measurements over a full bias range up to |V| ≤ 1.5 V for the case of graphene/arylalkane/graphene heterojunctions. Our findings clearly showed the demonstration of stable and reliable molecular junctions with graphene contacts and their intrinsic charge transport characteristics, as well as justifying the application of the voltage-induced barrier lowering approximation to the graphene-based molecular junction.

Keywords: molecular tunnel junction; Simmons model; barrier lowering; graphene

Citation: Im, K.; Seo, D.-H.; Song, H. Bias-Voltage Dependence of Tunneling Decay Coefficient and Barrier Height in Arylalkane Molecular Junctions with Graphene Contacts as a Protecting Interlayer. *Crystals* **2022**, *12*, 767. https:// doi.org/10.3390/cryst12060767

Academic Editors: Bo Chen, Rutao Wang and Nana Wang

Received: 26 April 2022
Accepted: 24 May 2022
Published: 26 May 2022

Publisher's Note: MDPI stays neutral with regard to jurisdictional claims in published maps and institutional affiliations.

Copyright: © 2022 by the authors. Licensee MDPI, Basel, Switzerland. This article is an open access article distributed under the terms and conditions of the Creative Commons Attribution (CC BY) license (https:// creativecommons.org/licenses/by/ 4.0/).

1. Introduction

The utilization of inherent molecular electronic functionality has been recognized as a fascinating idea to create the ultimate sub-nanoscale devices that are able to offer distinctive characteristics unavailable in the conventional semiconductor technologies [1–3]. However, the electronic properties of a molecular junction based on individual molecules directly rely upon the accurate control of molecular configurations in the junction, and the atomic precision essential for reproducibility cannot be accomplished with the top-down fabrication techniques of today [4–7]. On this account, the vertical junction architectures sandwiched with self-assembled monolayers (SAMs) are beneficial for their technological application, but the nondestructive fabrication of SAM electrode top contacts is required [8–14]. The process for top-contact formation should maintain the integrity and functionality of molecular layers, while being simultaneously compatible with the high-yield construction of molecular electronic junctions. It has been revealed that molecular junctions fabricated with the evaporation of top metal atoms indicate a very low device yield, often less than 1% [8,13]. Until now, a variety of methods have been suggested to protect molecular SAMs from the evaporated metal atoms. Most of these methods have employed non-evaporative electrode systems using a single or multilayer graphene [6,15,16], reduced graphene oxide [17], conducting polymer [11,14], metallic nanoparticle [9,10], or direct metal transfer [12,13] (see Table S1). In particular, graphene and its derivatives have received considerable attention as the contact materials of molecular junctions due to their excellent electronic and optical properties [5,6,15–17]. In this study, we investigated a tunnel junction with arylalkane SAMs as the insulating barriers, which are sandwiched between the top and bottom graphene contacts. The bottom graphene allows for robust covalent C−C bonds with the component molecules, and the top graphene serves as a protecting layer against the vapor-deposited metal atoms to prevent the formation of electrical shorts. Overall, the arrangement of molecular junctions based on a graphene heterostructure provides a stable device structure with a high yield (>80%) and makes cryogenic measurements possible.

To explore the charge transport characteristics of molecular junctions, one of the critical challenges is the demonstration of an easily accessible theoretical model, which is highly desirable to design novel electronic devices with predictable transport behaviors. The Simmons model has been recognized as the simplest approximation to tunneling via a constant rectangular barrier [18], but it seems to be inaccurate for experimental measurements on actual molecular junctions [19–21]. The original model with a constant barrier height was previously applied to a tunnel junction containing alkanethiol SAMs, where an additional fitting parameter α was required to achieve the best fit to the measurements [22,23]. The origin of the adjustable parameter α remains obscure. It may result from the non-rectangular barrier, the effective mass of electrons in SAMs, or a combination of those factors [19,22,24]. Furthermore, as noticed by earlier studies [18,22], the barrier lowering induced by the bias voltage has to be considered for a practical tunnel junction. This effect has been still overlooked in molecular junctions. In consequence, different parameters for a molecular tunnel barrier were often obtained from the same junctions, giving rise to a substantial variation in the fitting results [18]. In this regard, it is essential to demonstrate a more precise model that can be compared directly to experimental measurements. Here, we employed the model based on the bias-voltage dependence of a barrier height to describe the transport measurements for graphene-based molecular junctions. The barrier height is not demanded as a fitting parameter. Rather, it is straightforwardly determined by variable-length transport characterization and then is incorporated in the Simmons approximation to simulate the experiments. Agreement between the simulation and experiments is excellent for the cases of graphene/arylalkane/graphene heterojunctions. These results solidify the application of the barrier lowering model to quantitative transport analysis for molecular tunnel junctions.

2. Experimental Details

The schematic of a device structure is seen in Figure 1a, which was based on arylalkane monolayers anchored between two (top and bottom) graphene electrodes. To fabricate the device, a pre-patterned Au(100 nm)/Ti(3 nm) bottom contact pad was first deposited on oxidized Si wafers using an electron-beam evaporator at the low evaporation rate of ~0.1 Å/s. A single layer graphene (from Graphene Square), grown on a Cu foil by chemical vapor deposition, was transferred as a bottom electrode on the oxidized Si substrate. We employed a well-known poly(methyl methacrylate) (PMMA)-mediated graphene transfer method [25], in which a thermal-release tape was applied to the PMMA film spin-coated on the bottom graphene. The Cu foils were eliminated in the solution of 20 g/L ammonium persulfate in distilled water for more than 10 h, and the release tapes and PMMA were removed by warm acetone after the graphene transfer. Afterward, the bottom graphene was patterned through a shadow mask by O_2 plasma treatment. For molecular deposition, 20~50 mM diazonium compounds of arylalkanes were dissolved within dimethylformamide (DMF). The graphene-transferred substrates were submersed in the diazonium solution for a minimum of 24 h. Arylalkanes for three different lengths, denoted in the following as ArC8, ArC10, and ArC12 by the number of a C atom in the alkyl chain, were anchored to the bottom graphene via a dediazonization reaction [26]. This process allowed for a robust covalent bond between the bottom graphene and aryl diazonium compounds. The arylalkanes on the bottom graphene were rinsed with DMF to remove residual molecules and then dried completely in a vacuum chamber. The top graphene and Au (100 nm) contact pad were deposited on the molecular layers, the same as the procedures to create the bottom electrodes. As a final step, the redundant graphene and molecules out of the junction area were eliminated by O_2 plasma to avoid parasitic conducting paths. Figure 1b shows a top-view image of scanning electron microscopy (SEM) for the whole device fabricated, where the junction area, highlighted in a yellow square box, ranged from 4.0×10^{-8} to 6.3×10^{-8} m^2 (200–250 μm on a side). We used a Raman spectroscopy (NOST, FEX, Seongnam-si, Korea) and atomic force microscopy (AFM) (Park Systems, XE7, Suwon-si, Korea) to inspect the pristine and monolayer-covered graphenes and a Keithley 4200A-SCS parameter analyzer (Keithley Instruments, Solon, OH, USA) and Janis cryostat for transport measurements.

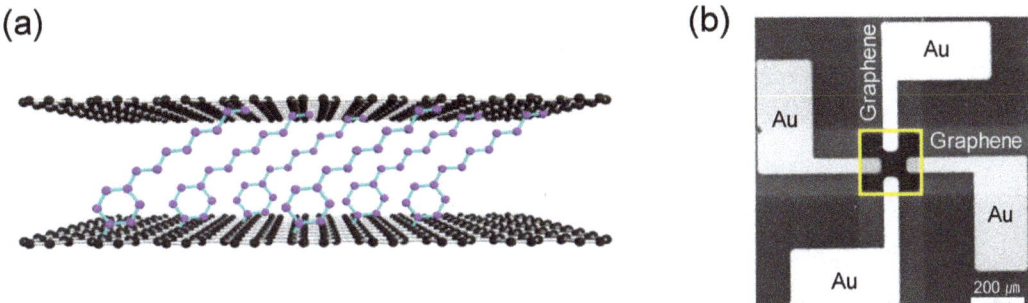

Figure 1. (**a**) Schematic of graphene/arylalkane/graphene heterojunctions. The arylalkane (ArC8) monolayer is sandwiched between top and bottom graphene contacts. (**b**) An SEM image of the fabricated device. The active area of a molecular junction is highlighted by a yellow square box.

3. Results and Discussion

The fabrication of a molecular junction based on SAMs mostly involves the deposition of the metal atoms for a top-side contact on a few nanometer-thick molecular layers. This process frequently results in a short circuit failure due to direct penetration of the evaporated metal atoms through SAMs, thus destroying the molecular junctions. To prevent this problem, here we employed the graphene film as a protecting interlayer for the stable formation of a molecule–electrode top contact. We also note that the graphene contacts did not take part in the molecular tunnel junctions as actively tunable components with the gate modulation [27,28], which was beyond the scope of the present study.

We measured the Raman spectra of the pristine and arylalkane (ArC10)-covered graphenes transferred on the substrates (see Figure S1). Two typical prominent peaks of the G band (1581 cm^{-1}) and 2D band (2674 cm^{-1}) and a large ratio (>~2.3) of Raman intensity between the G and 2D bands indicated that the transferred graphene film was a single layer [29]. The absence of the D peak (1349 cm^{-1}) in the pristine graphene implied that it was of good quality without substantial defects. The Raman peak corresponding to the D band appeared after the attachment of the arylalkane molecules to the graphene, which can be caused by the defects, resulting from the formation of a covalent C–C sp^3 bond between the molecules and graphene [26,29]. From the line profile analysis of the AFM images (see Figure S2), the height of the ArC10-grafted and pristine graphenes' layers transferred to the SiO$_2$ surface was measured as 2.5 and 0.5 nm, respectively. The offset of ~2 nm was nearly comparable to the molecular length. In total, the results for the Raman spectroscopy and AFM measurements consistently indicated the SAM formation onto the bottom graphene.

As the Fermi level (E_F) of electrodes is located within a large HOMO–LUMO gap of the short-length arylalkane series, tunneling via the molecular barrier can be rationally anticipated as the dominant transport mechanism of these junctions [22,23]. However, in the absence of variable-temperature characterization, other thermally activated mechanisms such as hopping conduction or thermionic emission cannot be ruled out. Figure 2a shows the representative semilogarithmic J–V characteristic of an ArC8 junction measured from the device structure described in Figure 1. It was obtained in a sufficiently wide temperature variation (from 298 to 78 K) and with 20 K steps to verify the conduction mechanism. The current density barely changed, which was in agreement with the results measured with similar metal/alkyl molecules/metal junctions. Figure 2b displays an Arrhenius analysis of the current density. The slope of ln(J) against 1/T at different biases exhibited no significant dependence, thus manifesting the absence of thermal activation. The measurements on ArC10 and ArC12 showed also little temperature dependence (see Figure S3). Hence, we concluded that charge tunneling is maintained as the conduction mechanism through arylalkane monolayers incorporated between graphene electrodes. Now that we established tunneling as the transport mechanism, theoretical calculations from a tunneling model could be used to simulate our molecular junctions based on a graphene heterostructure.

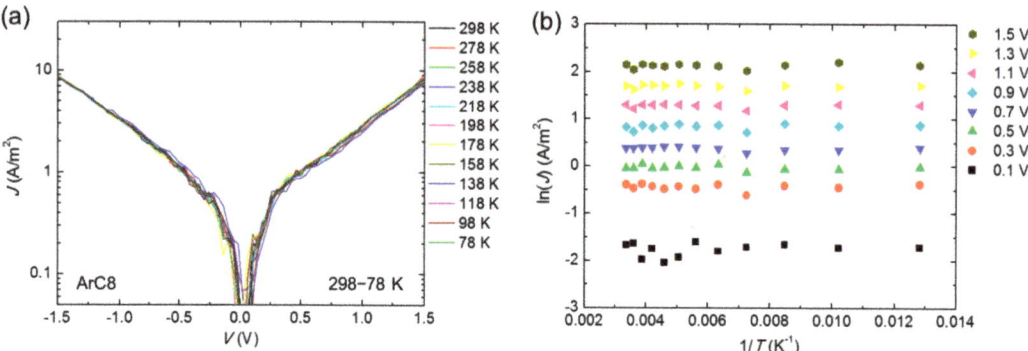

Figure 2. (a) Variable-temperature J–V characteristics with temperature variation from 298 to 78 K with 20 K steps, displayed in different color lines. (b) Arrhenius plot of ArC8 junctions at voltages from 0.1 to 1.5 V with 0.2 V steps.

One of the most essential parameters related to tunneling through a molecular junction is the decay coefficient (β), which can be determined by performing the length-dependent transport measurements based on the Simmons model [11,22,24]. Figure 3a shows ln(J) values plotted against the molecular length of arylalkanes at the different biases. By using ACD/Lab software, the molecular lengths of ArC8, ArC10, and ArC12 were estimated to be 13.3, 18.2, and 23.2 Å, respectively. The tunneling current densities were calculated from the average of approximately 80 devices per each molecule, indicating exponential dependency upon the molecular length (d) for a given bias. This length dependence represented a general feature of non-resonant tunneling, described by $J \propto \exp(-\beta d)$ [8,13,23]. The decay coefficient at each bias was determined from the slopes of linear fits and is displayed in Figure 3b. The uncertainty of the β values in Figure 3b denotes a linear fit error. The decay coefficients ranged from 1.08 to 0.92 Å^{-1}, dependent on the applied voltages. These values were similar to that of alkyl-based junctions with metallic contacts [13,24]. Our analysis of this length-dependent transport showed that the arylalkane monolayer not only maintained its molecular integrity in the junction but also predominated the charge transport properties and was not disturbed by the graphene–molecule contacts. Overall, temperature-independent and exponentially length-dependent J–V characteristics consistently showed the non-resonant tunneling transport for these junctions. These findings ruled out defect-mediated transport in the molecular tunnel barrier and graphene contacts, which is typically expected to show thermally activated transport characteristics [10,13].

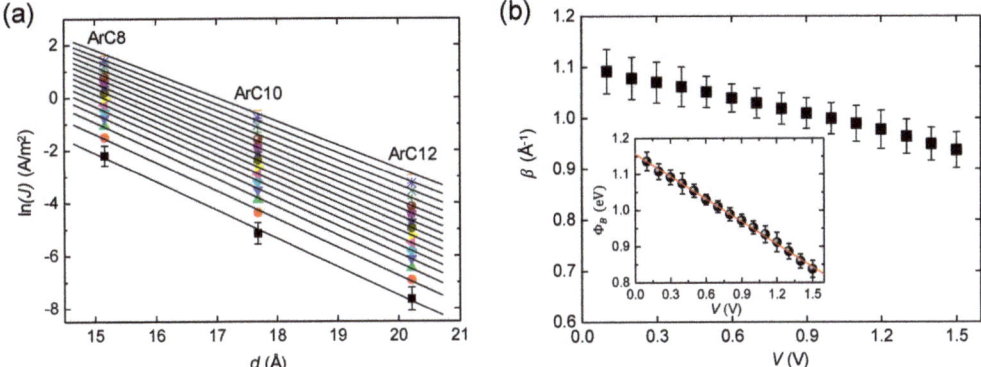

Figure 3. (**a**) Plot of ln(*J*) versus *d* for the bias range from 0.1 (bottom) to 1.5 V (top). For clarity, the data only at 0.1 V contain the error bar, indicating the standard deviation upon averaging the current density. The β values were determined from the slope of the linear fits (solid line). (**b**) Plot of β versus *V*. The inset shows the bias-voltage dependence of the barrier height. The solid line indicates the linear fit.

Within the Simmons model of a rectangular barrier (i.e., $\alpha = 1$), the relationship between the tunneling decay coefficient and barrier height (Φ_B) is given by [21,27]:

$$\beta = \frac{2\sqrt{2m}}{\hbar}\sqrt{\Phi_B} \qquad (1)$$

where *m* is the bare electron mass and h ($= 2\pi\hbar$) is the Planck constant. Using Equation (1), the Simmons barrier heights calculated from the β values at different biases are plotted in the inset of Figure 3b, where the barrier height linearly depends on the magnitude of the bias voltage. Our findings indicated that the increasing bias voltage led to lower tunneling barrier, thus decreasing the decay coefficient. This effect may be attributed to image potential, which has been known to reduce the barrier height and, therefore, enhances the probability of charge tunneling across molecular barriers [18,19]. The extraction of β and Φ_B was then straightforward by performing the length-dependent transport characterization, without adjustable fitting parameters. Based on the linear relation (red, solid line) observed in the inset of Figure 3b, the voltage dependence of the barrier height could be approximated as

$$\Phi_B(V) = -\gamma|eV| + \Phi_{B0} \qquad (2)$$

where $\Phi_B(V)$ and Φ_{B0} are defined as the bias-voltage dependent and zero-bias barrier height, respectively, in units of eV, and γ signifies the degree of the voltage dependency on the barrier height as a unitless constant. In this linear relationship, γ and Φ_{B0} can be determined from the slope and *y*-intercept, respectively, by which $\gamma = 0.21 \pm 0.01$ and $\Phi_{B0} = 1.16 \pm 0.01$ eV are found in the inset of Figure 3b. The voltage dependence of a barrier height was also previously reported for porphyrin molecular junctions formed by a conducting AFM technique without the length-dependent measurements [30].

The statistically representative *J*–*V* curves (data points) are presented in Figure 4 and were obtained by averaging all the data measured at 78 K for each arylalkane molecule with different lengths. The error bars indicate a 1σ standard error of the mean value. The traces of the *J*–*V* curves are almost symmetric in respect to the origin and display a nonlinearity that was shown to be more prominent as increasing the bias We initially tested the original Simmons model (see Equation (S1)) based on a constant rectangular barrier to simulate the representative *J*–*V* curves [18], where the barrier height was used as a fitting parameter. Apparently, there was inconsistency between the constant barrier approximation and the experimental *J*–*V* curves. As seen in Figure 4, the fitted current density (black, solid curves) with the constant barrier increased abruptly near ±1.1 V from an almost flat value, whereas the experimental J–V data rose gradually in a sigmoidal shape. This fitting result showed that the Simmons model based on the constant barrier height was not suitable to characterize the J–V curves of the graphene tunnel junctions without a further adjustable parameter.

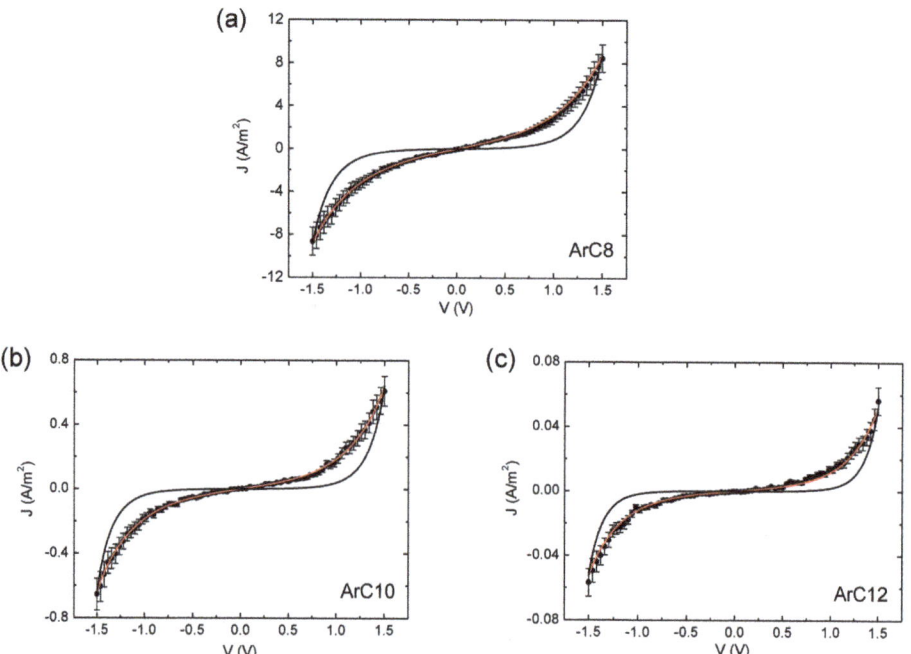

Figure 4. The statistically representative J–V curves (data points), obtained by averaging all the data measured at 78 K for (**a**) ArC8, (**b**) ArC10, and (**c**) ArC12 junctions. Black curves indicate the fitting results of the original Simmons model with a constant barrier height, and red curves denote the simulation of the VIBL-modified Simmons model.

Therefore, we presented an alternative method, where the Simmons model was modified with a voltage-induced barrier lowering (VIBL), quantitatively characterized by the length-dependent transport measurements. The significance of VIBL is that it allows for our experimental J–V curves to be simulated over a full bias range of $|V| \leq 1.5$ V by incorporating Equation (2) as the VIBL parameter in the Simmons model. Then, the modified Simmons equation expresses the current density through a molecular tunnel barrier as:

$$J = \frac{e}{4\pi^2 \hbar d^2} \left\{ \left(\Phi_B(V) - \frac{eV}{2} \right) \exp\left(-\frac{2d\sqrt{2m}}{\hbar} \sqrt{\Phi_B(V) - \frac{eV}{2}} \right) - \left(\Phi_B(V) + \frac{eV}{2} \right) \exp\left(-\frac{2d\sqrt{2m}}{\hbar} \sqrt{\Phi_B(V) + \frac{eV}{2}} \right) \right\} \quad (3)$$

Figure 4 demonstrates that the representative J–V characteristics averaged from extensive measurements on the tunnel junctions can be described with the modified Simmons simulation (red, solid curves) using Equation (3), where VIBL is included as $\Phi_B(V)$ in the model. As described above, the VIBL parameter $\Phi_B(V)$ was experimentally determined by the length-dependent analysis using Equations (1) and (2). It is noted that the J–V curves were not "fitted" and, thus, there was no need for the adjustable fitting parameters. The simulation with $\Phi_B(V)$ was accurate within the errors of measurements over the entire bias range, offering a better description for the measured J–V curves, as compared to the conventional constant barrier approximation. A clear consensus between the experimental and simulated transport characteristics demonstrated the validity for the use of the Simmons model combined with VIBL for the case of our graphene-based molecular junctions.

A critical verification of the VIBL approximation is if this model can exactly predict the experimental results. For a self-consistency check, we replotted the experimental and simulated data, as displayed in Figure 5, where the resistance R multiplied by the junction area A is shown on a logarithmic scale against the bias voltage for each molecule investigated [19]. Simmons earlier presented this analysis in the original paper to examine the effect of fitting parameters on the tunneling model [18]. We note that the log plot of RA versus V was beneficial to especially reveal the further details concerning the validity of the tunneling model. The simulations plotted using the VIBL model (red lines) precisely described the measurements (data points) in the whole bias region, as likewise observed in Figure 4. The fitting results with the constant barrier model are also plotted in Figure 5 (black lines), where the disagreement between the experiment and model is much more profound and the fits clearly seem to be not accurate over the entire voltage regime. With the barrier height alone as a fitting parameter, the analysis of RA versus V indicated that it was not possible to characterize the molecular tunnel junctions based on graphene electrodes.

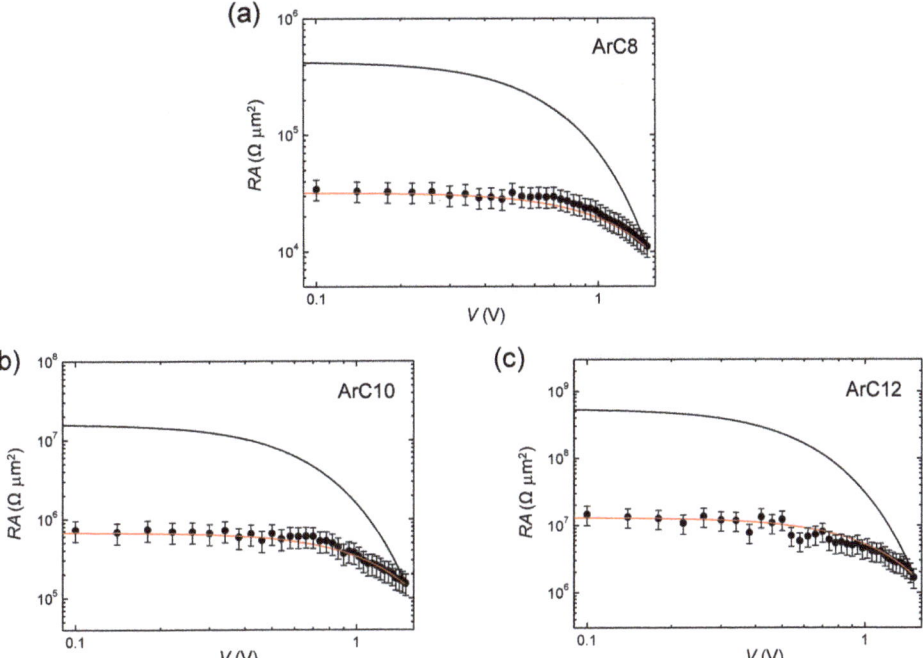

Figure 5. Log plots of RA versus V for (**a**) ArC8, (**b**) ArC10, and (**c**) ArC12 junctions. The fits with the original Simmons model for a constant barrier height (black curves) and the simulation of the VIBL-modified Simmons model (red curves) are shown.

4. Conclusions

The charge transport through vertical graphene/arylalkane/graphene heterojunctions was investigated by the Simmons model combined with the bias-voltage dependence of a tunneling barrier height. Both the temperature independence and exponential length dependence of the measured J–V characteristics clearly showed that non-resonant tunneling is the dominant conduction mechanism in these junctions. The tunneling decay coefficients at different bias voltages were determined by inspecting an exponential decrease in the current density with the molecular length, for an arylalkane series contacted between top and bottom graphene electrodes. The decay coefficient ranged from 1.08 to 0.92 Å^{-1} depending on the bias voltage. Based on the Simmons model, the length-dependent transport analysis indicated $\gamma = 0.21 \pm 0.01$ and $\Phi_{B0} = 1.16 \pm 0.01$ eV for the arylalkane tunnel junctions. The critical parameter $\Phi_B(V)$ of VIBL was experimentally extracted from the bias dependence of β. Then, we employed the Simmons model modified with $\Phi_B(V)$ to simulate the charge transport characteristics measured from graphene-based molecular junctions. Irrespective of the molecular length, the representative J–V data were in agreement with the simulation without adjustable fitting parameters over the whole range of applied voltage ($|V| \leq 1.5$ V), which was consistently revealed by the analysis of RA versus V plots on a logarithmic scale. The high quality of the simulations justifies the use of the VIBL-modified Simmons model for analysis of graphene-based molecular junctions. As presented in this study, the graphene contacts in molecular junctions effectively protect the molecular SAMs from penetration by the evaporated metal atoms, providing a stable test bed for the transport measurement. In addition, the incorporation of a gate electrode may become a fascinating future work to modulate the tunneling transport through graphene-based molecular devices.

Supplementary Materials: The following supporting information can be downloaded at: https://www.mdpi.com/article/10.3390/cryst12060767/s1, Figure S1: Raman spectrum for a pristine and ArC10-coverd graphene; Figure S2: AFM topographical images and line profile analyses of a pristine and ArC10-coverd graphene; Figure S3: Arrhenius plots of ArC10 and ArC12 junctions; Table S1: Comparison of various non-evaporative top contact molecular junctions [10].

Author Contributions: Conceptualization, H.S.; formal analysis, K.I. and D.-H.S.; investigation, K.I. and D.-H.S.; methodology, K.I.; supervision, H.S.; writing—original draft, H.S.; writing—review and editing, K.I., D.-H.S and H.S. All authors have read and agreed to the published version of the manuscript.

Funding: This work was supported by the National Research Foundation of Korea (grant No. 2020R1F1A1076107).

Institutional Review Board Statement: Not applicable.

Informed Consent Statement: Not applicable.

Data Availability Statement: Data is contained within the article or supplementary material.

Conflicts of Interest: The authors declare no conflict of interest.

References

1. Chen, H.; Stoddart, J.F. From molecular to supramolecular electronics. *Nat. Rev. Mater.* **2018**, *6*, 804–828. [CrossRef]
2. Bryce, M.R. A review of functional linear carbon chains (oligoynes, polyynes, cumulenes) and their applications as molecular wires in molecular electronics and optoelectronics. *J. Mater. Chem. C* **2021**, *9*, 10524–10546. [CrossRef]
3. Gupta, R.; Jash, P.; Sachan, P.; Bayat, A.; Singh, V.; Mondal, P.C. Electrochemical Potential-Driven High-Throughput Molecular Electronic and Spintronic Devices: From Molecules to Applications. *Angew. Chem. Int. Ed.* **2021**, *60*, 26904–26921. [CrossRef]
4. Xin, N.; Guan, J.; Zhou, C.; Chen, X.; Gu, C.; Li, Y.; Ratner, M.A.; Nitzan, A.; Stoddart, J.F.; Guo, X. Concepts in the design and engineering of single-molecule electronic devices. *Nat. Rev. Phys.* **2019**, *1*, 211–230. [CrossRef]
5. Fu, H.; Zhu, X.; Li, P.; Li, M.; Yang, L.; Jia, C.; Guo, X. Recent progress in single-molecule transistors: Their designs, mechanisms and applications. *J. Mater. Chem. C* **2022**, *10*, 2375–2389. [CrossRef]
6. Caneva, S.; Gehring, P.; García-Suárez, V.M.; García-Fuente, A.; Stefani, D.; Olavarria-Contreras, I.J.; Ferrer, J.; Dekker, C.; Van Der Zant, H.S.J. Mechanically controlled quantum interference in graphene break junctions. *Nat. Nanotechnol.* **2018**, *13*, 1126–1131. [CrossRef]
7. Zhang, J.L.; Zhong, J.Q.; Lin, J.D.; Hu, W.P.; Wu, K.; Xu, G.Q.; Wee, A.T.S.; Chen, W. Towards single molecule switches. *Chem. Soc. Rev.* **2015**, *44*, 2998–3022. [CrossRef] [PubMed]
8. Gorenskaia, E.; Turner, K.L.; Martín, S.; Cea, P.; Low, P.J. Fabrication of metallic and non-metallic top electrodes for large-area molecular junctions. *Nanoscale* **2021**, *13*, 9055–9074. [CrossRef] [PubMed]
9. Martín-Barreiro, A.; Soto, R.; Chiodini, S.; García-Serrano, A.; Martín, S.; Herrer, L.; Pérez-Murano, F.; Low, P.J.; Serrano, J.L.; Marcos, S.; et al. Uncapped Gold Nanoparticles for the Metallization of Organic Monolayers. *Adv. Mater. Interfaces* **2021**, *8*, 2100876. [CrossRef]

10. Puebla-Hellmann, G.; Venkatesan, K.; Mayor, M.; Lörtscher, E. Metallic nanoparticle contacts for high-yield, ambient-stable molecular-monolayer devices. *Nature* **2018**, *559*, 232–235. [CrossRef]
11. Akkerman, H.B.; Blom, P.W.M.; de Leeuw, D.M.; de Boer, B. Towards molecular electronics with large-area molecular junctions. *Nature* **2006**, *441*, 69–72. [CrossRef] [PubMed]
12. Loo, Y.-L.; Lang, D.V.; Rogers, A.J.A.; Hsu, J.W.P. Electrical Contacts to Molecular Layers by Nanotransfer Printing. *Nano Lett.* **2003**, *3*, 913–917. [CrossRef]
13. Jeong, H.; Kim, D.; Kwon, H.; Hwang, W.-T.; Jang, Y.; Min, M.; Char, K.; Xiang, D.; Jeong, H.; Lee, T. Statistical investigation of the length-dependent deviations in the electrical characteristics of molecular electronic junctions fabricated using the direct metal transfer method. *J. Phys. Condens. Matter* **2016**, *28*, 94003. [CrossRef] [PubMed]
14. Neuhausen, A.B.; Hosseini, A.; Sulpizio, J.A.; Chidsey, C.E.D.; Goldhaber-Gordon, D. Molecular Junctions of Self-Assembled Monolayers with Conducting Polymer Contacts. *ACS Nano* **2012**, *6*, 9920–9931. [CrossRef]
15. Wang, G.; Kim, Y.; Choe, M.; Kim, T.-W.; Lee, T. A New Approach for Molecular Electronic Junctions with a Multilayer Graphene Electrode. *Adv. Mater.* **2011**, *23*, 755–760. [CrossRef]
16. Seo, S.; Min, M.; Lee, S.M.; Lee, H. Photo-switchable molecular monolayer anchored between highly transparent and flexible graphene electrodes. *Nat. Commun.* **2013**, *4*, 1920. [CrossRef]
17. Kühnel, M.; Overgaard, M.H.; Hels, M.C.; Cui, A.; Vosch, T.; Nygård, J.; Li, T.; Laursen, B.W.; Nørgaard, K. High-Quality Reduced Graphene Oxide Electrodes for Sub-Kelvin Studies of Molecular Monolayer Junctions. *J. Phys. Chem. C* **2018**, *122*, 25102–25109. [CrossRef]
18. Simmons, J.G. Generalized Formula for the Electric Tunnel Effect between Similar Electrodes Separated by a Thin Insulating Film. *J. Appl. Phys.* **1963**, *34*, 1793–1803. [CrossRef]
19. Akkerman, H.B.; Naber, R.C.G.; Jongbloed, B.; van Hal, P.A.; Blom, P.W.M.; de Leeuw, D.M.; de Boer, B. Electron tunneling through alkanedithiol self-assembled monolayers in large-area molecular junctions. *Proc. Natl. Acad. Sci. USA* **2007**, *104*, 11161–11166. [CrossRef]
20. Huisman, E.H.; Guédon, C.M.; van Wees, B.J.; van der Molen, S.J. Interpretation of Transition Voltage Spectroscopy. *Nano Lett.* **2009**, *9*, 3909–3913. [CrossRef]
21. Xie, Z.; Bâldea, I.; Smith, C.E.; Wu, Y.; Frisbie, C.D. Experimental and Theoretical Analysis of Nanotransport in Oligophenylene Dithiol Junctions as a Function of Molecular Length and Contact Work Function. *ACS Nano* **2015**, *9*, 8022–8036. [CrossRef]
22. Wang, W.; Lee, T.; Reed, M.A. Mechanism of electron conduction in self-assembled alkanethiol monolayer devices. *Phys. Rev. B* **2003**, *68*, 035416. [CrossRef]
23. Holmlin, R.E.; Haag, R.; Chabinyc, M.L.; Ismagilov, R.F.; Cohen, A.E.; Terfort, A.; Rampi, M.A.; Whitesides, G.M. Electron Transport through Thin Organic Films in Metal−Insulator−Metal Junctions Based on Self-Assembled Monolayers. *J. Am. Chem. Soc.* **2001**, *123*, 5075–5085. [CrossRef] [PubMed]
24. Engelkes, V.B.; Beebe, A.J.M.; Frisbie, C.D. Length-Dependent Transport in Molecular Junctions Based on SAMs of Alkanethiols and Alkanedithiols: Effect of Metal Work Function and Applied Bias on Tunneling Efficiency and Contact Resistance. *J. Am. Chem. Soc.* **2004**, *126*, 14287–14296. [CrossRef]
25. Liang, X.; Sperling, B.A.; Calizo, I.; Cheng, G.; Hacker, C.; Zhang, Q.; Obeng, Y.; Yan, K.; Peng, H.; Li, Q.; et al. Toward Clean and Crackless Transfer of Graphene. *ACS Nano* **2011**, *5*, 9144–9153. [CrossRef] [PubMed]
26. MacLeod, J.M.; Rosei, F. Molecular Self-Assembly on Graphene. *Small* **2014**, *10*, 1038–1049. [CrossRef] [PubMed]
27. Kumar, S.B.; Seol, G.; Guo, J. Modeling of a vertical tunneling graphene heterojunction field-effect transistor. *Appl. Phys. Lett.* **2012**, *101*, 033503. [CrossRef]
28. Fallahazad, B.; Lee, K.; Kang, S.; Xue, J.; Larentis, S.; Corbet, C.; Kim, K.; Movva, H.C.P.; Taniguchi, T.; Watanabe, K.; et al. Gate-Tunable Resonant Tunneling in Double Bilayer Graphene Heterostructures. *Nano Lett.* **2015**, *15*, 428–433. [CrossRef]
29. Zhu, Y.; Murali, S.; Cai, W.; Li, X.; Suk, J.W.; Potts, J.R.; Ruoff, R.S. Graphene and Graphene Oxide: Synthesis, Properties, and Applications. *Adv. Mater.* **2010**, *22*, 3906–3924. [CrossRef]
30. Nawarat, P.; Beach, K.; Meunier, V.; Terrones, H.; Wang, G.-C.; Lewis, K.M. Voltage-Dependent Barrier Height of Electron Transport through Iron Porphyrin Molecular Junctions. *J. Phys. Chem. C* **2021**, *125*, 7350–7357. [CrossRef]

Article

Demonstration of Molecular Tunneling Junctions Based on Vertically Stacked Graphene Heterostructures

Seock-Hyeon Hong, Dong-Hyoup Seo and Hyunwook Song *

Department of Applied Physics, Kyung Hee University, Yongin 17104, Korea; gusbus97@khu.ac.kr (S.-H.H.); ehdguq1309@khu.ac.kr (D.-H.S.)
* Correspondence: hsong@khu.ac.kr

Abstract: We demonstrate the fabrication and complete characterization of vertical molecular tunneling junctions based on graphene heterostructures, which incorporate a control series of arylalkane molecules acting as charge transport barriers. Raman spectroscopy and atomic force microscopy were employed to identify the formation of the molecular monolayer via an electrophilic diazonium reaction on a pre-patterned bottom graphene electrode. The top graphene electrode was transferred to the deposited molecular layer to form a stable electrical connection without filamentary damage. Then, we showed proof of intrinsic charge carrier transport through the arylalkane molecule in the vertical tunneling junctions by carrying out multiprobe approaches combining complementary transport characterization methods, which included length- and temperature-dependent charge transport measurements and transition voltage spectroscopy. Interpretation of all the electrical characterizations was conducted on the basis of intact statistical analysis using a total of 294 fabricated devices. Our results and analysis can provide an objective criterion to validate molecular electronic devices fabricated with graphene electrodes and establish statistically representative junction properties. Since many of the experimental test beds used to examine molecular junctions have generated large variation in the measured data, such a statistical approach is advantageous to identify the meaningful parameters with the data population and describe how the results can be used to characterize the graphene-based molecular junctions.

Keywords: molecular junction; graphene electrode; charge tunneling; transition voltage spectroscopy

Citation: Hong, S.-H.; Seo, D.-H.; Song, H. Demonstration of Molecular Tunneling Junctions Based on Vertically Stacked Graphene Heterostructures. *Crystals* **2022**, *12*, 787. https://doi.org/10.3390/cryst12060787

Academic Editors: Bo Chen, Rutao Wang and Nana Wang

Received: 1 May 2022
Accepted: 24 May 2022
Published: 29 May 2022

Publisher's Note: MDPI stays neutral with regard to jurisdictional claims in published maps and institutional affiliations.

Copyright: © 2022 by the authors. Licensee MDPI, Basel, Switzerland. This article is an open access article distributed under the terms and conditions of the Creative Commons Attribution (CC BY) license (https://creativecommons.org/licenses/by/4.0/).

1. Introduction

Molecular electronics has the intention of creating a molecular junction device based on an individual molecule or its ensemble, whose current (I)−voltage (V) characteristics show the signatures of conventional electronic components or offer new electrical behaviors at the molecular level [1–3]. A variety of electronic functionalities demonstrated by molecular junctions, such as a diode [4], transistor [5], memory [6], photo-switching [7], and thermoelectric device [8], have been hitherto reported, which constitute a prospective component for future nanoscale electronic systems, as well as offer an ideal platform to explore new physical properties that occur in the charge transport through molecular systems. However, many challenges still have to be resolved for the technological applications and full understanding of molecular charge transport. The ensemble molecular junction (as opposed to a single-molecule junction) is typically built by sandwiching self-assembled monolayers between two thin metallic films. Such vertical junction arrangement most frequently involves the evaporative deposition of the top (second) metallic electrodes on vulnerable molecular monolayers, which can give rise to conductive filament formation, often leading to shorted circuit problems [9–12]. Comprehensive statistical studies on molecular junctions fabricated by the direct vapor deposition of top metal atoms have shown extremely low yields of working devices, frequently less than ~1% [12]. In this context, various methods have been proposed to prevent filamentary paths or related damage to the ultrathin molecular layer, arising from the evaporated metal atoms. Many of these methods take advantage of non-evaporative electrode systems using multilayer graphenes [13,14], reduced graphene oxides [15], conducting polymers [16], non-Newtonian liquid metals [9], or direct metal transfers [17]. Recently, the application of singe-layer graphenes (SLGs) to an electrical contact material in the junctions has gained substantial interest because of their outstanding mechanical, optical and electrical performance [8,18,19]. The SLG electrodes provide the possibility of using unique quantum transport phenomena with the control of Dirac points and high mobility [20]. They are also ultimately compatible with molecular self-assembled monolayers via both a chemical linkage and noncovalent interaction [21,22], offering a stable test bed to explore inherent molecular charge transport characteristics in the junctions.

In the present study, we demonstrate the fabrication and full characterization of vertical molecular tunneling junctions that incorporate a series of arylalkane monolayers inserted between two SLG interlayer electrodes. Raman spectra of the graphene sheets, deposited with aryl diazonium compounds and then modulated by the molecular doping effect, indicated that the component molecules were successfully grafted onto SLGs. The arylalkane monolayers can constitute prototypical molecular control series to corroborate the valid junction formation, because the alkyl-based molecules with different lengths have exhibited well-established charge transport pictures. The nearest molecular transport orbital (HOMO or LUMO) remains far above or below the Fermi level (E_F) of the electrode, and, accordingly, coherent non-resonant tunneling can be reasonably expected as the dominant conduction mechanism [23,24]. To investigate the transport behaviors of the vertical molecular junctions in detail, we employed various kinds of characterization techniques, including length- and temperature-dependent transport measurements and transition voltage spectroscopy, which were performed with follow-up statistical analysis for all the measured devices. Self-consistency for such multiprobe measurements validates the observation of intrinsic molecular electronic properties in the junctions fabricated with the graphene electrodes.

2. Experimental Details

Figure 1a illustrates a fabrication process for constructing a vertical molecular tunneling junction with two SLG interlayer electrodes. First, the bottom 50 nm-thick Au contact lead with an adhesion layer of 5 nm-thick Ti was patterned on Si/SiO$_2$ substrates using a shadow mask and electron-gun evaporator at a deposition rate of ~0.2 Å/s (step 1). After the treatment of O$_2$ plasma to improve the surface wettability, chemical vapor deposition (CVD)-grown SLGs (from Graphene Square, Pohang, Korea) on a Cu foil were transferred to the substrates using a polymethyl methacrylate (PMMA)-mediated method [25]. Thermal release tapes were attached to spin-coated PMMA films on the transferred graphene sheet. The copper foil layers were removed with 20 g/L ammonium persulfate solution in distilled water for 12 h. The remaining supporting tapes and PMMA films were eliminated using warm acetone after the transfer process. Then, the bottom SLGs were patterned by O$_2$ plasma etching with a shadow mask (step 2). The arylalkane molecule of three different lengths, indicated as C8, C10 and C12, respectively, according to the number of carbons in the alkyl chain, was covalently grafted to the bottom SLGs (step 3). For molecular self-assembly, the samples were immersed in 10 mM arylalkane diazonium solution in dimethylformamide (DMF) for over 12 h in a N$_2$ glove box (less than O$_2$ level of 10 ppm). Before further progress, they were thoroughly cleansed using DMF and dried inside the glove box. Thereafter, the top SLGs and the contact lead of 50 nm-thick Au were formed on the monolayers (steps 4 and 5) by repeating the same process as described previously when constructing the bottom electrodes. Finally, the residual graphene layers and molecules outside the active area of the junctions were removed using O$_2$ plasma treatment with a shadow mask to prevent the formation of a direct conducing pathway between the top and bottom electrodes (step 6). Figure 1b shows an optical image of the fabricated device, where the junction area, highlighted with a red dashed line, was estimated to be 250 × 250 µm^2.

The graphene layers in pristine condition and with arylalkane molecules were characterized using Raman spectroscopy (NOST, FEX, Seongnam, Korea) and atomic force microscopy (AFM) (Park Systems, XE7, Suwon, Korea). The electrical characterization was carried out using parameter analyzers (Keithley, 4200A-SCS, Solon, OH, USA) and cryogenic probe stations (Lake Shore Cryotronics, Model TTPX, Westerville, OH, USA).

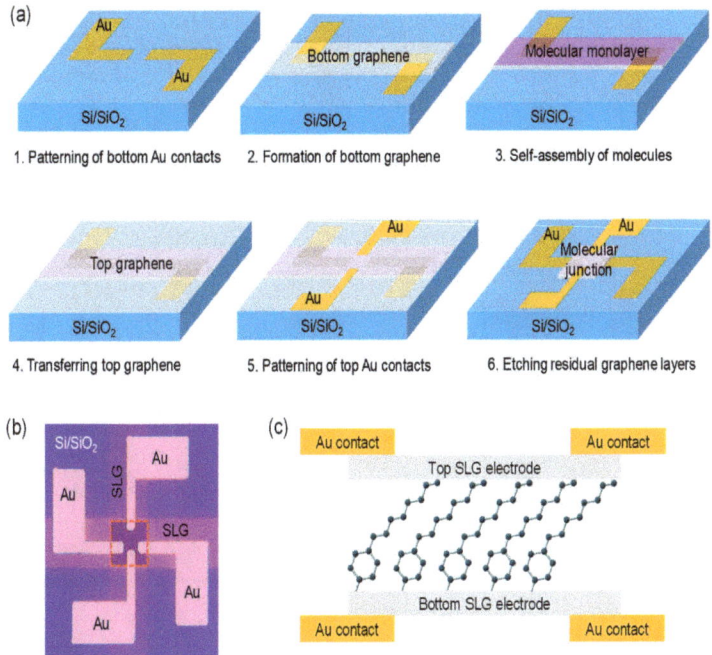

Figure 1. (**a**) Illustration of the fabrication process (step 1 to step 6) of the vertically stacked graphene/arylalkane/graphene device. (**b**) Optical image of a complete device. Red dashed square denotes the area of an active junction, which was estimated to be 250 × 250 μm^2. (**c**) Schematic of the device structure in a cross-sectional view. The component molecule in the junction is C8.

3. Results and Discussion

The arylalkane molecules were self-assembled on the patterned bottom SLGs by a dediazonization process of sp^2 hybridization carbon networks, which can form a robust covalent C−C bond between the molecules and graphene basal planes [21,22]. contrary, the top end of the molecular monolayers was contacted with the top SLGs by means of van der Waals interactions. Figure 1c illustrates the schematic of the complete device structure in a cross-sectional view. The arrangement based on vertically stacked graphene heterostructures showed excellent stability for high-yield devices (>80%) and successive electrical measurements. By performing a line profile analysis of the AFM topographical images (Figure S1 in the Supplementary Materials), the heights of a pristine and C10-grafted graphene layer transferred to the SiO$_2$ surface were measured as 0.5 nm and 2.5 nm, respectively. The height difference of ~2 nm reasonably confirmed the monolayer formation on SLGs, as compared to the molecular length estimated from ChemDraw (CambridgeSoft, Cambridge, MA, USA).

Raman spectroscopy has been extensively used as an analytical technique for the non-destructive investigation of SLGs and their derivatives decorated with the organic molecules [26,27]. We measured the Raman spectra of the pristine and arylalkane-grafted graphenes transferred on the substrates, as shown in Figure 2, which were obtained at 10 different spots with a 532 nm laser. The assigned peak position and intensity in each spectrum were determined by Lorentzian function. The spectrum of the pristine graphene sheet entirely concurred with previously reported results [26], where two prominent peaks of G band (1582 cm^{-1}) and 2D band (2675 cm^{-1}) appeared. The omission of D band (1350 cm^{-1}), typically induced by disorder or defects in the graphene basal plane [27], implied that the CVD graphene was of good quality. A large Raman intensity ratio between 2D and G bands ($I_{2D}/I_G > 2.4$) indicated that the transferred graphene film was a single layer [28]. Noticeably, the Raman peak corresponding to D band appeared after the attachment of the arylalkane molecules to SLGs. The D peak can be caused by increased defects resulting from the formation of a covalent C–C sp^3 bond between the molecules and SLGs [22]. After molecular deposition, we also observed that the intensity ratio of the 2D band against the G band (I_{2D}/I_G) decreased, and the position of the G peak was downshifted (Figure S2 in the Supplementary Materials). It has been reported that such a decrease in the ratio of I_{2D}/I_G results from a molecular doping effect and G peak's shift to lower frequency, namely, softening of the G band indicates electron donating on graphene [29]. Collectively, our observation in the Raman spectra consistently showed the signatures of molecular layer formation on SLGs.

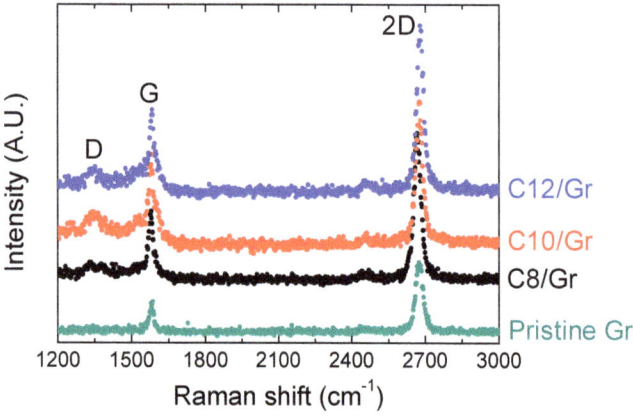

Figure 2. Raman spectra of pristine and arylalkane (C8, C10 and C12)-grafted graphenes transferred to SiO$_2$ surface.

The alkyl chains with different lengths constitute an important control series in molecular junctions, because a coherent picture has been clearly established for the non-resonant tunneling mechanism and length-dependent transport characteristics with the alkyl-based molecules [2,12,23]. We performed electrical measurements on a total of 294 fabricated devices (78 for C8, 105 for C10, and 111 C for 12), excluding open- or short-circuit failures (38 devices). The current density (J)–voltage (V) data collected by measuring enough devices provided the statistical picture of molecular electronic properties in the vertical junctions. Figure 3a displays the current density histograms of C8, C10 and C12 junctions measured at 2 V, which fitted well with the Gaussian curves. The intact statistics, without any device selection, showed that the J values were log-normally distributed. The peak positions of Gaussian curves in the histograms represented the most probable value of the current density for each junction, manifesting distinct molecular length dependency. We note that the log-normal distribution derives from the primary factor exponentially affecting the current density. It would probably be the variation in the tunneling distance that can be influenced by the detailed microscopic configurations, such as a molecular binding site and conformation in the junctions [2]. Such a result was not monitored in the graphene–graphene structure with the absence of component molecules, but rather its histogram indicated Gaussian distribution at the linear scale (not shown here). Figure 3b displays the statistically representative J–V curve on a log scale, which was generated by averaging the junctions within three sigma (3σ) regimes of the Gaussian distribution in Figure 3a [30]. The error bar indicates standard deviation from the averaged J value. The current density of the molecular junctions appeared to be much lower than that of the direct graphene–graphene (Gr–Gr) contact (top data in Figure 3b). For the shortest C8 junction, it was reduced by a factor of ~10^3 in A/m^2 unit, denoting the formation of a transport barrier sandwiched between two SLGs. In addition, Figure 3b shows that the current density of the molecular junctions rapidly decreased as the number of carbons in the alkyl chains increased, which accorded with the characteristic of tunneling [23].

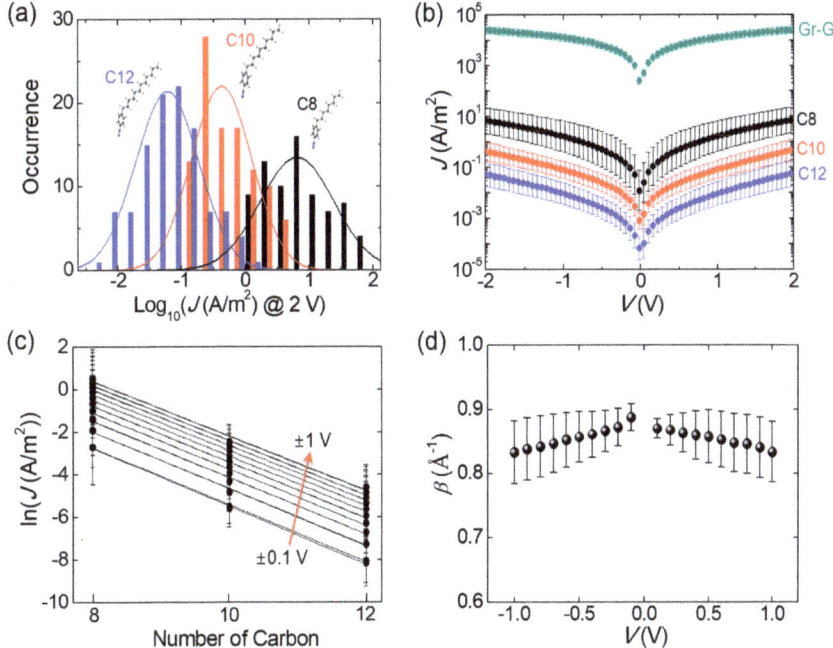

Figure 3. (a) Statistical histograms of the current density at 2 V for C8, C10 and C12 junctions. The peak position of Gaussian distribution indicates the most probable value for each junction. (b) Statistical representative J–V curves on a log scale, obtained from the devices within 3σ regimes of Gaussian distribution. The error bars denote the standard deviation for averaging. (c) Plots of ln(J) versus the number of carbons at different voltages. $β$ is determined by the slopes of the linear fits. (d) Plots of the $β$ values from −1 V to 1 V. Error bars denote the uncertainty about the linear fits.

To investigate the length-dependent charge transport of the junctions in more detail, we presented the semi-log plot of current density (−1 to +1 V) as a function of the molecular length, as shown in Figure 3c. Within the simplified Simmons model for trapezoidal barrier approximation, the molecular tunneling junction is typically described by $J \propto \exp(-βd)$, where $β$ is the tunneling decay coefficient and d is the molecular length. $β$ mathematically scales with the square root of the barrier height and quantifies the decay of the tunneling probability with increasing d [23], depending on the molecular structure. Furthermore, the decay coefficient is reproducible across various experimental platforms [24], which can accordingly provide a valuable benchmark for the formation of a valid molecular junction. Figure 3c reveals an apparent exponential relationship existing between J and d, corresponding to $J \propto \exp(-βd)$, where the tunneling decay parameter was obtained from the slopes of the linear fits in Figure 3c. As plotted in Figure 3d, the $β$ values were estimated to be 1.04 to 1.12 per carbon (equal to 0.83 to 0.89 $Å^{-1}$). The error bar denotes uncertainty about the linear fits. These $β$ values were reasonably consistent with those observed in conventional metal/alkyl-containing monolayer/metal junctions [16,23,24]. Despite still being within the range of uncertainty, a subtle reduction in β with increasing voltage may be ascribed to a large electric field-induced barrier lowering in the tunneling junctions [31]. It is also noteworthy that a few recent studies on the rough topographic condition of the bottom electrodes have shown a significantly lower β value (~0.5 Å^{-1}), even for alkyl-based junctions [7,32]. In this context, we checked the surface roughness of a graphene layer transferred on the oxidized Si substrate using AFM (Figure S3 in the Supplementary Materials). The root mean square (r.m.s.) roughness was estimated to be ~0.14 nm. Such a smooth graphene surface would be desirable to obtain an accurate β value.

The independence of $J-V$ characteristics on temperature variation is a well-known verification of charge tunneling, because it can eliminate many of the other thermally activated transport mechanisms, for example, thermionic emission or hopping conduction [11,23]. The temperature-variable measurement (80 to 280 K) of $J-V$ curves on C8 junctions is shown in Figure 4a. Figure 4b displays corresponding Arrhenius plots and logarithmic J versus inverse temperatures, transformed from the data in Figure 4a. No temperature dependency of the $J-V$ characteristics was clearly confirmed. This result demonstrates that the charge transport occurred by tunneling via the molecular monolayer incorporated into two SLGs. A coherent and clear picture of non-resonant tunneling has, so far, emerged for alkyl-based molecular junctions, because the E_F easily lies in the large HOMO−LUMO gap (8~10 eV) of the very short molecule [2,24].

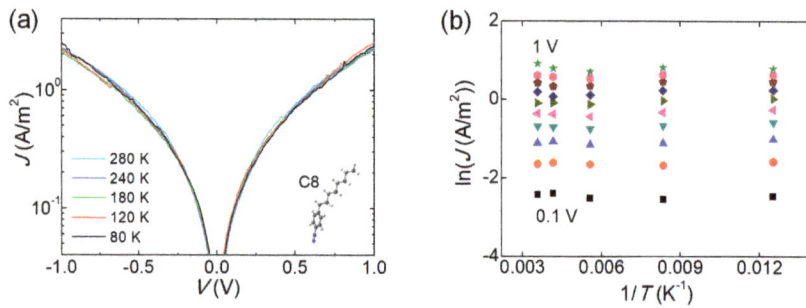

Figure 4. (a) Representative $J-V$ curves of the C8 junction on a semi-log scale as the temperature varies from 80 to 280 K. (b) Arrhenius plots of the current density at different voltages from 0.1 to 1 V.

Transition voltage spectroscopy (TVS) facilitates an evaluation of the difference ($|\varepsilon_0 - E_F|$) between the nearest molecular energy level (ε_0) and E_F [33,34], which refers to the height of a transport barrier in the junctions, by estimating the transition voltage (V_t) to produce an inflection point of the nonlinear Fowler—Nordheim (FN) plot, namely, $\ln(J/V^2)$ versus $1/V$ (see the Supplementary Materials for details). Since its first introduction [33], TVS has become a prevailing analytical technique to study the energy level alignment in molecular junctions, due to its simplicity and validity. Interpretation of the FN curves based on the Landauer transport model showed that such inflection can take place when ε_0 (HOMO for this case) is quite close to a resonance position by the bias voltage (Figure 5a) [35], where it seems to be in the Lorentzian shape broadened by the coupling with electrodes [30]. Accordingly, a measurement of V_t offers experimental approximation to $|\varepsilon_0 - E_F|$. Figure 5b shows the representative TVS analysis of C8, C10 and C12 junctions. The minimum point on the FN curves refers to V_t (as marked by arrows). A TVS histogram of the inflection events was constructed from these minimum values (Figure 6a–c), in which the Gaussian distribution showed no significant asymmetry for the bias polarity. This effect was observed in the molecular junctions, which have comparable coupling strengths for both molecule–electrode contacts [34], thus indicating that the graphene-based vertical molecular junctions can provide good stability for the top physical contacts via van der

Waals interactions [36]. A graphical summary of the TVS measurement on the arylalkane junctions is shown in Figure 6d. It can be clearly observed that the average of V_t (point data with error bar) fell within the standard deviation (dashed lines) of the values measured in each junction. This result demonstrated that V_t was not dependent on the molecular length. Similarly, a study of ultraviolet photoelectron spectroscopy using alkyl thiol monolayers on the Au surface showed that the energy offset ($|\varepsilon_{HOMO} - E_F|$) of HOMO and E_F was independent of the alkyl chain's length [33]. The constancy of V_t in alkyl-based molecular junctions with variable lengths exactly coincided with the Landauer transport model of TVS, where it was invariant for constant $|\varepsilon_0 - E_F|$ [37], whereas it decreased as the tunneling gap in a molecule-free vacuum junction increased, complying with the Simmons model [38]. Consequently, the findings of the TVS analysis presented further evidence of molecular barrier formation in the vertical tunneling junctions.

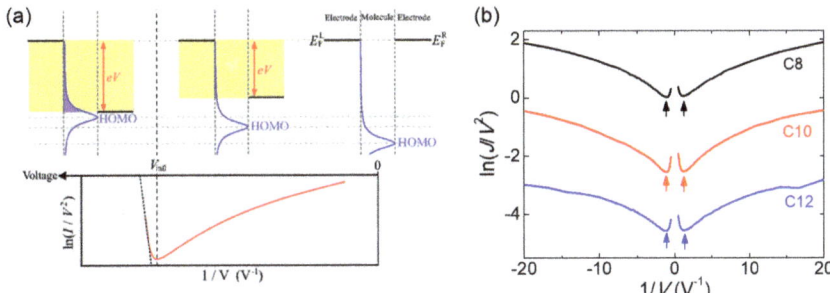

Figure 5. (**a**) Illustration of the inflection behavior in the FN curve and corresponding energy band diagrams. The inflection voltage V_{infl} (vertically dashed line) in the FN curve denotes the transition voltage. Reproduced from Ref. [29], with permission from American Physical Society. (**b**) Representative TVS analysis of C8, C10 and C12 junctions. Each arrow indicates the transition voltage.

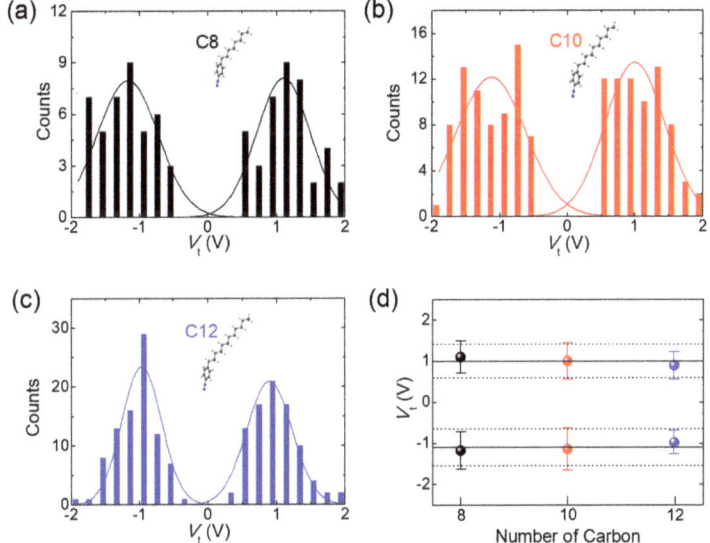

Figure 6. Statistical TVS histograms of (**a**) C8, (**b**) C10 and (**c**) C12 junctions. The solid curves indicate Gaussian distribution, where no significant asymmetry was observed. (**d**) Graphical summary of TVS measurement on the arylalkane junctions. The transition voltage of the junctions was independent of the molecular length.

4. Conclusions

In summary, we reported an alternative molecular device architecture using graphene heterostructures. Our prototype devices were the SLG electrode-based molecular junctions with arylalkane molecules acting as a vertical tunneling barrier. They, indeed, exhibited intrinsic molecular charge transport characteristics with a working device yield of more than 80% among all the fabricated devices. Such devices have considerable potential for a versatile test platform in molecular electronics. We showed inherent molecular contribution to the charge tunneling in the junctions by accomplishing various characterization techniques: (1) intact statistical analysis, (2) $J-V$ curves independent of temperature variation, (3) accurate exponential decay of the current density with the length of alkyl chains, and (4) the TVS analysis coincided with the Landauer transport model, which fully demonstrated that arylalkane molecules acted as a controllable transport barrier in the vertical junctions, by which one can constitute a valid molecular electronic device with SLG electrodes. The intrinsic tunneling characteristics indicate that the integrity of the component molecular layer as a tunnel barrier is preserved in the fabricated device, and the charge transport is not dominated by defects and imperfections. As demonstrated by our findings, the graphene contacts effectively protect the molecular layer from penetration of the evaporated metal atoms and preserve its integrity in the junction, offering a reliable test platform for molecular charge transport characterization. In addition, the application of a back-gate electrode to the graphene-based molecular junctions, to actively control the tunneling transport, suggests promising avenues for future studies.

Supplementary Materials: The following supporting information can be downloaded at: https://www.mdpi.com/article/10.3390/cryst12060787/s1, Figure S1: AFM topographical images and line profile analyses; Figure S2: Intensity ratio and shift of Raman peak; Figure S3: Surface roughness; Figure S4: Representative $J-V$ curves of C10 and C12 junctions.

Author Contributions: Conceptualization, H.S.; Investigation, S.-H.H. and D.-H.S.; Methodology, S.-H.H. and D.-H.S.; Supervision, H.S.; Writing–original draft, H.S.; Writing–review & editing, S.-H.H., D.-H.S. and H.S. All authors have read and agreed to the published version of the manuscript.

Funding: This work was supported by National Research Foundation of Korea (grant No. 2020R1F1A1076107).

Institutional Review Board Statement: Not applicable.

Informed Consent Statement: Not applicable.

Data Availability Statement: Data is contained within the article or supplementary material.

Conflicts of Interest: The authors declare no conflict of interest.

References

1. Evers, F.; Korytár, R.; Tewari, S.; van Ruitenbeek, J.M. Advances and challenges in single-molecule electron transport. *Rev. Mod. Phys.* **2020**, *92*, 035001. [CrossRef]
2. Vilan, A.; Aswal, D.; Cahen, D. Large-Area, Ensemble Molecular Electronics: Motivation and Challenges. *Chem. Rev.* **2017**, *117*, 4248–4286. [CrossRef] [PubMed]
3. Han, Y.; Nickle, C.; Zhang, Z.; Astier, H.P.A.G.; Duffin, T.J.; Qi, D.; Wang, Z.; del Barco, E.; Thompson, D.; Nijhuis, C.A. Electric-field-driven dual-functional molecular switches in tunnel junctions. *Nat. Mater.* **2020**, *19*, 843–848. [CrossRef] [PubMed]
4. Song, P.; Guerin, S.; Tan, S.J.R.; Annadata, H.V.; Yu, X.J.; Scully, M.; Han, Y.M.; Roemer, M.; Loh, K.P.; Thompson, D.; et al. Stable Molecular Diodes Based on π–π Interactions of the Molecular Frontier Orbitals with Graphene Electrodes. *Adv. Mater.* **2018**, *30*, 1706322. [CrossRef] [PubMed]
5. Song, H. Electrostatic Gate Control in Molecular Transistors. *Top. Curr. Chem.* **2018**, *376*, 37. [CrossRef] [PubMed]
6. Qiu, X.; Chiechi, R.C. Large-Area Molecular Junctions: Synthesizing Integrated Circuits for Next-Generation Nonvolatile Memory. *Trends Chem.* **2020**, *2*, 869–872. [CrossRef]
7. Seo, S.; Min, M.; Lee, S.M.; Lee, H. Photo-switchable molecular monolayer anchored between highly transparent and flexible graphene electrodes. *Nat. Commun.* **2013**, *4*, 1920. [CrossRef]
8. Park, S.; Kim, H.R.; Kim, J.; Hong, B.-H.; Yoon, H.J. Enhanced Thermopower of Saturated Molecules by Noncovalent Anchor-Induced Electron Doping of Single-Layer Graphene Electrode. *Adv. Mater.* **2021**, *33*, 2103177. [CrossRef]

9. Wan, A.; Jiang, L.; Sangeeth, C.S.S.; Nijhuis, C.A. Reversible Soft Top-Contacts to Yield Molecular Junctions with Precise and Reproducible Electrical Characteristics. *Adv. Funct. Mat.* **2014**, *24*, 4442–4456. [CrossRef]
10. Jie, Y.; Wang, D.; Huang, J.; Feng, Y.; Yang, J.; Fang, J.; Chen, R. Metal–Molecule–Metal Junctions on Self-Assembled Monolayers Made with Selective Electroless Deposition. *ACS Appl. Mater. Inter.* **2022**, *14*, 1609–1614. [CrossRef]
11. Jeong, H.; Kim, D.; Xiang, D.; Lee, T. High-Yield Functional Molecular Electronic Devices. *ACS Nano* **2017**, *11*, 6511–6548. [CrossRef] [PubMed]
12. Song, H.; Lee, T.; Choi, N.-J.; Lee, H. A statistical method for determining intrinsic electronic transport properties of self-assembled alkanethiol monolayer devices. *Appl. Phys. Lett.* **2007**, *91*, 253116. [CrossRef]
13. Wang, G.; Kim, Y.; Choe, M.; Kim, T.-W.; Lee, T. A New Approach for Molecular Electronic Junctions with a Multilayer Graphene Electrode. *Adv. Mater.* **2011**, *23*, 755–760. [CrossRef]
14. Koo, J.; Jang, Y.; Martin, L.; Kim, D.; Jeong, H.; Kang, K.; Lee, W.; Kim, J.; Hwang, W.-T.; Xiang, D.; et al. Unidirectional Real-Time Photoswitching of Diarylethene Molecular Monolayer Junctions with Multilayer Graphene Electrodes. *ACS Appl. Mater. Inter.* **2019**, *11*, 11645–11653. [CrossRef] [PubMed]
15. Min, M.; Seo, S.; Lee, S.M.; Lee, H. Voltage-Controlled Nonvolatile Molecular Memory of an Azobenzene Monolayer through Solution-Processed Reduced Graphene Oxide Contacts. *Adv. Mater.* **2013**, *25*, 7045. [CrossRef]
16. Akkerman, H.B.; Blom, P.W.M.; de Leeuw, D.M.; de Boer, B. Towards molecular electronics with large-area molecular junctions. *Nature* **2006**, *441*, 69–72. [CrossRef] [PubMed]
17. Jeong, H.; Kim, D.; Kim, P.; Cho, M.R.; Hwang, W.-T.; Jang, Y.; Cho, K.; Min, M.; Xiang, D.; Park, Y.D.; et al. A new approach for high-yield metal–molecule–metal junctions by direct metal transfer method. *Nanotechnology* **2015**, *26*, 025601. [CrossRef]
18. Jang, Y.; Kwon, S.-J.; Shin, J.; Jeong, H.; Hwang, W.-T.; Kim, J.; Koo, J.; Ko, T.Y.; Ryu, S.; Wang, G.; et al. Interface-Engineered Charge-Transport Properties in Benzenedithiol Molecular Electronic Junctions via Chemically p-Doped Graphene Electrodes. *ACS Appl. Mater. Inter.* **2017**, *9*, 42043–42049. [CrossRef]
19. Chen, S.; Su, D.; Jia, C.; Li, Y.; Li, X.; Guo, X.; Leigh, D.A.; Zhang, L. Real-time observation of the dynamics of an individual rotaxane molecular shuttle using a single-molecule junction. *Chem* **2022**, *8*, 243–252. [CrossRef]
20. Novoselov, K.S.; Fal'ko, V.I.; Colombo, L.; Gellert, P.R.; Schwab, M.G.; Kim, K. A roadmap for graphene. *Nature* **2012**, *490*, 192–200. [CrossRef]
21. Liu, Y.C.; McCreery, R.L. Reactions of Organic Monolayers on Carbon Surfaces Observed with Unenhanced Raman Spectroscopy. *J. Am. Chem. Soc.* **1995**, *117*, 11254–11259. [CrossRef]
22. MacLeod, J.M.; Rosei, F. Molecular Self-Assembly on Graphene. *Small* **2014**, *10*, 1038–1049. [CrossRef] [PubMed]
23. Wang, W.; Lee, T.; Reed, M.A. Mechanism of electron conduction in self-assembled alkanethiol monolayer devices. *Phys. Rev. B* **2003**, *68*, 035416. [CrossRef]
24. Xie, Z.; Bâldea, I.; Frisbie, C.D. Energy Level Alignment in Molecular Tunnel Junctions by Transport and Spectroscopy: Self-Consistency for the Case of Alkyl Thiols and Dithiols on Ag, Au, and Pt Electrodes. *J. Am. Chem. Soc.* **2019**, *141*, 18182–18192. [CrossRef] [PubMed]
25. Liang, X.; Sperling, B.A.; Calizo, I.; Cheng, G.; Hacker, C.A.; Zhang, Q.; Obeng, Y.; Yan, K.; Peng, H.; Li, Q.; et al. Toward Clean and Crackless Transfer of Graphene. *ACS Nano* **2011**, *5*, 9144–9153. [CrossRef] [PubMed]
26. Malard, L.M.; Pimenta, M.A.; Dresselhaus, G.; Dresselhaus, M.S. Raman spectroscopy in graphene. *Phys. Rep.* **2009**, *473*, 51–87. [CrossRef]
27. Wu, J.-B.; Lin, M.-L.; Cong, X.; Liu, H.-N.; Tan, P.-H. Raman spectroscopy of graphene-based materials and its applications in related devices. *Chem. Soc. Rev.* **2018**, *47*, 1822–1873. [CrossRef]
28. Graf, D.; Molitor, F.; Ensslin, K.; Stampfer, C.; Jungen, A.; Hierold, C.; Wirtz, L. Spatially Resolved Raman Spectroscopy of Single- and Few-Layer Graphene. *Nano Lett.* **2007**, *7*, 238. [CrossRef]
29. Dong, X.; Fu, D.; Fang, W.; Shi, Y.; Chen, P.; Li, L.-J. Doping Single-Layer Graphene with Aromatic Molecules. *Small* **2009**, *5*, 1422–1426. [CrossRef]
30. Reus, W.F.; Nijhuis, C.A.; Barber, J.R.; Thuo, M.M.; Tricard, S.; Whitesides, G.W. Statistical Tools for Analyzing Measurements of Charge Transport. *J. Phys. Chem. C* **2012**, *116*, 6714–6733. [CrossRef]
31. Tung, R.T. The physics and chemistry of the Schottky barrier height. *Appl. Phys. Rev.* **2014**, *1*, 011304.
32. Yuan, L.; Jiang, L.; Zhang, B.; Nijhuis, C.A. Dependency of the Tunneling Decay Coefficient in Molecular Tunneling Junctions on the Topography of the Bottom Electrodes. *Angew. Chem. Int. Ed.* **2014**, *53*, 3377–3381. [CrossRef] [PubMed]
33. Beebe, J.M.; Kim, B.; Gadzuk, J.W.; Frisbie, C.D.; Kushmerick, J.G. Transition from Direct Tunneling to Field Emission in Metal-Molecule-Metal Junctions. *Phys. Rev. Lett.* **2006**, *97*, 026801. [CrossRef]
34. Bâldea, I. Ambipolar transition voltage spectroscopy: Analytical results and experimental agreement. *Phys. Rev. B* **2012**, *85*, 035442. [CrossRef]
35. Araidai, M.; Tsukada, M. Theoretical calculations of electron transport in molecular junctions: Inflection behavior in Fowler-Nordheim plot and its origin. *Phys. Rev. B* **2010**, *81*, 235114. [CrossRef]
36. Song, P.; Sangeeth, C.S.S.; Thompson, D.; Du, W.; Loh, K.P.; Nijhuis, C.A. Noncovalent Self-Assembled Monolayers on Graphene as a Highly Stable Platform for Molecular Tunnel Junctions. *Adv. Mater.* **2016**, *28*, 631–639. [CrossRef] [PubMed]

37. Huisman, E.H.; Guédon, C.M.; van Wees, B.J.; van der Molen, S.J. Interpretation of Transition Voltage Spectroscopy. *Nano Lett.* **2009**, *9*, 3909–3913. [CrossRef]
38. Trouwborst, M.L.; Martin, C.A.; Smit, R.H.M.; Guédon, C.M.; Baart, T.V.; van der Molen, S.J.; van Ruitenbeek, J.M. Transition Voltage Spectroscopy and the Nature of Vacuum Tunneling. *Nano Lett.* **2011**, *11*, 614–617. [CrossRef]

Article

High-Temperature Electronic Transport Properties of PEDOT:PSS Top-Contact Molecular Junctions with Oligophenylene Dithiols

Dong-Hyoup Seo, Kyungjin Im and Hyunwook Song *

Department of Applied Physics, Kyung Hee University, Yongin 17104, Korea; ehdguq1309@khu.ac.kr (D.-H.S.); limkj0512@khu.ac.kr (K.I.)
* Correspondence: hsong@khu.ac.kr

Abstract: In this study, we investigated the high-temperature electronic transport behavior of spin-coated PEDOT:PSS top-contact molecular ensemble junctions based on self-assembled monolayers (SAMs) of oligophenylene dithiols. We observed irreversible temperature-dependent charge transport at the high-temperature regime over 320 K. The effective contact resistance and normalized resistance decreased with increasing temperature (320 to 400 K), whereas the tunneling attenuation factor was nearly constant irrespective of temperature change. These findings demonstrate that the high-temperature transport properties are not dominated by the integrity of SAMs in molecular junctions, but rather the PEDOT:PSS/SAMs contact. Transition voltage spectroscopy measurements indicated that the contact barrier height of the PEDOT:PSS/SAMs is lowered at elevated temperatures, which gives rise to a decrease in the contact resistance and normalized resistance. The high-temperature charge transport through these junctions is also related to an increase in the grain area of PEDOT cores after thermal treatment. Moreover, it was found that there was no significant change in either the current density or normalized resistance of the annealed junctions after 60 days of storage in ambient conditions.

Keywords: molecular junction; PEDOT:PSS; off-resonant tunneling; transition voltage spectroscopy

Citation: Seo, D.-H.; Im, K.; Song, H. High-Temperature Electronic Transport Properties of PEDOT:PSS Top-Contact Molecular Junctions with Oligophenylene Dithiols. *Crystals* **2022**, *12*, 962. https://doi.org/10.3390/cryst12070962

Academic Editors: Bo Chen, Rutao Wang and Nana Wang

Received: 17 June 2022
Accepted: 8 July 2022
Published: 10 July 2022

Publisher's Note: MDPI stays neutral with regard to jurisdictional claims in published maps and institutional affiliations.

Copyright: © 2022 by the authors. Licensee MDPI, Basel, Switzerland. This article is an open access article distributed under the terms and conditions of the Creative Commons Attribution (CC BY) license (https://creativecommons.org/licenses/by/4.0/).

1. Introduction

Molecular ensemble junctions have been considered a key element in molecular electronics in which self-assembled monolayers (SAMs) of individual molecules are vertically sandwiched between top and bottom electrodes and the constituent molecules span the two electrodes [1–3]. This junction arrangement typically prefers quantum mechanical tunneling as a charge transport mechanism through SAMs because of their large HOMO–LUMO (the highest occupied and lowest unoccupied molecular orbitals) gap with a very short molecular length [4–6]. Over the past few decades, a variety of pioneering techniques, such as a conducting atomic force microcopy [7], Hg drop [8], eutectic Ga-In [9], nanopore [6], micro-via-hole [5], nanoparticle top-contact [10], metal transfer printing [11], and graphene interlayer [12], have been demonstrated to fabricate and characterize the molecular ensemble junctions based on SAMs. In particular, the highly conducting polymer poly(3,4-ethylenedioxythiophene):ploy(styrenesulfonate) (PEDOT:PSS) top-contact molecular ensemble junction is one of the most successful methods to achieve the high yield of working devices as well as the outstanding stability and reproducibility of large area junctions (up to several hundreds of μm^2 in area) [2,13,14], which are crucial for any technological applications of molecular electronic devices. However, the physisorbed contact properties between the spin-coated PEDOT:PSS top-contact and SAMs have been not thoroughly investigated in the test device platform of molecular junctions. Moreover, most of the charge transport studies on the molecular junction have been performed at low (cryogenic) or room temperature due to its thermal instability [3,13,15].

In this work, we studied the electronic transport properties of PEDOT:PSS top-contact molecular junctions containing oligophenylene dithiols with three different lengths. The PEDOT:PSS top electrodes spin-coated on the SAMs of oligophenylene dithiols effectively prevent electrical shorts from vapor-deposited top metal contacts. The active region of a molecular junction is generated by via-holes photolithographically pattered in photoresist to eliminate parasitical current paths and protect the junction from ambient conditions. The oligophenylene dithiols are one of the simplest prototype molecules with a π-conjugated backbone, which provide a control series to systemically investigate the charge transport through molecular junctions because the off-resonant tunneling mechanism has been coherently demonstrated so far for these molecules [16,17]. Here, our primary focus is on high-temperature charge transport behavior through the annealed PEDOT:PSS molecular junctions. Specifically, we observed that the low-bias junction resistance is temperature-dependent at elevated temperatures above 320 K, which is highly related to total area of the PEDOT grains and the conductivity of spin-coated PEDOT:PSS film. The high-bias nonlinear transport is also examined by transition voltage spectroscopy, indicating that the barrier height at the PEDOT:PSS/SAMs contact is lowered with increasing annealing temperature and therefore the effective contact resistance of these junctions is decreased. Finally, we demonstrate the long-term stability of PEDOT:PSS molecular junctions after thermal treatment.

2. Experimental Details

The device structure of a molecular ensemble junction with the PEDOT:PSS top-contact is schematically shown in Figure 1a. Three different oligophenylene dithiols, abbreviated as OPD-n, where n denotes the number of a phenyl ring (see Figure 1b), are sandwiched between the PEDOT:PSS (top) and Au (bottom) electrodes. To fabricate the device, a 2 nm thick Ti adhesion layer and a 60 nm thick Au bottom electrode are vapor-deposited on a thermally oxidized Si substrate using an e-beam evaporator. As seen in Figure 1c, a photoresist layer is spin-coated and via-hole arrays of diameter 20–200 μm are created by a typical photolithographic method. For the SAM formation on Au bottom electrodes, the substrate is immersed into an ethanol solution of OPD-n molecules at a concentration of 1–2 mM for a minimum of 24 h. Before the next process, it is thoroughly rinsed with clean ethanol and then dried in a vacuum desiccator. The conducting polymer PEDOT:PSS (Clevios™ PH1000) modified with 5% dimethyl sulfoxide (DMSO) is spin-coated on top of the OPD-n SAMs, producing a film thickness of 90–100 nm with a conductivity of about 300 S cm^{-1} at room temperature. It has been known that the conductivity of DMSO-modified PEDOT:PSS film is 2–3 orders of magnitude greater than that of pure PEDOT:PSS [13]. Moreover, the pure PEDOT:PSS makes short molecular junctions indistinguishable due to its low conductivity and high contact resistance [14]. In order to secure low contact resistance, a 50 nm Au top-contact pad is vapor-deposited through a shadow mask. Lastly, residual SAMs and PEDOT:PSS film outside the effective region of a molecular junction are eliminated using reactive ion etching, in which the top Au layer acts as an etching mask. All the electrical measurements were performed in a probe station equipped with a built-in temperature controller (MS tech, M5VC model) using a semiconductor parameter analyzer (Keithley 4200A-SCS), where the sample stage cools using a liquid nitrogen. The experimental conditions (e.g., spin-coating and deposition rate and solvent concentrations) are consistently preserved to examine the effect of anneal temperature on charge transport through the molecular junctions. To investigate the long-term stability of annealed molecular junctions, the samples are stored in ambient conditions (usually temperatures of 293–302 K and the relative humidity of 40–50%) without light shielding, humidity control, or any other precautions, and then are remeasured after 60 days in air.

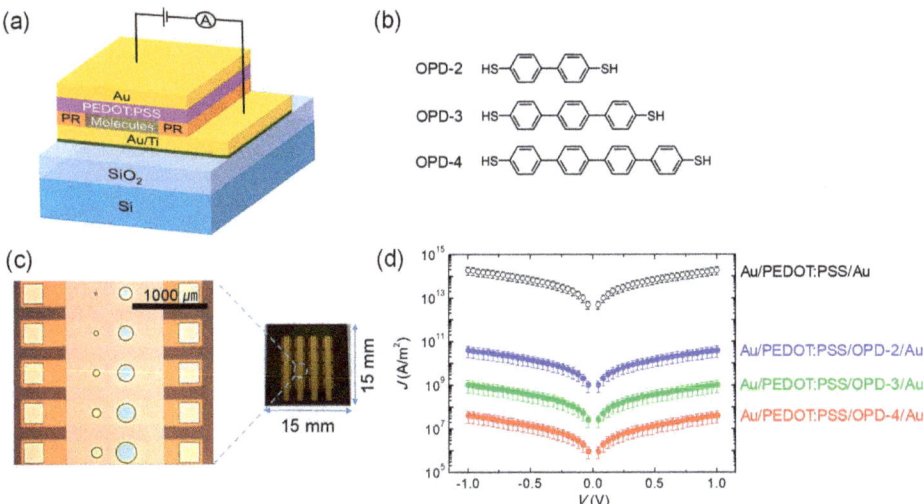

Figure 1. (**a**) Schematic image of PEDOT:PSS top-contact molecular ensemble junctions. (**b**) Chemical structure of oligophenylene dithiols with three different lengths (n = 2, 3 and 4). (**c**) Optical images of fabricated devices, showing via-hole arrays with diameter 20–200 μm. (**d**) Room temperature $J(V)$ curves of OPD-2, OPD-3, OPD-4, and PEDOT:PSS only junctions.

3. Results and Discussion

The oligophenylene dithiols with different lengths constitute an important control series in molecular junctions because it has been unambiguously determined that off-resonant tunneling is the dominant conduction mechanism for these molecules and the length-dependent transport behavior with various electrode systems is extensively reported in studies [1–3], providing a crucial comparative study to validate the demonstration of molecular junctions based on PEDOT:PSS top-contact. Moreover, oligophenylene forms the conjugated backbone of a molecular wire, and the application of such conductive organic molecules as an active component in molecular junctions is highly desirable in the field of molecular electronics. Figure 1d presents the current density (J) of OPD-n (n = 2, 3 and 4) junctions on a logarithmic scale as a function of the bias voltage (V), which was measured at room temperature. The $J(V)$ curves were produced by an average over 20 devices per molecule, and the error bars denote a standard deviation. The ODP-n junction shows a much lower J value than that of a PEDOT:PSS-only device (that is, without ODP-n SAMs). For the case of OPD-2 (the shortest molecule), it was decreased by a factor of ~10^4 in a unit of A/m^2, indicating that a molecular tunnel barrier was formed in the vertically sandwiched junctions. Figure 1d also exhibits that the current density exponentially decreases with increasing n or a molecular length, which can be considered as a typical characteristic of off-resonant tunneling [6,14]. The most remarkable feature observed in the PEDOT:PSS top-contact molecular junctions is that the charge transport exhibits temperature-dependent behavior at a high temperature (typically over room temperature). Figure 2a shows the normalized resistance RA (where R is the resistance and A is the junction area) plotted against the annealing temperature. R is determined by the linear fits of $J(V)$ curves in a low bias region ranging from −0.1 to +0.1 V and the plots are displayed for different molecular lengths of OPD-n in Figure 2a. The RA values of all junctions exponentially decrease with increasing temperature at more than 320 K, whereas those are nearly constant at low temperatures (80–300 K). As clearly shown in Figure 2b, the $J(V)$ characteristics of OPD-n junctions indicate an irreversible change for temperature variation. This result implies that the charge transport properties might be not caused by thermally activated conduction mechanisms or thermal broadening of the Fermi-Dirac distribution at the contacts.

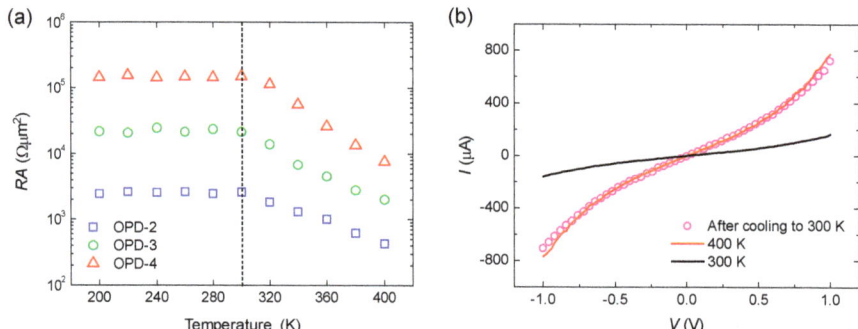

Figure 2. (a) Semilog plots of RA versus temperature for OPD-n junctions at different annealing temperatures (300, 340, and 380 K). A vertically dashed line indicates 300 K. (b) Irreversible $I(V)$ characteristic of OPD-4 junction after thermal annealing at 400 K.

To understand the charge transport properties of PEDOT:PSS top-contact molecular junctions at the high-temperature regime ($\geq$320 K), we performed a variety of transport characterizations, including the measurements of a tunneling attenuation factor β, effective contact resistance R_c, and transition voltage V_{trans} for OPD-n molecular series. Figure 3a displays the normalized resistance RA against the number of a phenyl ring (n = 2, 3, 4) at different annealing temperatures (300, 340, and 380 K). It was observed that the RA values exponentially increase with n. The low-bias resistance R of a molecular tunnel junction is typically expressed as [6,9]:

$$R = R_c \exp(\beta n) \quad (1)$$

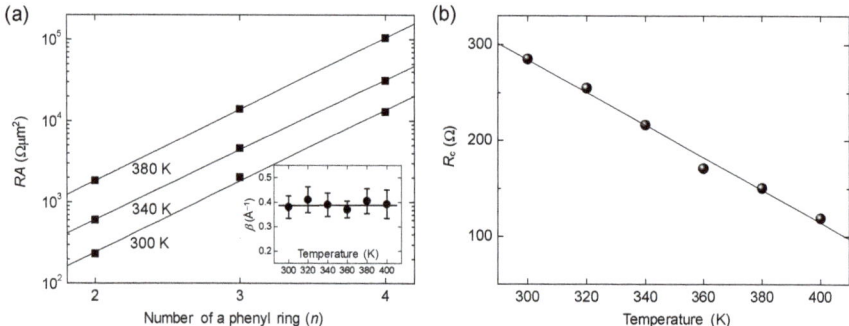

Figure 3. (a) Semilog plots of RA versus n. Inset shows β values as a function of temperature. Solid lines indicate linear fits. Error bars denote fitting errors. (b) Plot of R_c versus temperature. Solid lines indicate a linear fit.

Equation (1) is a general feature of off-resonant tunneling and also convenient because it distinguishes the exponential length dependence of R from the molecule–electrode contact contribution (R_c). β is determined from the slope of linear fits (solid lines) in Figure 3a. It has been well-known that β is a parameter that depends on the molecular backbone structure in the junctions, reflecting the extent to which the electron wave function decays with a molecular length [3,6]. The β value obtained at 300 K is 1.61 per a phenyl ring (=0.38 Å^{-1}), which reasonably agrees with previous studies for OPD-n junctions [16,17]. Noticeably, β is nearly constant across a range of annealing temperatures as shown in the inset of Figure 3a. This finding clearly indicates that the efficiency of the tunneling process through the conjugated molecular backbone of OPD-n is maintained, thus implying that the integrity of SAMs in the molecular junctions is saved from considerable damage within the temperature range investigated (up to 400 K). On the contrary, we observed that R_c monotonously decreases with elevating temperature (Figure 3b), which is determined from an extrapolation at $n = 0$ in Figure 3a using Equation (1). These results indicate that the high-temperature charge transport behavior is not dominated by the integrity of SAMs themselves in the junctions, but rather electrical contacts between PEDOT:PSS top electrodes and OPD-n SAMs.

It has been recognized that the morphology of spin-coated PEDOT:PSS films has a core–shell structure consisting of highly-conductive PEDOT cores enclosed by shells with non-conductive PSS chains [18]. The electrical conduction in PEDOT:PSS occurs mainly by charge hopping through PEDOT grains, but not PSS shells. Therefore, the total area of PEDOT segments occupied in the conducting polymer is a crucial factor in determining its conductivity. The grain area of PEDOT conductive regions in the spin-coated PEDOT:PSS film can be examined by its phase image obtained from an atomic force microscopy (AFM) [18]. Figure 4a displays representative AFM topographical and phase images of a PEDOT:PSS film obtained at 300 K and after being annealed at 400 K. A bright (in positive degree) and dark (in negative degree) region in the phase images corresponds to the grains of a PEDOT core and PSS shell, respectively. We obtained all the AFM images after cooling the samples to room temperature and found that the total area of PEDOT grains increases with thermal treatment. For comparison, the PEDOT grain area was calculated to be 0.452 μm^2 at 300 K and 0.538 μm^2 at 400 K in the phase images of 1 μm^2 using a built-in analysis software (see Figure 4a, bottom). Figure 4b shows a direct correlation between the entire grain area of PEDOT cores and four-point-probe conductivity of PEDOT:PSS films with an increase in the annealing temperature from 300 to 400 K. This result demonstrates that high temperatures over 320 K give rise to an increase in the total grain area of conductive PEDOT cores and thus the enhancement of PEDOT:PSS electrode's conductivity. In addition, residual solvent and moisture, acting as a tunnel barrier in the spin-coated PEDOT:PSS top-contact molecular junctions, can be evaporated during an annealing process. These effects generate better electrical contact between the PEDOT:PSS film and OPD-n SAMs, leading to a decrease in the normalized resistance (Figure 2a) and effective contact resistance (Figure 3b) of the junctions.

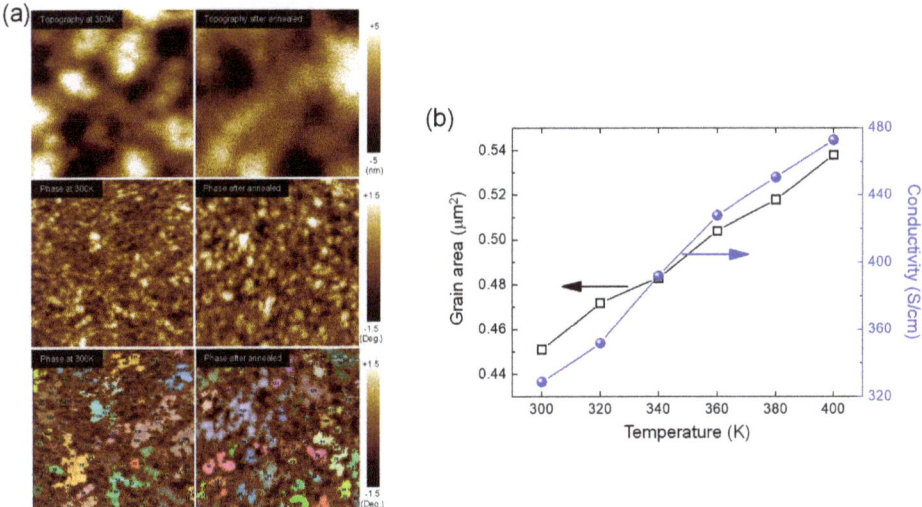

Figure 4. (a) AFM topography image (top), phase images (middle), and grain area calculation (bottom) of PEDOT:PSS film in of 1×1 μm^2 area obtained at 300 K and after being annealed at 400 K. (b) Plots of the total grain area of PEDOT cores and the conductivity of PEDOT:PSS film against temperature.

For a more comprehensive investigation of charge transport behavior through PEDOT:PSS molecular junctions, we also examined the high-bias (nonlinear) regime of the current (I)−voltage characteristics at different temperatures. In a situation of off-resonant tunneling observed in the OPD-n junctions, a tunnel barrier can be approximately estimated from the inflection (or transition) voltage $V = V_{trans}$ in a Fowler–Nordheim plot of the same $I(V)$ characteristics [19]. This method is so-called transition voltage spectroscopy (TVS) [20]. Owing to its simplicity and reproducibility, TVS have been widely utilized as a tool to quantify the barrier height for nonlinear transport through molecular tunnel junctions [20–22]. As representatively illustrated in Figure 5a, we examined the shape of nonlinear $I(V)$ curves obtained from OPD-n junctions by recasting the curves as Fowler–Nordheim coordinates, that is, $\ln|I^2/V|$ versus $1/V$. These plots produce well-defined minima at V_{trans} (as denoted by arrows in Figure 5a), which is the key quantity of TVS and gives an approximation to the height of tunneling barrier in the junctions. To investigate the influence of thermal treatment on charge transport through the junctions, we plot V_{trans} as a function of temperature in Figure 5b. It was found that V_{trans} consistently decreases for OPD-n junctions as temperature increases. A decease in V_{trans} indicates the lowering of a barrier height with high temperature, which can be attributed to a decrease in the effective contact resistance at the interface between PEDOT:PSS and OPD-n due to an increase in the grain area of PEDOT cores (as described in Figure 4) and the removal of residual solvent and water in the spin-coated polymer films.

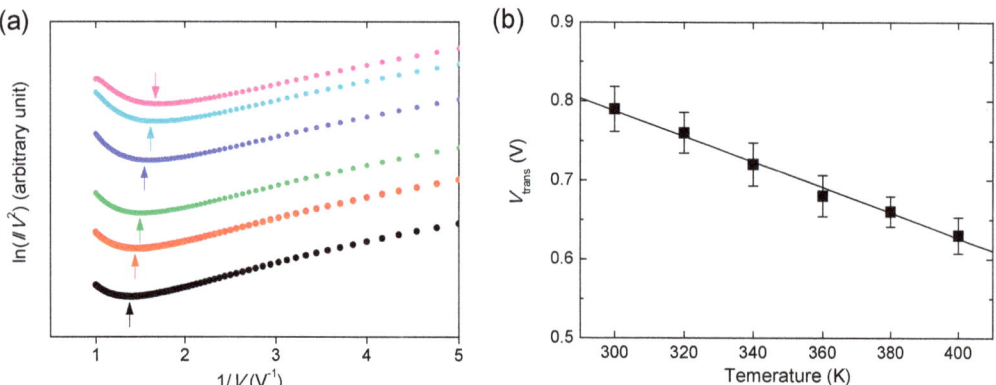

Figure 5. (a) Fowler–Nordheim plots of OPD-n junctions. Arrows indicate V_{trans}. (b) Plot of V_{trans} versus temperature. Solid line indicates a linear fit and error bars denote a standard deviation.

Furthermore, to examine the long-term stability of PEDOT:PSS molecular junctions with high temperature, the samples were annealed at 400 K and then stored in ambient conditions for subsequent electrical measurements. We remeasured the junctions after 60 days. As shown in Figure 6a, there is no significant change in the $J(V)$ curves. A slight decrease in the current density is probably due to an increase in the hopping distance of PEDOT:PSS films or an additional contact barrier resulting from water molecules in air [18]. Figure 6b presents the RA values of eight distinct junctions after 2 months of storage in ambient conditions, where the only small increase in RA values was observed for all measured devices.

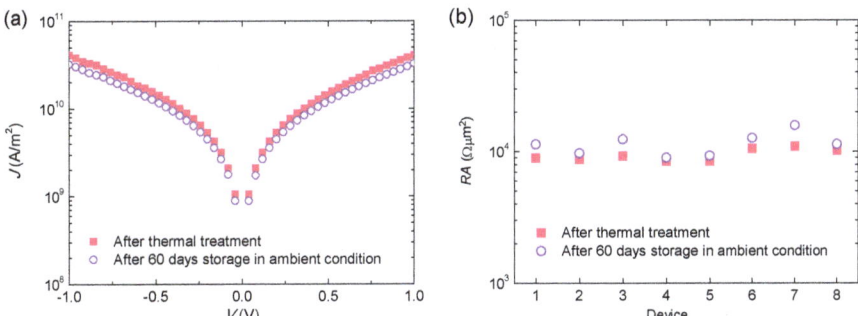

Figure 6. (**a**) The $J(V)$ characteristics of OPD-2 junctions measured after being annealed at 400 K and 60 days of storage in ambient conditions. (**b**) RA values for 8 OPD-2 junctions measured after being annealed at 400 K and 60 days of storage in ambient conditions.

4. Conclusions

We investigated the charge transport properties of spin-coated PEDOT:PSS top-contact molecular ensemble junctions with OPD-n SAMs. The current density exponentially decreased with molecular length, indicating a general feature of off-resonant tunneling. The charge transport exhibited irreversible temperature-dependent behavior at the high-temperature regime. The normalized resistance rapidly decreased with increasing temperature above 320 K. β showed a constant value of 0.38 Å^{-1}, whereas R_c linearly decreased as temperature increases from 320 to 400 K. These results indicate that the high-temperature transport behavior was dominated by the PEDOT:PSS/SAMs contact, but not the integrity of SAMs in the junctions. A decease in V_{trans} can be attributed to the contact barrier lowering at the PEDOT:PSS/SAMs interface, which is most probably due to an increase in the grain area of PEDOT cores and the removal of residual solvent and water in the spin-coated polymer films after thermal annealing. In addition, no significant change was found in the current density and normalized resistance of the annealed molecular junctions after 60 days of storage in ambient conditions. Recent studies also indicate that PEDOT:PSS electrodes can provide various approaches to create molecular junctions [23–25].

Author Contributions: Conceptualization, H.S.; Investigation, D.-H.S. and K.I.; Methodology, D.-H.S. and K.I.; Supervision, H.S.; Writing–original draft, H.S.; Writing–review & editing, D.-H.S., K.I. and H.S. All authors have read and agreed to the published version of the manuscript.

Funding: This work was supported by the National Research Foundation of Korea (grant No. 2020R1F1A1076107).

Institutional Review Board Statement: Not applicable.

Informed Consent Statement: Not applicable.

Data Availability Statement: Data is contained within the article.

Conflicts of Interest: The authors declare no conflict of interest.

References

1. Liu, Y.; Qiu, X.; Soni, S.; Chiechi, R.C. Charge Transport through Molecular Ensembles: Recent Progress in Molecular Electronics. *Chem. Phys. Rev.* **2021**, *2*, 021303. [CrossRef]
2. Vilan, A.; Aswal, D.; Cahen, D. Large-Area, Ensemble Molecular Electronics: Motivation and Challenges. *Chem. Rev.* **2017**, *117*, 4248–4286. [CrossRef]
3. Xiang, D.; Wang, X.; Jia, C.; Lee, T.; Guo, X. Molecular-Scale Electronics: From Concept to Function. *Chem. Rev.* **2016**, *116*, 4318–4440. [CrossRef]
4. Jeong, H.; Kim, D.; Xiang, D.; Lee, T. High-Yield Functional Molecular Electronic Devices. *ACS Nano* **2017**, *11*, 6511–6548. [CrossRef]

5. Song, H.; Lee, T.; Choi, N.-J.; Lee, H. A statistical method for determining intrinsic electronic transport properties of self-assembled alkanethiol monolayer devices. *Appl. Phys. Lett.* **2007**, *91*, 253116. [CrossRef]
6. Wang, W.; Lee, T.; Reed, M.A. Mechanism of electron conduction in self-assembled alkanethiol monolayer devices. *Phys. Rev. B* **2003**, *68*, 035416. [CrossRef]
7. Wold, D.J.; Frisbie, C.D. Formation of Metal-Molecule-Metal Tunnel Junctions: Microcontacts to Alkanethiol Monolayers with a Conducting AFM Tip. *J. Am. Chem. Soc.* **2000**, *122*, 2970–2971. [CrossRef]
8. Holmlin, R.E.; Haag, R.; Chabinyc, M.L.; Ismagilov, R.F.; Cohen, A.E.; Terfort, A.; Rampi, M.A.; Whitesides, G.M. Electron Transport through Thin Organic Films in Metal−Insulator−Metal Junctions Based on Self-Assembled Monolayers. *J. Am. Chem. Soc.* **2001**, *123*, 5075–5085. [CrossRef]
9. Yuan, L.; Jiang, L.; Zhang, B.; Nijhuis, C.A. Dependency of the Tunneling Decay Coefficient in Molecular Tunneling Junctions on the Topography of the Bottom Electrodes. *Angew. Chem. Int. Ed.* **2014**, *53*, 3377–3381. [CrossRef]
10. Puebla-Hellmann, G.; Venkatesan, K.; Mayor, M.; Lörtscher, E. Metallic nanoparticle contacts for high-yield, ambient-stable molecular-monolayer devices. *Nature* **2018**, *559*, 232–235. [CrossRef]
11. Jeong, H.; Kim, D.; Kim, P.; Cho, M.R.; Hwang, W.-T.; Jang, Y.; Cho, K.; Min, M.; Xiang, D.; Park, Y.D.; et al. A new approach for high-yield metal–molecule–metal junctions by direct metal transfer method. *Nanotechnology* **2015**, *26*, 025601. [CrossRef]
12. Park, S.; Kim, H.R.; Kim, J.; Hong, B.-H.; Yoon, H.J. Enhanced Thermopower of Saturated Molecules by Noncovalent Anchor-Induced Electron Doping of Single-Layer Graphene Electrode. *Adv. Mater.* **2021**, *33*, 2103177. [CrossRef]
13. Akkerman, H.B.; Blom, P.W.M.; de Leeuw, D.M.; de Boer, B. Towards molecular electronics with large-area molecular junctions. *Nature* **2006**, *441*, 69–72. [CrossRef]
14. Akkerman, H.B.; Naber, R.C.G.; Jongbloed, B.; van Hal, P.A.; Blom, P.W.M.; de Leeuw, D.M.; de Boer, B. Electron tunneling through alkanedithiol self-assembled monolayers in large-area molecular junctions. *Proc. Natl. Acad. Sci. USA* **2007**, *104*, 11161–11166. [CrossRef]
15. Kühnel, M.; Overgaard, M.H.; Hels, M.C.; Cui, A.; Vosch, T.; Nygård, J.; Li, T.; Laursen, B.W.; Nørgaard, K. High-Quality Reduced Graphene Oxide Electrodes for Sub-Kelvin Studies of Molecular Monolayer Junctions. *J. Phys. Chem. C* **2018**, *122*, 25102–25109. [CrossRef]
16. Nguyen, Q.V.; Xie, Z.; Frisbie, C.D. Quantifying Molecular Structure-Tunneling Conductance Relationships: Oligophenylene Dimethanethiol vs Oligophenylene Dithiol Molecular Junctions. *J. Phys. Chem. C* **2021**, *125*, 4292–4298. [CrossRef]
17. Xie, Z.; Bâldea, I.; Frisbie, C.D. Determination of Energy-Level Alignment in Molecular Tunnel Junctions by Transport and Spectroscopy: Self-Consistency for the Case of Oligophenylene Thiols and Dithiols on Ag, Au, and Pt Electrodes. *J. Am. Chem. Soc.* **2019**, *141*, 3670–3681. [CrossRef]
18. Nardes, A.M.; Janssen, R.A.J.; Kemerink, M. A Morphological Model for the Solvent-Enhanced Conductivity of PEDOT:PSS Thin Films. *Adv. Funct. Mater.* **2008**, *18*, 865–871. [CrossRef]
19. Beebe, J.M.; Kim, B.; Gadzuk, J.W.; Frisbie, C.D.; Kushmerick, J.G. Transition from Direct Tunneling to Field Emission in Metal-Molecule-Metal Junctions. *Phys. Rev. Lett.* **2006**, *97*, 026801. [CrossRef]
20. Beebe, J.M.; Kim, B.; Frisbie, C.D.; Kushmerick, J.G. Measuring Relative Barrier Heights in Molecular Electronic Junctions with Transition Voltage Spectroscopy. *ACS Nano* **2008**, *2*, 827–832. [CrossRef]
21. Chen, J.; Markussen, T.; Thygesen, K.S. Quantifying Transition Voltage Spectroscopy of Molecular Junctions: Ab Initio Calculations. *Phys. Rev. B* **2010**, *82*, 121412. [CrossRef]
22. Nose, D.; Dote, K.; Sato, T.; Yamamoto, M.; Ishii, H.; Noguchi, Y. Effects of Interface Electronic Structures on Transition Voltage Spectroscopy of Alkanethiol Molecular Junctions. *J. Phys. Chem. C* **2015**, *119*, 12765–12771. [CrossRef]
23. Bruce, J.P.; Oliver, D.R.; Lewis, N.S.; Freund, M.S. Electrical Characteristics of the Junction between PEDOT:PSS and Thiophene-Functionalized Silicon Microwires. *ACS Appl. Mater. Interfaces* **2015**, *7*, 27160–27166. [CrossRef]
24. Yildirim, E.; Zhu, Q.; Wu, G.; Tan, T.L.; Xu, J.; Yang, S.-W. Self-Organization of PEDOT:PSS Induced by Green and Water-Soluble Organic Molecules. *J. Phys. Chem. C* **2019**, *123*, 9745–9755. [CrossRef]
25. Zhu, Q.; Yildirim, E.; Wang, X.; Soo, X.Y.D.; Zheng, Y.; Tan, T.L.; Wu, G.; Yang, S.-W.; Xu, J. Improved Alignment of PEDOT:PSS Induced by in-situ Crystallization of "Green" Dimethylsulfone Molecules to Enhance the Polymer Thermoelectric Performance. *Front. Chem.* **2019**, *7*, 783. [CrossRef]

Review

Research Progress on MXene-Based Flexible Supercapacitors: A Review

Baoshou Shen [1,2,*], Rong Hao [1,2], Yuting Huang [1,2], Zhongming Guo [1,2] and Xiaoli Zhu [1,2,*]

1. Institute of Earth Surface System and Hazards, College of Urban and Environmental Sciences, Northwest University, Xi'an 710127, China
2. Shaanxi Key Laboratory of Earth Surface System and Environmental Carrying Capacity, Xi'an 710127, China
* Correspondence: bsshen@nwu.edu.cn (B.S.); xiaolizhu@nwu.edu.cn (X.Z.); Tel.: +86-188-5119-8960 (B.S.); +86-135-7256-2258 (X.Z.)

Abstract: The increasing demands for portable, intelligent, and wearable electronics have significantly promoted the development of flexible supercapacitors (SCs) with features such as a long lifespan, a high degree of flexibility, and safety. MXenes, a class of unique two-dimensional materials with excellent physical and chemical properties, have been extensively studied as electrode materials for SCs. However, there is little literature that systematically summarizes MXene-based flexible SCs according to different flexible electrode construction methods. Recent progress in flexible electrode fabrication and its application to SCs is reviewed according to different flexible electrode construction methods based on MXenes and their composite electrodes, with or without substrate support. The fabrication methods of flexible electrodes, electrochemical performance, and the related influencing factors of MXene-based flexible SCs are summarized and discussed in detail. In addition, the future possibilities of flexible SCs based on MXene are explored and presented.

Keywords: MXenes; two-dimensional materials; flexible electrode; flexible supercapacitors

1. Introduction

With the emergence of more foldable, wearable, and flexible electronic devices, it is imperative to develop high-performing, safe, and cost-effective flexible energy storage devices that are compact and flexible enough to match these electronic components [1,2]. Due to the benefits of supercapacitors (SCs), such as long cycling, fast charging–discharging rates, and higher power density [3,4], flexible SCs are seen as advantageous candidates for powering flexible devices due to their being space efficient, lightweight, easy to handle, reliable, and compatible with other flexible electronic components [5–7]. They are usually made with thin-film electrode materials in the shape of fibers or flat sheets, with or without soft-matter substrates for support. Active materials are firmly integrated with flexible substrates, which allows the device to be robust and flexible mechanically. Free-standing films, fibers, or papers can serve as flexible electrodes generally without the need for current collectors and insulating binders, providing superior volumetric and gravimetric capacitance. Additionally, they can be easily fabricated into any desired shape or structural form [8]. Many nanostructured materials are frequently used in flexible SCs as electrode materials with improved performance, including carbon nanotubes, hierarchically porous carbon, and hollow metal oxides/sulfides [9–16]. However, there are few electrode materials that can have both high volumetric and gravimetric specific capacitance.

MXenes, a specific class of 2D transition metal carbides and nitrides, are obtained by selectively etching the A layer in their precursor MAX phase, where A is mainly group-13 or group-14 elements [17,18]. Since $Ti_3C_2T_x$ nanosheets were successfully synthesized with a selective etching method in 2011, MXenes have attracted great attention from scientists [19].

In the past decade, although more than thirty MXenes have been synthesized, including Ti_2CT_x, Ti_2NT_x, $Ti_3C_2F_x$, $Ti_4N_3T_x$, $Mo_{1.33}CT_x$, Mo_2CT_x, Nb_2CT_x, $Nb_4C_3T_x$, $Hf_3C_2T_x$, $Zr_3C_2T_x$, Cr_2CT_x, and V_2AlC, MXene-based SC electrodes mainly focus on $Ti_3C_2T_x$, Ti_2CT_x, Mo_2CT_x, V_2CT_x, and $Mo_{1.33}CT_x$ [20–39]. In addition to having the advantages of numerous electrochemically active sites, intact electron transport channels, and a two-dimensional structure for the diffusion of electrolytes, MXenes also exhibit metal conductivity and contain functional groups, including oxygen and fluorine, which are conducive to current charge–discharge and rate performance [23]. Furthermore, MXenes contribute most capacitance in the form of intercalation pseudocapacitance through intercalation processes with better reversibility and reaction kinetics, showing excellent cycling stability when employed as electrode materials [40]. Since the crystal layers of MXenes are flexible, their large gaps can accommodate more electrolyte ions, and MXenes with high packing density theoretically have higher volumetric capacity and energy density. For example, MXene hydrogels provide a volumetric capacitance of 1500 F cm^{-3}, which surpasses the hitherto unequaled value of RuO_2, and its rate characteristic is superior to that of carbon [24]. Hence, MXenes, as a layered 2D material, have demonstrated the most promising application potential in flexible SCs because of their higher conductivity and excellent dispersibility, which are beneficial to making films due to their hydrophilic surface. There is also a lot of reported research about MXenes as flexible SC electrode materials.

Several reviews on the synthesis and applications of MXenes have been published, particularly in energy storage and conversion [41–57]. For example, Ma et al. summarized the latest developments in flexible MXene-based composites for wearable devices, emphasizing preparation processes, working mechanisms, performance, and a vast array of applications, including sensors, SCs, and electromagnetic interference (EMI)-shielding materials [58]. Huang et al. reviewed the synthetic processes and fundamental features of functional 2D MXene nanostructures, highlighting their applications in EMI shielding, sensors, photodetectors, and catalysis [59]. Liu et al. overviewed the synthesis methods and the associated mechanisms of the Ti_3C_2 MXene/graphene composite, highlighting their potential application as energy storage materials, such as lithium–sulfur batteries, lithium-ion batteries, SCs, etc. [53]. Jiang et al. presented the most recent developments in MXene-based microsupercapacitors (MSCs), including device architecture, electrode material design, and different methods of depositing and patterning [41]. In 2021, Xu et al. reviewed recent research and breakthroughs in the chemical and physical synthesis of 2D MXenes and their applications in different flexible devices [50]. Zhang et al. reviewed the most typical flexible electrode materials at this point of development in terms of the recent advancements and challenges of flexible SCs [60]. Ma et al. comprehensively reviewed the most recent developments in $Ti_3C_2T_x$-based SC electrodes, paying special emphasis to the crucial role played by $Ti_3C_2T_x$ MXene in the exceptional electrochemical performance as well as the underlying mechanisms [49]. Yang et al. summarized the recent advances in MXene-based electrochemical immunosensors, emphasizing the roles played by MXenes in various types of electrochemical immunosensors [61]. Vasyukova et al. reviewed methods for synthesizing MXenes as well as their potential medical and environmental applications [62]. Yang et al. reviewed the recent research advancements in the structure, construction, and application of MXene-based heterostructures such as SCs, sensors, batteries, and photocatalysts [63]. However, for all that, the above-reported reviews are not specifically for the application of MXene-based electrode materials in flexible SCs.

Despite the numerous reviews that have referred to MXenes for their electrochemical energy storage capabilities, there have been a limited number of reviews about the different construction methods of electrodes for MXene-based flexible SCs. Here, we present the most recent developments in flexible electrode manufacturing and their applications in SCs according to the different construction methods of flexible electrodes based on MXenes and their composite electrodes, with or without substrate support. Firstly, since the distinctive physical and chemical properties of $Ti_3C_2T_x$ MXene are directly related to the process of preparation, a brief description of the synthesis strategy of $Ti_3C_2T_x$ MXene and its impact on the electrochemical properties will be presented. Secondly, construction methods such as self-supporting, PET-supported, fabric fiber-supported, and other substrate-supported MXene-based films as flexible electrodes are reviewed. An overview of the fabrication methods, electrode structures, working mechanisms, electrochemical performance, and related influencing factors of $Ti_3C_2T_x$-based SC electrodes is provided. In addition, the future possibilities of SC materials based on $Ti_3C_2T_x$ are outlined to encourage more research and development on MXenes in this fast-growing field.

2. Synthesis of $Ti_3C_2T_x$ MXene

The outstanding properties of $Ti_3C_2T_x$ are highly dependent upon its synthesis processes, which determine its chemical composition, electrical conductivity, lateral size, etching efficiency, surface terminations, and defects. Since the first preparation of $Ti_3C_2T_x$ in 2011, researchers have conducted extensive research on the new MAX phase and on the etching method. At present, many types of etchants are being explored for the production of $Ti_3C_2T_x$ MXene, including fluoride etching [26,30,64–68], fluoride-based salt etching [69,70], and fluoride-free etching, which have a significant impact on the electrochemical performance.

2.1. HF Etching

MXene is typically prepared by selectively etching the A layer of the MAX phase, and the mechanism can be described as follows [71,72]:

$$M_{(n+1)} AX_n + 3HF = AlF_3 + 1.5H_2 + M_{(n+1)} X_n \tag{1}$$

$$M_{(n+1)} X_n + 2H_2O = M_{(n+1)} X_n (OH)_2 + H_2 \tag{2}$$

$$M_{(n+1)} X_n + 2HF = M_{(n+1)} X_n F_2 + H_2 \tag{3}$$

In reaction (1), the A elements are separated from the MAX phase, resulting in the $M_{n+1}X_n$ phase. The functional groups of -F and/or -OH originate from reactions (2) and (3). Figure 1a,b show the schematic of the exfoliation process and characterization of structure morphology for Ti_3AlC_2. Naguib et al. prepared $Ti_3C_2T_x$ MXene with an accordion-like shape (Figure 1b) by etching Ti_3AlC_2 powders for 2 h in a 50% concentrated HF solution [73]. Mashtalir et al. investigated the influence of process parameters and particle size on the etching of Al from Ti_3AlC_2 in a 50% HF solution. The results showed that reducing the initial MAX particle size, prolonging reaction time, and increasing the immersion temperature were advantageous for the phase transformation of bulky Ti_3AlC_2 into $Ti_3C_2T_x$ [74]. During the etching procedure, etching duration, temperature, and HF concentration significantly impact the products. Al can be removed from the Ti_3AlC_2 MAX phase by HF with concentrations as high as 5%. However, an accordion-like particle form is commonly observed when HF concentrations exceed 10%. Moreover, the higher the HF percentage, the more defects there are in the $Ti_3C_2T_x$ flakes, affecting the quality, stability in the environment, and properties of the MXene obtained [75,76]. As the HF method has a low reaction temperature and is easy to operate, it is ideal for etching Al-containing MAX phases and portions of non-MAX phases. The HF etchant, however, is highly corrosive, toxic, poses operational risks, and has adverse environmental effects.

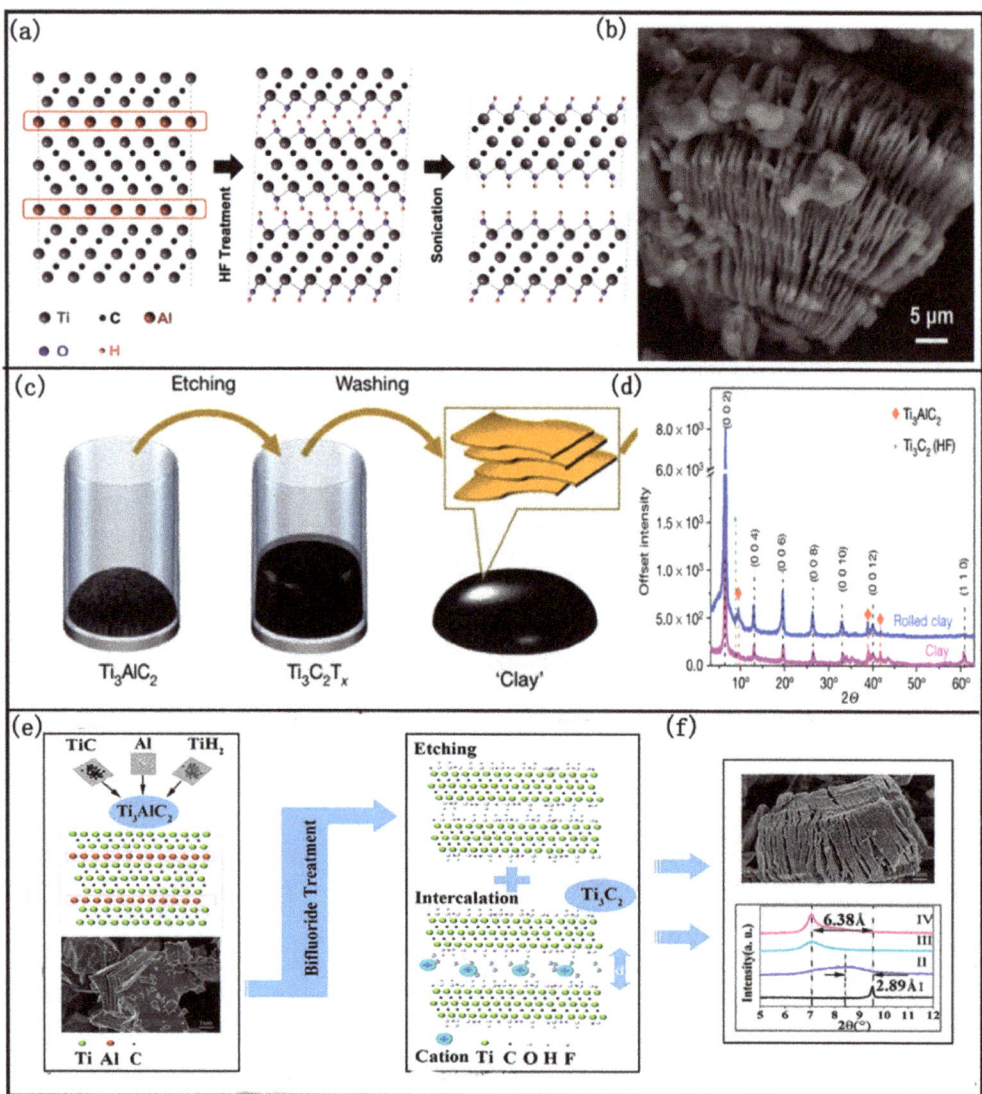

Figure 1. (a) Schematic of the exfoliation process for Ti₃AlC₂. (b) SEM image of a sample after HF treatment. Reprinted with permission from Ref. [73]. Copyright 2011 WILEY-VCH Verlag GmbH & Co., KGaA. (c) Ti₃AlC₂ etched in a solution of HCl + LiF and then washed with water to obtain Ti₃C₂Tₓ; the resulting Ti₃C₂Tₓ behaves like a clay. (d) XRD patterns of samples produced by etching in LiF + HCl solution. Reprinted with permission from Ref. [77]. Copyright 2014 *Nature*. (e) The schematic illustration of a reaction between Ti₃AlC₂ and bifluorides. (f) SEM images of samples exfoliated by NH₄HF₂ and the XRD patterns of Ti₃AlC₂ and different Ti₃C₂ samples: (I) Ti₃C₂ from etching Ti₃AlC₂ with HF; (II) Ti₃C₂ from etching Ti₃AlC₂ with NaHF₂; (III) Ti₃C₂ from etching Ti₃AlC₂ with KHF₂; (IV) Ti₃C₂ from etching Ti₃AlC₂ with NH₄HF₂. Reprinted with permission from Ref. [78]. Copyright 2017 *Materials & Design*.

2.2. Fluoride-Based Salt Etching

To develop significantly safer and gentler methods to manufacture $Ti_3C_2T_x$ flakes, researchers attempted to use hydrochloric acid (HCl) and fluoride salts as etchants to dissolve the Al element and generate 2D transition metal carbides. Ghidiu et al. prepared MXenes by dissolving LiF in 6 M HCl (Figure 1c) [77]. Compared to the HF etchant, this method can produce MXene flakes with bigger lateral dimensions, higher yields, and better quality. Furthermore, compared with the lattice parameter of HF-produced $Ti_3C_2T_x$ (c < 20 Å), the value in this study was 27–28. (Figure 1d). The increased interlayer spacing allows for the creation of more electrochemically active surfaces as well as shorter electrolyte ion transport routes. Furthermore, the milder etchant of LiF + HCl produces the $Ti_3C_2T_x$ flakes with wider lateral dimensions and fewer nanosized flaws. TEM also showed that the majority of the $Ti_3C_2T_x$ flakes had diameters of 500–1500 nm. The amount of HCl and LiF during the synthesis of fluoride-based salts affects the size, processing capacity, and quality of $Ti_3C_2T_x$. For example, LiF:Ti_3AlC_2 molar ratios increase from 5 to 7.5 when the HCl concentration is increased from 6 to 9 M, improving the quality and size of the $Ti_3C_2T_x$ flakes [79].

In addition to LiF, various fluoride salts such as NH_4HF_2, KF, NaF, FeF_3, NH_4F, etc., have been utilized to produce $Ti_3C_2T_x$ MXene. In 2014, it was reported that NH_4HF_2 could be used to etch sputter-deposited epitaxial Ti_3AlC_2 films at room temperature [80]. In contrast to the films etched with HF, the films intercalated with NH_3 and NH^{4+} species showed 25% larger c lattice parameters. Feng et al. described the effect of etching duration and temperature on the synthesis of $Ti_3C_2T_x$ in 1 M of different bifluoride solutions ($NaHF_2$, KHF_2, NH_4HF) (Figure 1e) [78]. In 1 M of bifluoride solution at 60 °C, the minimum etching duration for the onset of exfoliation of Ti_3AlC_2 was 8 h. Using bifluoride, KHF_2, or NH_4HF_2 as an etchant allowed the formation of Ti_3C_2 with greater interplanar spacing in a single-stage process and the retention of the 2D flake structure (Figure 1f). Wang et al. proposed using iron fluoride (FeF_3) and hydrogen chloride (HCl) as an etching for the production of $Ti_{n+1}C_nT_x$ from $Ti_{n+1}AlC_n$ (n = 1 or 2) [81]. Compared to the HF etching method, the fluorine content of Ti_3C_2 made with FeF_3/HCl is lower. By adjusting the immersion time in the water, it was possible to tune the partial oxidation of $Ti_3C_2T_x$, which enabled the preparation of a composite of anatase and $Ti_3C_2T_x$.

The fluoride salts used in synthesizing $Ti_3C_2T_x$ are less poisonous and milder than HF. This $Ti_3C_2T_x$ has a relatively large size, few flaws, a low fluorine concentration, and large interlayer spacing, allowing for further structural modification.

2.3. Fluoride-Free Etching

Even though fluorine-containing etching produces MXenes with a good layer-sheet structure, long etching times can cause defects in the product, and impurity groups (-F, -OH) can change the properties of the MXenes. The specific capacitance of the material can also be affected when it is used as an SC electrode. Researchers have developed a variety of fluorine-free MXene etching techniques in response to these problems [82–89].

Electrochemical etching is a method for preparing 2D $Ti_3C_2T_x$ with good capacitive performance. This method can be carried out in electrolytes devoid of fluorides in order to produce $Ti_3C_2T_x$ MXene devoid of fluorine terminations. Al layers can be selectively etched by applying a steady voltage, allowing chloride ions (Cl^-) that have a strong affinity for Al to break the Ti-Al bonds. Feng et al. proposed an electrochemical method for layering Ti_3C_2 using binary aqueous electrolytes [89]. The anodic etching of Al is facilitated by chloride ions, which allow Ti-Al bonds to be broken quickly. Then, ammonium hydroxide (NH_4OH) is added to make it easier to etch below the surface of the anode that has already been etched. More than 90% of the Ti_3AlC_2 etched in a short period of time is a single layer or double layer, and the average lateral dimension exceeds 2 µm. In addition, an all-solid-state SC fabricated from exfoliated sheets exhibits excellent volumetric and areal capacitances of 439 F cm^{-3} and 220 mF cm^{-2}, respectively, at a scan rate of 10 mV s^{-1}, which is larger than for the classical HF etching process [90].

Alkali is also anticipated to achieve selective etching of the MAX phase. Li et al. successfully prepared multilayer MXene with 92 wt% purity based on the Bayer process using only alkali-assisted hydrothermal methods [91]. Initially, a solution of Ti_3AlC_2 was oxidized by NaOH and then dissolved into Al $(OH)^{4-}$, resulting in the surface termination of MXene with various functional groups of Al atoms, such as -OH and/or -O. After that, the inner Al atoms began to oxidize, producing new Al hydroxides ($(Al(OH)_3)$) and dehydrated oxide hydroxides (AlO(OH)). These Ti layers provided lattice confinement, preventing these insoluble compounds from reacting readily with -OH to produce dissolvable Al $(OH)_4^-$, which interfered with MXene synthesis and had to be eliminated. The schematic diagram of the reaction between Ti_3AlC_2 and the NaOH aqueous solution under different conditions is shown in Figure 2a. The $Ti_3C_2T_x$ film electrode was prepared (52 μm thick) in 1 M H_2SO_4 with a gravimetric capacitance of 314 F g^{-1} at 2 mV s^{-1}, which is 28.2% higher than the LiF + HCl-$Ti_3C_2T_x$ clay (75 μm thick) [48] and 214% higher than the HF-$Ti_3C_2T_x$ [92]. Similarly, Li et al. etched 0.1 g Ti_3AlC_2 by using 0.35 g of KOH in a hydrothermal reactor at 180 °C for 24 h [93]. Al atoms are replaced with -OH groups, resulting in nanosheets of $Ti_3C_2(OH)_2$ with significant lateral dimensions. When the MAX phase is etched with concentrated alkali, highly hydrophilic products with F-free terminations can be achieved. The use of high alkali concentrations and high temperatures limits its applications for preparing MXene on a large scale.

Due to their electron acceptor, transition metal halides in the molten state are capable of reacting with the A layer of the MAX phase. As shown in Figure 2b, Li et al. presented a method for etching MAX phases based on direct redox coupling between A and a Lewis acid molten salt cation [94]. This general Lewis acid etching procedure also expanded the range of MAX-phase precursors, which can be used to prepare new MXenes (Figure 2c). These MXenes exhibited increased storage capacity for Li^+ and high current in nonaqueous electrolytes, making them suitable electrode materials for Li-ion batteries and multifunctional devices, including capacitors [95,96]. Using the one-step molten salt reaction of $SnCl_2$ in situ, Wu et al. synthesized $Ti_3C_2T_x$ MXene/Sn composites directly from Ti_3AlC_2 MAX phase precursors in Figure 2d [97]. The $SnCl_2$ is etched as a Lewis acid during this process to etch the Ti_3AlC_2 MAX phase and obtain $Ti_3C_2T_x$ MXenes. The structure of $Ti_3C_2T_x$ MXene displays a typical accordion design. It was found that the interlayer spacing of $Ti_3C_2T_x$ MXene was 1.15 nm (Figure 2e), a much greater spacing than that obtained by acid etching (0.96–0.98 nm), widely used at the time. Although the nonaqueous molten salt etching method offers a broader range of etching and chemical safety, it is still in its infancy, which requires further investigation into the physicochemical properties of the MXenes produced.

It is more difficult to produce MXenes using water as a solvent at room temperature. For example, the presence of water adversely affects the synthesis of polymeric nanocomposites with MXenes reinforced by means of in situ polymerization [98,99]. The residual water may have an impact on the successful loading of specific quantum dots onto MXene sheets [100,101]. Moreover, when organic electrolytes are employed, the presence of water could decrease the stability of the electrolyte and reduce the electrochemical voltage window, resulting in the performance degradation of lithium-ion and sodium-ion batteries. In recent years, researchers have made great efforts to find a breakthrough. Using organic substances and deep eutectic solvents (DES) as etching solvents, they have succeeded in preparing MXene products with good electrochemical properties.

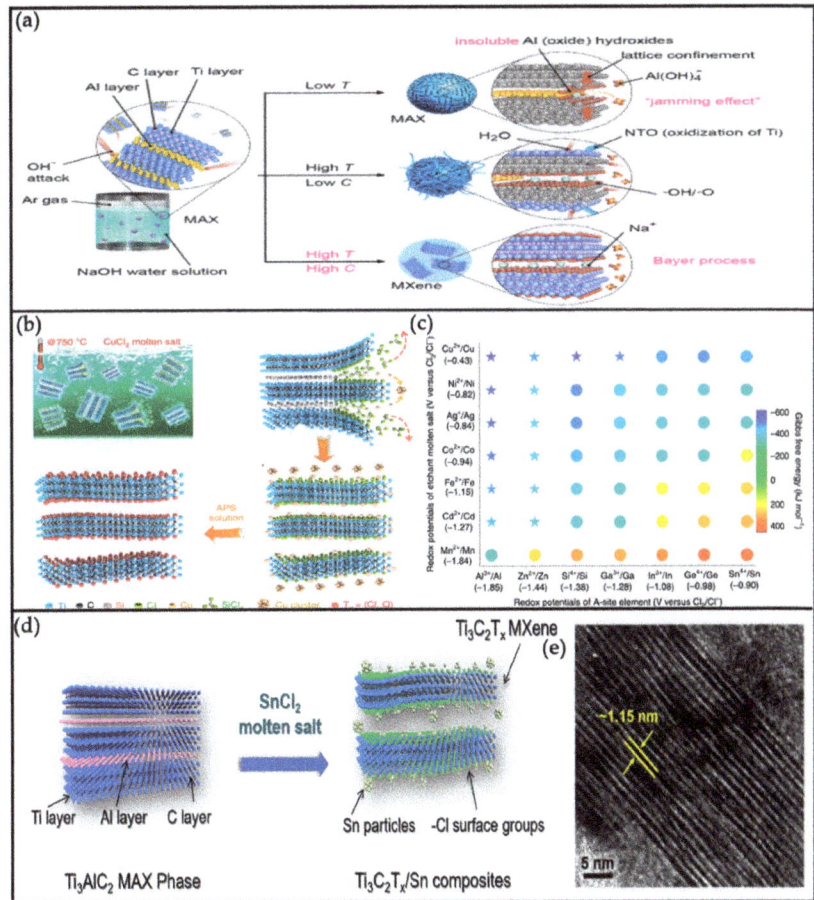

Figure 2. (a) The reaction between Ti$_3$AlC$_2$ and NaOH water solution under different conditions. Reprinted with permission from Ref. [91]. Copyright 2018 WILEY-VCH. (b) Schematic of Ti$_3$C$_2$T$_x$ MXene preparation. (c) Generalization of the Lewis acid etching route to a large family of MAX phases. Reprinted with permission from Ref. [94]. Copyright 2020 *Natural Materials*. (d) Schematic of the synthesis of Ti$_3$C$_2$T$_x$ and Ti$_3$C$_2$T$_x$/Sn composites by SnCl$_2$ molten salt reaction. (e) HRTEM image of Ti$_3$C$_2$T$_x$ MXene. Reprinted with permission from Ref. [97]. Copyright 2022 *Electrochimica Acta*.

As shown in Figure 3a, Wu et al. reported a highly reliable and water-free ion thermal method for synthesizing Ti$_3$C$_2$ MXene in deep eutectic solvents (DES) [102]. The DES used in the production of Ti$_3$C$_2$ MXene offers the following special advantages over earlier high-risk processes: (i) The processing of solid precursors and products at room temperature was highly safe, rather than making use of hazardous solutions; (ii) the low vapor pressure of DES and its excellent solvation properties enable the etching process to generate HF in situ through a reaction between H$_2$C$_2$O$_4$ and NH$_4$F in a mild environment; (iii) the cations of choline were intercalated into the layers of Ti$_3$C$_2$, resulting in a larger interlayer spacing of 1.35 nm in comparison with HF-Ti$_3$C$_2$ (0.98 nm); and (iv) DES can be recycled and reutilized throughout the etching process, which is promising for the industrial preparation of MXene at a low cost. Shi et al. developed a new iodine-assisted nonaqueous etching strategy [103]. MAX powders were immersed in an I$_2$-CH$_3$CN mixture with a 1:3 molar ratio of Ti$_3$AlC$_2$ to I$_2$, as shown in Figure 3b. Iodine can remove Al layers from Ti$_3$AlC$_2$ because Ti−Al bonds are more reactive than Ti−C bonds. Then, manual shaking in an HCl solution was sufficient for separating 2D MXene sheets. Because of the benefits of the nonaqueous etching process, 2D MXene sheets exhibit good structural stability. MXene films have a higher conductivity of 1250 S cm^{-1} compared to films made with fluoride etchants. The exfoliated MXene sheets, through iodine etching containing extensive oxygen surface groups, can be fabricated into SCs with gravimetric capacitances and cycling stability, which surpasses the performance of most MXene materials previously reported [34,104,105].

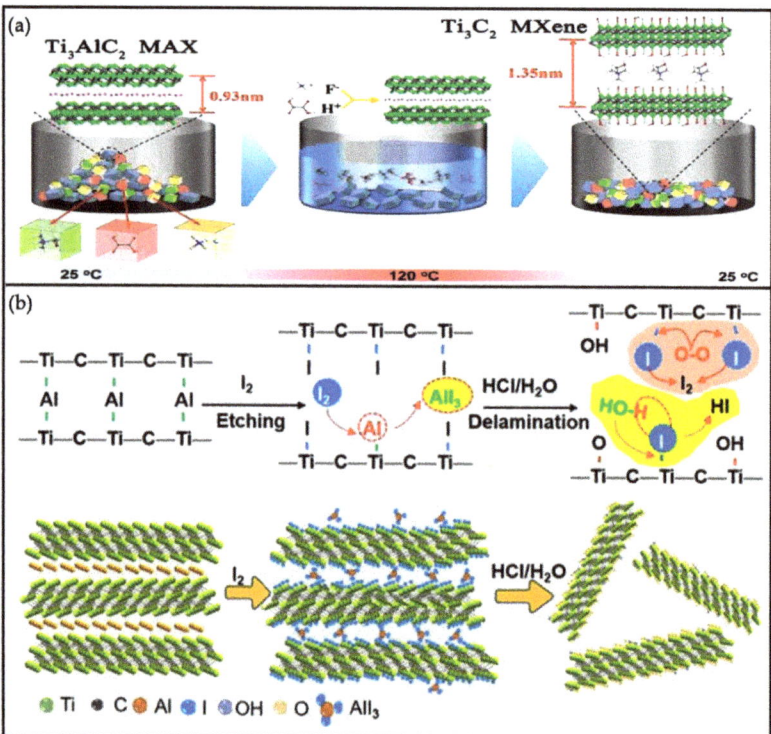

Figure 3. (**a**) Scheme of the ionothermal synthesis of DES-Ti$_3$C$_2$ MXene. Reprinted with permission from Ref. [102]. Copyright 2019 *Journal of Energy Chemistry*. (**b**) The iodine-assisted etching and delamination of Ti$_3$AlC$_2$ towards 2D MXene sheets. Reprinted with permission from Ref. [103]. Copyright 2021 Wiley-VCH GmbH.

3. MXene-Based Flexible Electrode Materials

As electrode materials, MXenes should exhibit not only excellent electrochemical performance but also excellent properties such as hydrophilicity, malleability, and two-dimensional structure (atomic layer thickness and micrometer-scale lateral dimensions), which make them suitable for the formation of the thin film serving as a flexible electrode. As a result, the electrochemical performance of SCs is largely determined by MXene electrode material structure design, such as electrode architecture, surface terminations, interlayer spacing, and composites (Figure 4) [55]. Recently, a lot of research has explored the application of MXenes and their composites for fabricating SCs on different substrates, including self-supporting, PET-supported, carbon-cloth-fiber-supported, and so on. In this section, we provide a report on recent developments in MXene-based flexible electrodes.

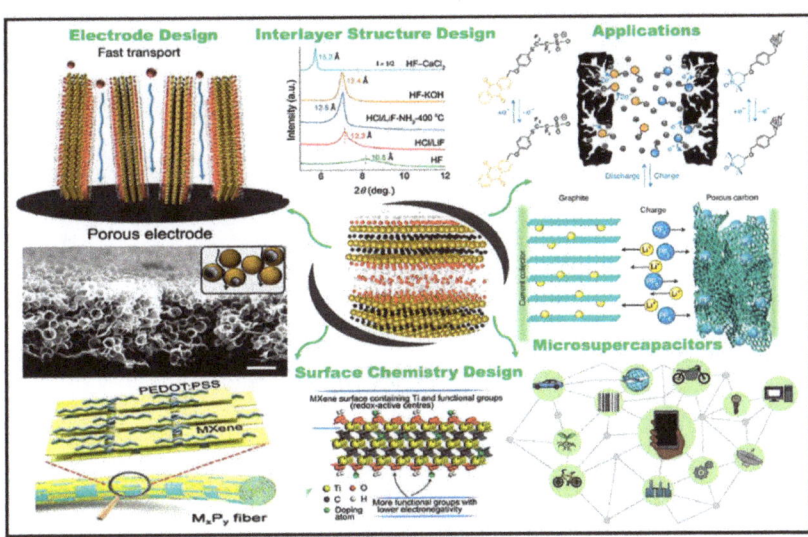

Figure 4. Modification design strategies for MXenes and applications, including interlayer structure design, surface chemistry design, electrode architecture design, and composites. Reprinted with permission from Ref. [55]. Copyright 2021 *Journal of Electroanalytical Chemistry*.

3.1. Self-Support MXene-Based Films as Flexible Electrodes

3.1.1. Pure MXene

MXene-based electrodes, especially freestanding MXene films, have immense potential for SCs and flexible electronics [44]. In addition, the many terminations endow MXenes with excellent hydrophilicity and a rich surface charge, allowing MXene nanosheets to be disseminated uniformly in aqueous solutions.

By using vacuum filtration (VAF) [106], the MXene dispersions can be easily turned into MXene films. These films can be directly charged as electrodes for flexible SCs to achieve high specific capacitance and good cycling performance. Ghidiu et al. rolled hydrophilic MXene (Ti_3C_2) into thin films using LiF and HCl etching [77]. The volumetric capacitance of the Ti_3C_2 electrode is up to 900 F cm^{-3}, which is at least twice that of MXene (300 F cm^{-3}) generated through hydrofluoric acid etching and is characterized by exceptional rate performance and excellent cyclability [23]. Furthermore, the synthetic method allows film production to be much faster while avoiding the handling of hazardous concentrated hydrofluoric acid. It is necessary for the delaminated Ti_3C_2 (d−Ti_3C_2) films to have a sufficiently high stacking density and electrical conductivity for them to be capable of producing high volumetric performances. As shown in Figure 5a, Que et al. developed a much thinner, more flexible MXene-film electrode obtained through vacuum filtration by applying external pressure to the membrane [107]. The application of external pressures could increase the density of the delaminated Ti_3C_2 (d−Ti_3C_2) films, resulting in good wettability, a comparatively high electrical conductivity, and high surface activity, thereby facilitating effective ion transport. The d−Ti_3C_2 film, pressed at a pressure of 40 MPa, exhibits an extraordinarily high capacitance of 633 F cm^{-3}, a high energy density, and outstanding cyclical stability (Figure 5c,d). Furthermore, the corresponding SC in the organic electrolyte has a volumetric energy density of 41 Wh L^{-1} (Figure 5b).

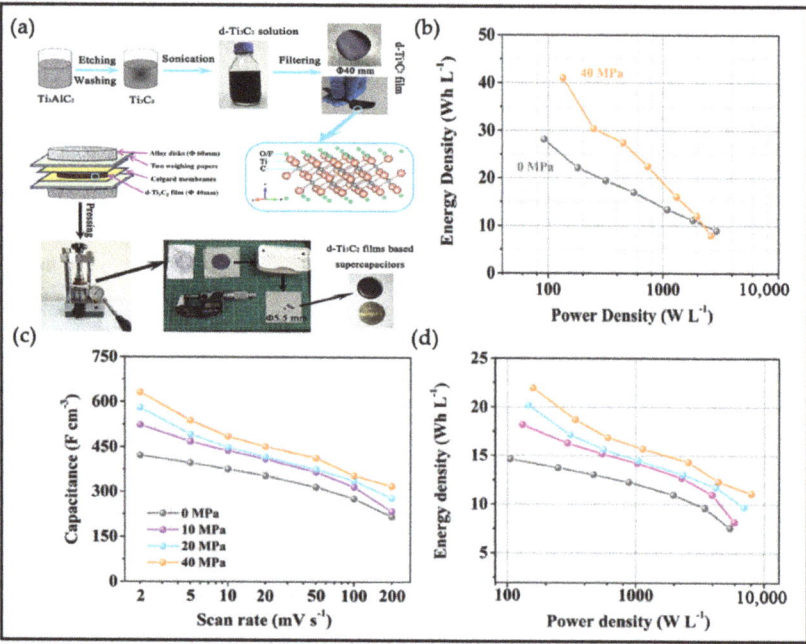

Figure 5. (a) Schematic illustration of the high-pressure d−Ti$_3$C$_2$ film synthesis and electrode preparation. (b) Ragone plots of volumetric energy and power densities. (c) Volumetric capacitances of the d−Ti$_3$C$_2$ film electrodes in different symmetric SCs. (d) Ragone plots of volumetric energy and power densities obtained from the symmetric SCs based on the d−Ti$_3$C$_2$ films under different pressures. Reprinted with permission from Ref. [107]. Copyright 2018 WILEY-VCH Verlag GmbH & Co., KGaA.

Heat treatment is an efficient approach for eliminating the terminals of MXene and improving its electrochemical performance [108]. A method for enhancing the capacitance performance of Ti$_3$C$_2$T$_x$ film by annealing at a low temperature in inert gas was presented by Zhang et al. [109]. Due to more C−Ti−O active sites and greater interlayer voids, the annealed film at 200 °C in an Ar atmosphere has an energy density of 29.2 Wh Kg^{-1} and a capacitance of 429 F g^{-1} in 1 M H$_2$SO$_4$ electrolyte. Subsequently, Zhao et al. investigated the high-temperature annealing of Ti$_3$C$_2$T$_x$ films to enhance capacitance performance [110]. In addition to its gravimetric value of 442 F g^{-1}, the film also has a high volumetric capacitance and an excellent rate capability after being annealed at 650 °C in an Ar atmosphere. As a simple and environmentally friendly process, alkalinizing followed by annealing has been certified to increase the gravimetric capacitance of MXenes. By alkalinizing and annealing Ti$_3$C$_2$T$_x$ film (Figure 6a), Zhang et al. synthesized a flexible and binderless MXene film (named ak−Ti$_3$C$_2$T$_x$ film−A) [106]. As a result of the alkalizing and annealing processes, more oxygen-containing groups are exposed to the aqueous electrolyte, increasing the pseudocapacitance during the charge–discharge process [111,112]. Furthermore, the annealing treatment also increases the crystalline order, which enhances the conductivity of the MXene film. In addition to extremely high volumetric capacitance (Figure 6b), the film electrode has remarkable cycling stability. The symmetric SC with ak−Ti$_3$C$_2$T$_x$ film−A also showed a volumetric energy density of 45.2 Wh L^{-1}.

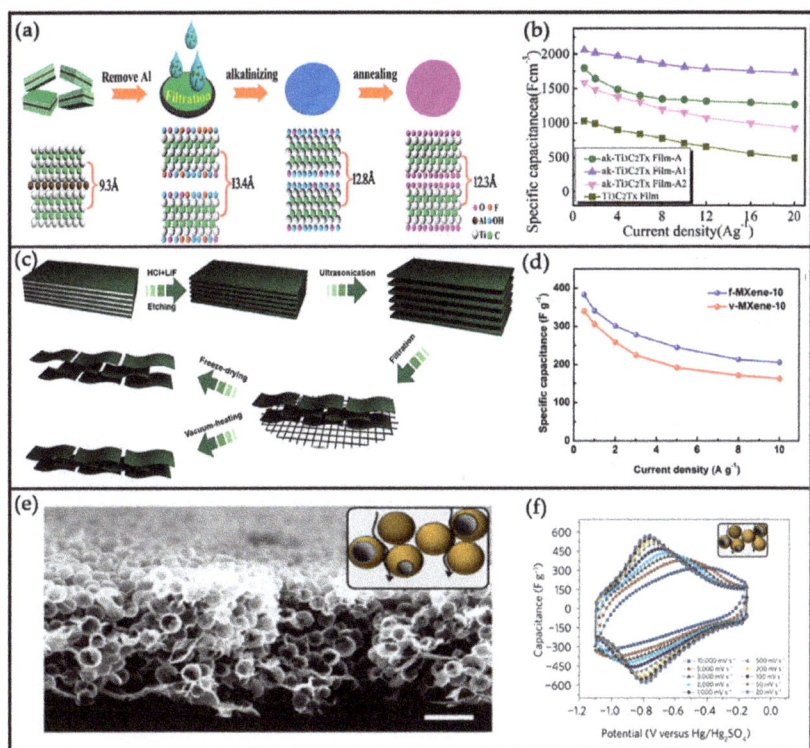

Figure 6. (a) Schematic diagram showing the synthesis process of ak−Ti$_3$C$_2$T$_x$ film−A. (b) The specific capacitance as a function of current density. Reprinted with permission from Ref. [106]. Copyright 2018 *Electrochimica Acta*. (c) Schematic illustration for the fabrication of f-MXene and v-MXene films. (d) Specific capacitance of f−MXene−10 and v−MXene−10 at different current densities. Reprinted with permission from Ref. [113]. Copyright 2020 *Applied Surface Science*. (e) SEM image of macroporous templated Ti$_3$C$_2$T$_x$ electrode cross-section. Insets show schematically the ionic current pathway in electrodes of different architectures. (f) Cyclic voltammetry profiles of a macroporous 13−μm−thick film with a 0.43 mg cm^{-2} loading collected in 3 M H$_2$SO$_4$ at scan rates from 20 to 10,000 mV s^{-1}; the inset shows a schematically macroporous electrode architecture and the ionic current pathways in it. Reprinted with permission from Ref. [24]. Copyright 2017 *Nature Energy*.

Although MXenes can be easily built into a film using a simple vacuum filtration process, this results in the horizontal restacking of the delaminated MXene nanosheets, slowing ion transport as well as inadequate active site exposure, hence reducing capacitance and rate performance. Many ways have been reported for enhancing ion accessibility to active sites. For example, the freeze-drying treatment is an efficient way of producing 2D materials with extremely complex structures since some of the metastable designs can be preserved during the process [114]. During the process, the solvent molecules function as pore creators, which prevent the flakes from stacking, resulting in an increase in the specific surface area. Additionally, MXene films with suitable porosity architectures can be produced by modifying the freeze-drying procedure. Xia et al. presented a method for fabricating a Ti$_3$C$_2$T$_x$ film electrode with malleable, freestanding, and vertically aligned properties by mechanically shearing a liquid–crystalline phase of MXene nanosheets and then freeze-drying the nanosheets to remove ethanol [115]. Ran et al. fabricated freestanding and flexible MXene films with vacuum-filtering and freeze-drying techniques (Figure 6c) [113]. The frozen solvent molecules were eliminated by sublimation throughout the freeze-drying process, mitigating the detrimental effect of van der Waals forces and enhancing layer spacing. In comparison to the dense stacking of vacuum-heated MXene (v−MXene) film, the freeze-dried MXene (f−MXene) film exhibited a porous structure that enhanced electrolyte ion shuttling, thus enhancing electrochemical performance. Therefore, the f−MXene film electrode has a maximum specific capacitance of 341.5 F g^{-1} at 1 A g^{-1} and 206.2 F g^{-1} when the current density reaches 10 A g^{-1}, which is a significant improvement over the v-MXene film electrode (Figure 6d).

The templating method is also widely used to produce porous 2D materials by putting a template material into the nanosheet interlayers of 2D materials and then removing it [115–117]. Typically, polymers are utilized as templates for designing 3D macroporous electrode films by ordered assembly. Lukatskaya et al. created a flexible $Ti_3C_2T_x$ electrode with an open, porous architecture using microspheres of polymethylmethacrylate (PMMA) as a sacrificial template and then removed the template via annealing (Figure 6e) [24]. The $Ti_3C_2T_x$ electrode exhibited a capacitance of 200 F g^{-1} at scan rates as high as 10 V s^{-1} (Figure 6f).

3.1.2. MXene/Graphene

Integrating MXenes with graphene is a promising method of fabricating composite films. The irregular $Ti_3C_2T_x$ acts as an intercalator and dispersant within the graphene layer, lessening graphene agglomeration and increasing specific surface area. The $Ti_3C_2T_x$ with superior electrical conductivity and hydrophilicity will enhance the electrochemical properties of the composites and their capacitive deionization characteristics [118–120]. Thus, the synergistic effect enabled by the bilayer effect of graphene and the pseudocapacitive characteristics of $Ti_3C_2T_x$ may enhance the energy storage performance of the composite electrode [121–125].

Most of the time, MXenes and graphene are made by mixing solutions of MXenes with reduced graphene oxide (rGO) or graphene oxide nanosheets and then vacuum-filtering to form the composite films [126–129]. Yan et al. came up with a way to make MXene/rGO SC electrodes using the electrostatic self-assembly of negatively charged MXene and positively charged, chemically oxidized rGO, as shown in Figure 7a [130]. The MXene/rGO composite effectively prevents the self-restacking of both rGO and MXene while maintaining extremely high electrical conductivity (2261 S cm^{-1}) and large density (3.1 g cm^{-3}). The MXene/rGO-5 wt% composite electrode has an excellent volumetric capacitance at 2 mV s^{-1}, a capacitance retention capacity of 61% at 1 V s^{-1}, and an extended cycle life. In addition, this binder-free symmetric SC displays an extremely high volumetric energy density of 32.6 Wh L^{-1}. Fan et al. prepared modified MXene/holey graphene films by filtering alkalized MXene and holey graphene oxide (HGO) dispersions and annealing them in Figure 7b [128]. Alkali is capable of leading not only to the destruction of charge balance in holey graphene oxide and MXene dispersions but also of causing the transition of the -F group into a -OH group. Furthermore, annealing may also remove most of the -OH groups and increase the number of Ti atoms, which could lead to greater pseudocapacitive reactions. It can provide extremely high capacitances (1445 F cm^{-3} at 2 mV s^{-1}), high mass loading capacities, and excellent rate performance as an electrode material for SCs. Furthermore, the assembled symmetric SC exhibits a tremendous volumetric energy density (38.5 Wh L^{-1}).

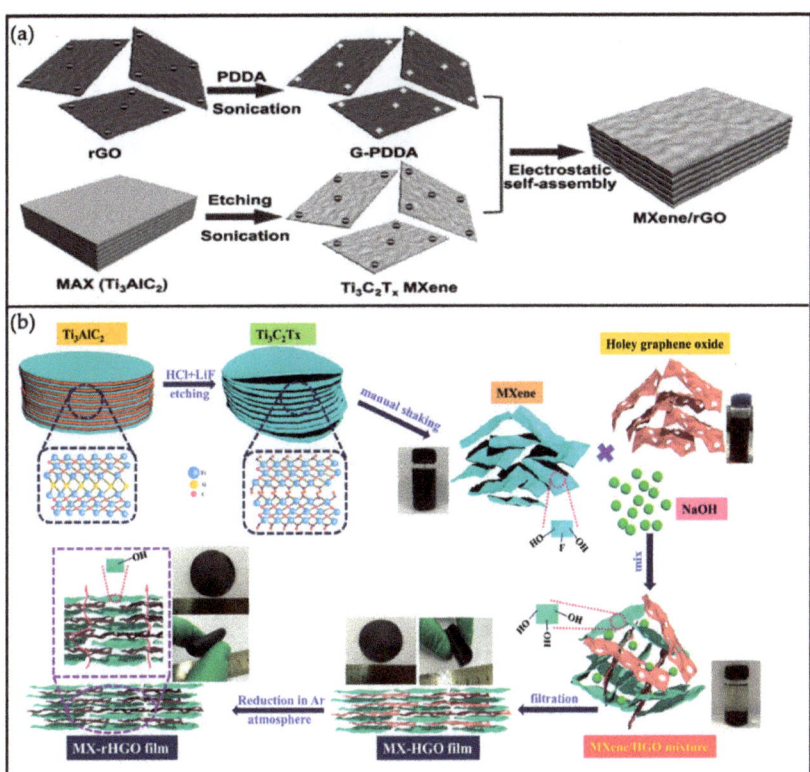

Figure 7. (a) Schematic illustration for the synthesis of the MXene/rGO hybrids. Reprinted with permission from Ref. [130]. Copyright 2017 WILEY-VCH Verlag GmbH & Co., KGaA. (b) Illustration of synthesis of the modified MXene/holey graphene film. Reprinted with permission from Ref. [128]. Copyright 2018 WILEY-VCH Verlag GmbH & Co., KGaA.

These heterostructured films were manufactured via a vacuum-assisted filtration process, which is both time-consuming and size-constrained, making it unsuitable for large-scale production. Therefore, for a satisfactory stacking of each component, as well as for high-speed processing [131], Miao et al. developed a simple and effective method for fabricating a 3D porous MXene film using self-propagating reduction. The process can be completed in 1.25 s, resulting in a 3D porous framework via the immediate release of substantial amounts of gas. MXene/rGO films have a higher capacitance and rate performance because the 3D porous structure provides abundant ion-accessible active sites and allows rapid ion transport [132]. Yang et al. developed an effective and rapid self-assembly method for creating a 3D porous oxidation-resistant MXene/graphene (PMG) composite using the template in Figure 8a [133]. The 3D porous design could successfully prevent the oxidation of MXene layers, ensuring superior electrical conductivity and an adequate number of electrochemically active sites. Therefore, the PMG−5 electrode has an exceptional cycling stability, excellent rate performance, and a remarkable specific capacitance (Figure 8b,c). In addition, the as-assembled asymmetric SC (ASC) has excellent cycling stability with a specific capacitance degradation of 4.3% after 10,000 cycles and a notable energy density of 50.8 Wh kg^{-1} (Figure 8d,e).

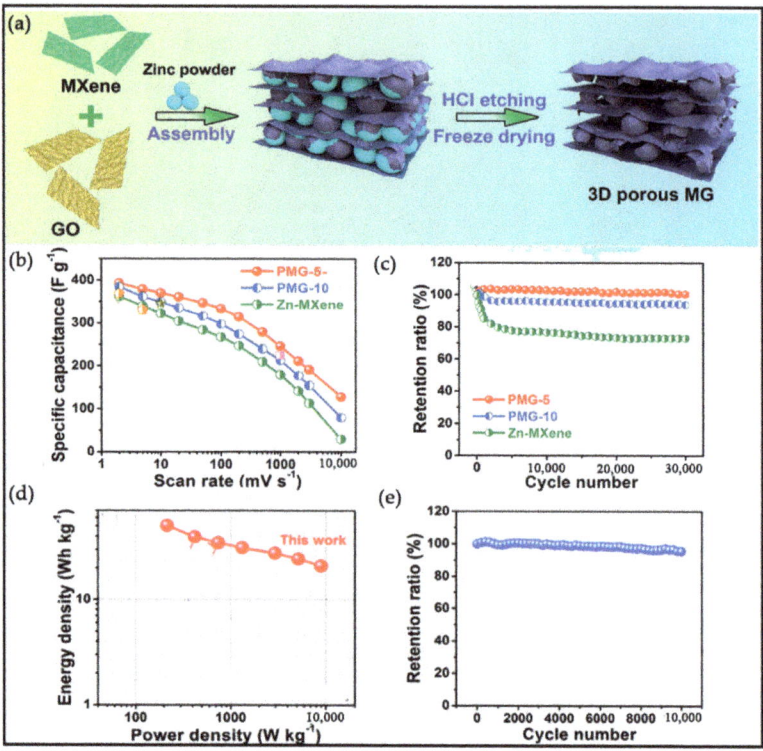

Figure 8. (**a**) Schematic illustration of the synthesis of 3D porous MG nanocomposite. (**b**) Specific capacitance at different scan rates. (**c**) Cycling stability of the Zn−MXene, PMG−5, and PMG−10 electrode measured at 200 mV s^{-1}. (**d**) Energy and power density for PMG−5//NHRGO ASCs. (**e**) Capacitance retention ratio of ASC at a scan rate of 100 mV s^{-1} for 10,000 cycles. Reprinted with permission from Ref. [133]. Copyright 2021 Wiley-VCH GmbH.

3.1.3. MXene/Carbon Nanotubes

Carbon nanotubes (CNTs), a common and well-studied type of 1D carbon nanomaterial, are also used to make SC electrodes by combining them with $Ti_3C_2T_x$ MXene. CNTs can increase the performance of energy storage by enlarging the specific surface area, regulating the interlayer gap, enabling ion diffusion, and improving electrical conductivity [134–139].

The layer-by-layer assembly method is a well-established method of constructing microstructures [140,141]. Zhao et al. have exploited sandwich-like, flexible MXene/CNT film electrodes for SCs using the alternate filtration of MXene and CNTs from aqueous solutions (Figure 9a) [142]. In comparison to pure MXene and randomly mixed MXene/CNT paper electrodes, these electrodes are highly flexible and freestanding, with highly significant volumetric capacitances and excellent rate performance. At a scan rate of 2 mV s^{-1}, the MXene/SWCNT paper electrode showed a high capacitance of 390 F cm^{-3}. It also displayed a volumetric capacitance of 350 F cm^{-3} at 5 A g^{-1}, which did not degrade after 10,000 cycles (Figure 9b,c).

Self-assembly technology, also known as electrostatic assembly, is an easy method of synthesizing hybrid materials. In self-assembly technology, one material is constructed on the surface of another, thus forming composites by utilizing the electrostatic interaction between distinct charges. Dall'Agnese et al. produced a flexible $Ti_3C_2T_x$/CNT film by using the self-assembly approach and studied the electrochemical behavior of Ti_3C_2 MXene in different organic electrolytes [143]. This electrode exhibited excellent cycling stability and rate performance. By means of vacuum filtration, Xu et al. also produced a Ti$_3$C$_2$T$_x$/SCNT self-assembled composite electrode, which exhibited a capacitance of 220 mF cm^{-2} (314 F cm^{-3}) and retained 95% of its capacitance after 10,000 cycles [144]. The enhanced capacitance could be attributed to the increase in the interlayer spacing of MXene and the improved ion accessibility brought about by the utilization of SWCNTs as spacers.

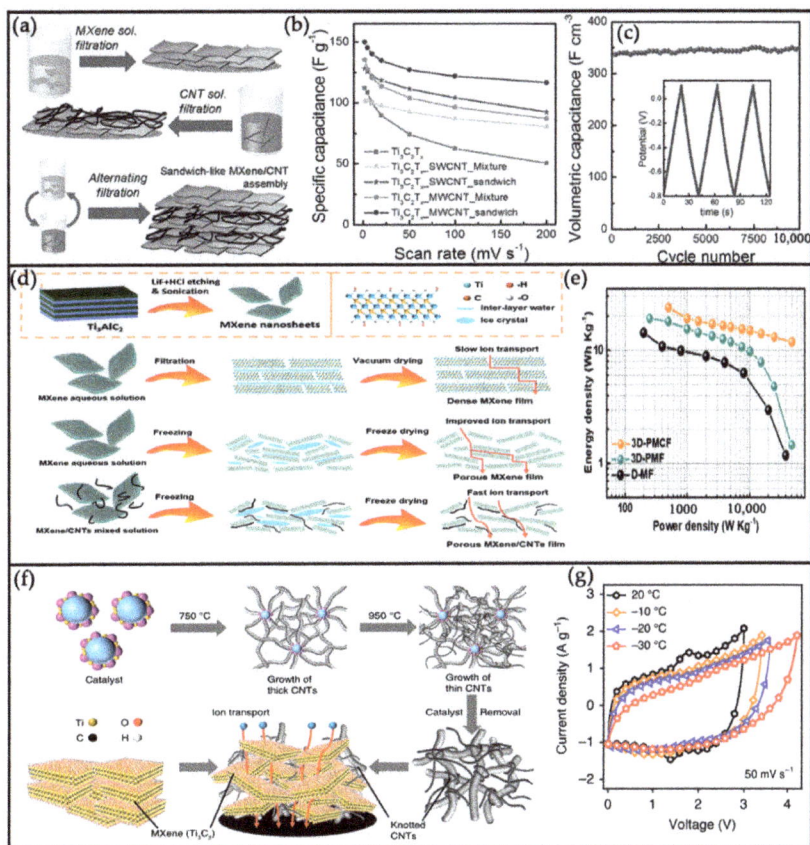

Figure 9. (**a**) The fabrication of MXene/CNT papers. (**b**) Gravimetric capacitances of Ti$_3$C$_2$T$_x$ and Ti$_3$C$_2$T$_x$/CNT electrodes at different scan rates. (**c**) Cycling stability of a sandwich-like Ti$_3$C$_2$T$_x$/SWCNT electrode at 5 A g^{-1}. Reprinted with permission from Ref. [142]. Copyright 2014 WILEY-VCH Verlag GmbH & Co., KGaA. (**d**) Schematic illustration of the fabrication process of MXene nanosheets, vacuum-dried D−MF film, freeze-dried 3D−PMF film, and freeze-dried 3D−PMCF film. (**e**) Gravimetric energy and power density profiles for 3D−PMCF, 3D−PMF, and D−MF. Reprinted with permission from Ref. [144]. Copyright 2020 WILEY-VCH Verlag GmbH & Co., KGaA. (**f**) Design of the MXene-knotted CNT composite electrodes for efficient ion transportation. (**g**) Cyclic voltammograms with larger voltage windows at low temperatures for the full cell using a MXene-knotted CNT composite electrode with a CNT content of 17% as the negative electrode. Reprinted with permission from Ref. [145]. Copyright 2020 *Nature Communications*.

Even though the MXene/MWCNT composite electrodes made with these methods seem to have a larger gap between layers compared to unmodified MXenes, the 2D layers continue to be horizontally stacked, indicating that the stacking problem persists, which restricts ion accessibility and slows ion kinetics. As shown in Figure 9d, Zhang fabricated a flexible 3D porous Ti$_3$C$_2$T$_x$/CNTs film (3D−PMCF) using an in situ ice template strategy [144]. After freeze-drying, the resulting Ti$_3$C$_2$T$_x$/CNTs film possessed a 3D structural network with a highly porous structure, which was templated by interlayered ice in conjunction with CNTs as functional spacers. In addition to exposing several active sites, 3D-PMCF facilitates rapid ion transport, resulting in superior electrochemical performance. The symmetric SCs based on 3D−PMCF achieved a high energy density of 23.9 Wh kg^{-1}, demonstrating their potential as flexible electrodes for supercapacitors (Figure 9e). In order to achieve improved ion transport at low temperatures, Gao applied knotted CNTs, which broke the traditional horizontal alignment of the 2D layers of MXene Ti$_3$C$_2$ [145]. As a result of knot-like structures, the Ti$_3$C$_2$ flakes are prevented from restacking, providing fast pathways for ion transport, which results in the improved low-temperature operation of Ti$_3$C$_2$ MXene-based SCs (Figure 9f,g).

3.1.4. MXene/Polymer

Since polymers are simple to produce, are inexpensive, and have tunable functionalities, they have been widely employed to prepare MXene-based composites [146,147]. MXene and polymer-formed composite films have also been increasingly applied to flexible devices over the past few years.

Solvent processing is the most common method of production. In most cases, MXene is added to a polymer solution in the colloidal form (often aqueous). Then, the solvent is removed from the solution using evaporation, vacuum filtration, or precipitation into a nonsolvent. Ling et al. fabricated Ti$_3$C$_2$/polymer membranes by applying charged polydiallyldimethylammonium chloride (PDDA) and polyvinyl alcohol (PVA) via a VAF method (Figure 10a) [148]. Compared to pure Ti$_3$C$_2$T$_x$ film (2.4 × 10^5 S m^{-1}), the conductivity of Ti$_3$C$_2$T$_x$/PVA composite film is 2.2 × 10^4 S m^{-1}. However, the composite Ti$_3$C$_2$T$_x$/PVA films displayed a significantly higher tensile strength than the pure PVA or Ti$_3$C$_2$T$_x$ films (Figure 10b). Intercalating and confining the polymer between the MXene flakes helped increase both cationic intercalation and flexibility, resulting in an outstanding volumetric capacitance (Figure 10c). The volumetric capacitance was still quite respectable after 10,000 cycles, indicating satisfactory cyclic stability (Figure 10d). In addition, as shown in Figure 10d, Boota et al. fabricated a Ti$_3$C$_2$/polypyrrole (PPy) flexible film by using the oxidant-free polymerization of PPy and a subsequent VAF approach [149]. By intercalating homogeneous polymer chains, the interlayer spacing is widened, and the orderly alignment of the polymer chains facilitates charge transport and ion diffusion within the electrolyte, significantly enhancing the pseudocapacitive. As SC electrodes, the PPy/Ti$_3$C$_2$T$_x$ film retained a capacitance of 92% after 25,000 cycles and showed an excellent volumetric capacitance of 1000 F cm^{-3} (Figure 10e). In Figure 10f, Luo et al. presented the simple physical mixing of MXene nanosheets with PANI nanofibers followed by a suction filtration procedure to create MXene/PANI films [150]. In addition to offering a channel for charge carriers, PANI nanofibers can enhance MXene layer spacing, which is advantageous for electrolyte ion infiltration. The assembled device exhibited a specific capacitance of 272.5 F g^{-1} at 1 A g^{-1} (Figure 10g).

It is more convenient and less expensive in industrial production to directly combine a Ti$_3$C$_2$T$_x$ supernatant with PEDOT aqueous solution than to polymerize a monomer in situ on the sheet surface. Li et al. proposed an SC constructed from a Ti$_3$C$_2$/poly (3,4-ethylenedioxythiophene):poly (styrene sulfonate) (PEDOT:PSS) membrane treated with sulfuric acid (H$_2$SO$_4$), with the hybrid film serving as the negative electrode [151]. H$_2$SO$_4$ can remove a portion of the insulating PSS, which results in an increased conductivity in the composite. As well as providing electroactive surfaces, PEDOT chains create electronic transport pathways that accelerate electrochemical reactions. In comparison to pure Ti$_3$C$_2$, the hybrid film has an increase in specific surface area of 4.5 times, as well as exceptional volumetric capacitance (1065 F cm^{-3} at 2 mV s^{-1}).

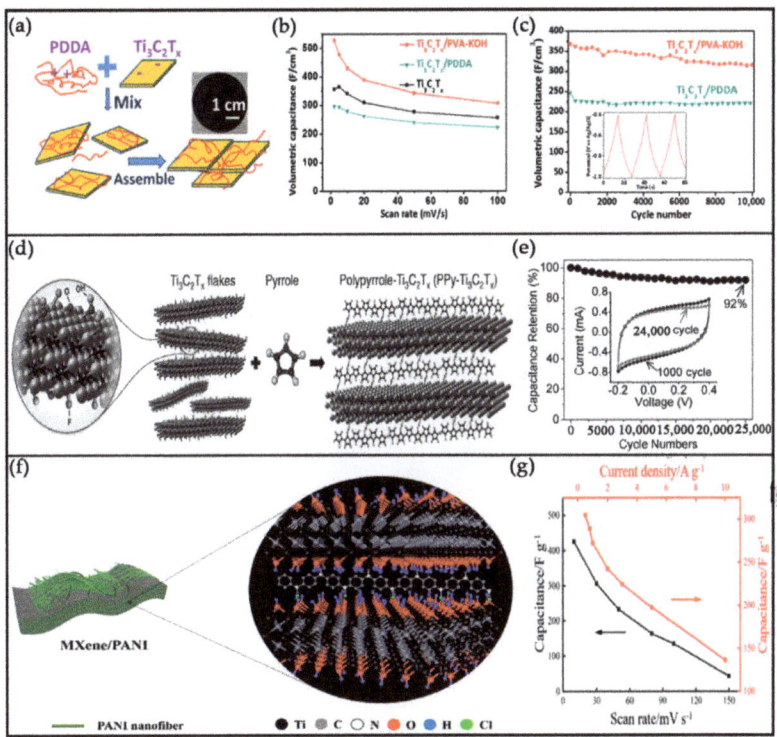

Figure 10. (**a**) Schematic illustration of the preparation of $Ti_3C_2T_x$/PDDA films. (**b**) Volumetric capacitances at different scan rates for $Ti_3C_2T_x$, $Ti_3C_2T_x$/PDDA, and $Ti_3C_2T_x$/PVA–KOH films. (**c**) Cyclic stability of $Ti_3C_2T_x$/PDDA and $Ti_3C_2T_x$/PVA–KOH electrodes at a current density of 5 A g^{-1}; the inset shows the last three cycles of a $Ti_3C_2T_x$/PVA–KOH capacitor. Reprinted with permission from Ref. [148]. Copyright 2014 *PNAS*. (**d**) Schematic illustration of pyrrole polymerization using MXene. The terminating groups on the latter contribute to the polymerization process. (**e**) Cycle life performance showing high capacitance retention of the PPy/$Ti_3C_2T_x$ (1:2) film after 25,000 cycles at 100 mV s^{-1}. Inset shows that the shape of the CV was retained after cycling, confirming the high electrochemical stability. Reprinted with permission from Ref. [149]. Copyright 2015 WILEY-VCH Verlag GmbH & Co., KGaA. (**f**) Schematic diagram of binding mechanism between MXene nanosheets and PANI nanofibers of the PPy confined between the MXene layers. (**g**) The plot of specific capacitance versus scan rates and current densities for MP5. Reprinted with permission from Ref. [150]. Copyright 2022 *Electrochimica Acta*.

3.1.5. MXene/Metal Oxides, Metal Hydroxide Composites

Transition metal compounds have large specific capacitances in theory (RuO_2 (720 F g^{-1}), MnO_2 (1370 F g^{-1}), and MoS_2 (811 F g^{-1})). However, the rate performance and cycle stability are not good when using these compounds alone [152–155]. The conductivity of MXenes is high, while the capacitance is relatively low in comparison with transition metal compounds. The integration of pseudocapacitive materials and MXenes will enhance the pseudocapacitance. As a pseudocapacitive material, oxide/hydroxide nanoparticles are used as intercalation materials to prevent the restacking of MXene sheets [156–161].

As a typical pseudocapacitive material, MnO_2 possesses plentiful resources, low toxicity, low cost, and high capacitance in theory. In addition to enhancing its conducting properties, the MnO_x/MXene composite can achieve higher specific capacitances. Tian et al. exploited freestanding and flexible MnO_x-Ti_3C_2 films using a simple in situ wet-chemistry synthesis approach [162]. In comparison to random mixing approaches or layer-by-layer assembly, this method ensures a strong connection between the components, thus reducing contact resistance and improving electrochemical performance. The MnO$_x$-Ti$_3$C$_2$ film electrodes exhibited outstanding electrochemical properties. In addition to a volumetric capacity of 602.0 F cm^{-3}, they also show good rate capability. The MnO$_x$-Ti$_3$C$_2$ film-based symmetric SC has an energy density of 13.64 mWh cm^{-3} at 2 mV s^{-1}, a power density of 3755.61 mW cm^{-3} at 100 mV s^{-1}, and remarkable cycling stability. As shown in Figure 11a, Zhou et al. made a highly flexible, all-pseudocapacitive electrode by combining Ti$_3$C$_2$T$_x$ with ultralong MnO$_2$ NWs [158]. MnO$_2$ nanosheets can be useful as electrochemically active materials and interlayers for preventing MXene restacking and improving pseudo-capacitance, as well as retaining outstanding flexibility. When used as an electrode for SCs, the resulting film (Ti$_3$C$_2$T$_x$/MnO$_2$ = 6) has excellent volumetric and a specific areal capacitance of 1025 F cm^{-3} and 205 mF cm^{-2}, respectively (Figure 11b). It also retains its capacitance after 10,000 cycles at 98.38% and has high capacitance retention, outperforming the previously reported Ti$_3$C$_2$T$_x$ MXene-based flexible electrodes.

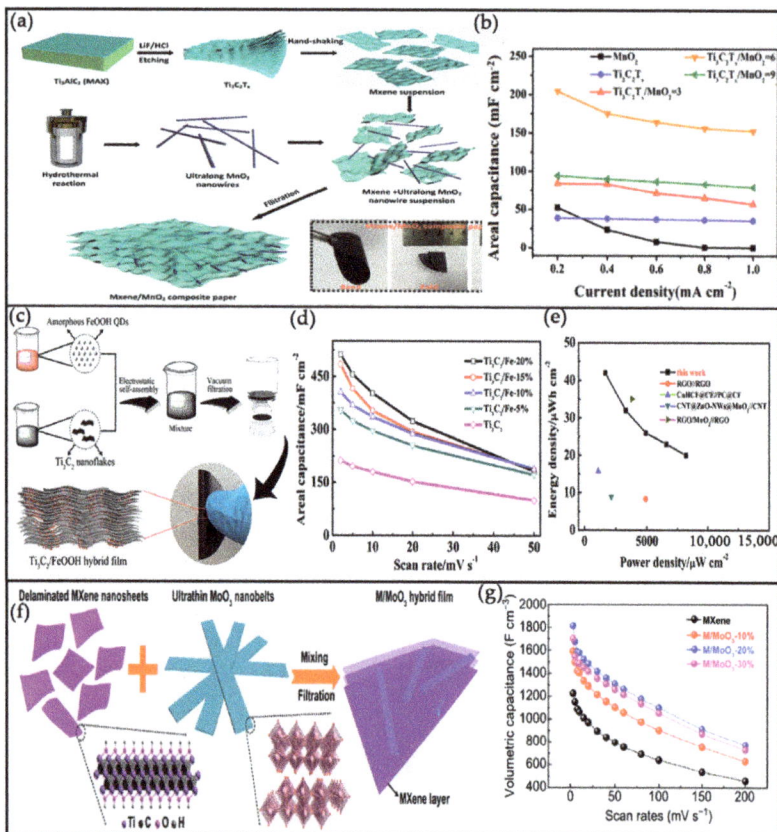

Figure 11. (a) A schematic representation of the fabrication process. (b) Specific areal capacitance of different samples versus current density. Reprinted with permission from Ref. [158]. Copyright 2018 WILEY-VCH Verlag GmbH & Co., KGaA. (c) Schematic illustration of the synthesis of Ti$_3$C$_2$/FeOOH hybrid films. (d) The areal capacitance as a function of scan rates. (e) Ragone plots of the Ti$_3$C$_2$/Fe−15%//MnO$_2$/CC device in comparison with the other reported ASCs. Reprinted with permission from Ref. [163]. Copyright 2019 *Electrochimica Acta*. (f) Schematic illustration of the fabrication process of M/MoO$_3$ hybrid films. (g) Corresponding volumetric specific capacitance of various electrodes. Reprinted with permission from Ref. [164]. Copyright 2020 Nano-Micro Lett.

Zhao et al. synthesized a freestanding Ti_3C_2/FeOOH quantum dots (QDs) hybrid film by electrostatic self-assembly in Figure 11c [163]. Amorphous FeOOH QDs anchored on Ti_3C_2 nanosheets can serve as both pillars to prevent the nanosheets from being restacked as well as active materials to provide considerable capacitance. Ti_3C_2 nanosheets as conductive layers were used to make up for the low conductivity of FeOOH. In addition to possessing a capacitance that is 2.3 times higher than the traditional Ti_3C_2 film, the hybrid Ti_3C_2/FeOOH QDs film shows excellent cycle stability with neutral electrolytes (Figure 11d). An ASC was created by combing the hybrid film with MnO_2/CC. The ASC provided a maximum power density of 8.2 mW cm^{-2} and an energy density of 42 µWh cm^{-2} when working in a wide potential window of 1.6 V (Figure 11e). Simple, flexible devices made from Ti_3C_2/Fe−15%//MnO_2/CC demonstrated outstanding flexibility. These results show that the Ti_3C_2/Fe−15% hybrid film has a lot of potential for use in flexible ASCs, where it will allow for a larger applied potential difference window and a higher energy density.

MoO_3 nanobelts, as pseudocapacitive materials, exhibit excellent potential for MXene films, such as mechanical stability, simple preparation procedure, strong electrochemical reaction activity, and high pseudocapacitance in acidic conditions. As depicted in Figure 11f, Wang et al. manufactured all-pseudocapacitive and highly malleable MXene/MoO_3 hybrid films using a vacuum-assisted technique. [164]. As a result of the excellent synergetic effect, the MXene nanosheets exhibit the highest pseudocapacitance in an acidic electrolyte. The as-prepared freestanding MXene/MoO_3-20% hybrid film exhibits an extremely high volumetric capacitance of 1817 F cm^{-3}, which is over 1.5 times greater than the capacitance of pure MXene film (Figure 11g).

3.2. PET (Terephthalic Acid Glycol Ester) as Flexible Substrate

Due to their good flexibility and stability, polymers are commonly used as substrate materials for flexible SCs [165,166]. Flexible SCs are typically constructed by combining conductive materials with PET flexible substrates through deposition, spraying, printing, and coating processes.

As shown in Figure 12a, Rosen et al. fabricated a high-performance solid-state SC from $Mo_{1.33}$C MXene/PEDOT:PSS-aligned polymer films [167]. This process involved vacuum-filtering the composite, followed by acid treating the as-obtained film for 24 h before preparing the all-solid SCs using PET as a flexible substrate. PEDOT nanofibers are aligned and confined between layers of high-conducting $Mo_{1.33}$C, allowing rapidly reversible oxidation reactions as well as short diffusion paths to facilitate ion transport. Thus, these flexible solid-state SCs have a maximal capacitance of 568 F cm^{-3}, a power density of 19,470 mW cm^{-3} (Figure 12b), an extremely high energy density of 33.2 mWh cm^{-3}, and a capacitive retention of 90% after 10,000 cycles. As shown in Figure 12c, a flexible hybrid film electrode composed of 3D cubic Ni-Fe oxide and 2D $Ti_3C_2T_x$ layers was developed by Zhang et al. [157], and it was made to adhere to a PET flexible substrate for the purposes of electrochemical measurements. MXene layers were utilized as binders and conductive additives to assist charge transfer in the electrode, thereby preventing a substantial loss in conductivity. As a result of the cubic Ni-Fe oxide being used as a spacer between the MXene layers, more interlayer space was created, which improved the diffusion of electrolytes. Based on the flexible composite film electrode, a solid-state flexible SC was fabricated that displays exceptionally robust cycling stability, retaining 90% of its capacitance after 10,000 charging–discharging cycles and maintaining steady energy storage capability following 50 cycles of mechanical bending.

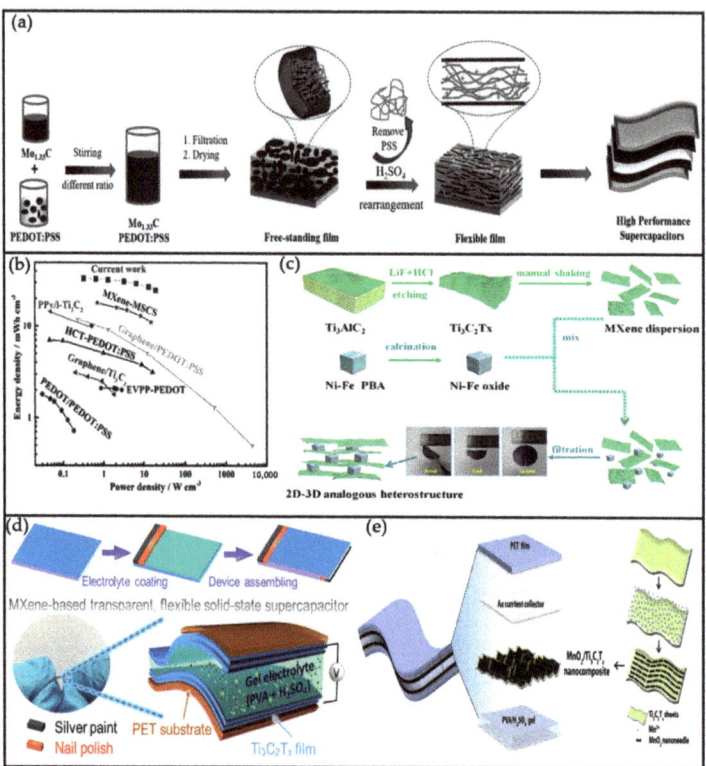

Figure 12. (a) Schematic illustration of the preparation of composite films and the fabrication of a solid-state SC. (b) Energy and power density of the M:P = 10:1−24 h device compared with previously reported devices. Reprinted with permission from Ref. [167]. Copyright 2017 WILEY-VCH Verlag GmbH & Co., KGaA. (c) Schematic illustration of the fabrication process for the composite film electrode. Reprinted with permission from Ref. [157]. Copyright 2020 *Chemical Engineering Journal*. (d) Schematic demonstration of $Ti_3C_2T_x$ MXene-based transparent, flexible solid-state supercapacitor fabrication. Reprinted with permission from Ref. [168]. Copyright 2017 WILEY-VCH Verlag GmbH & Co., KGaA. (e) Schematic illustration of the $MnO_2/Ti_3C_2T_x$ nanocomposite-based flexible supercapacitor preparation. Reprinted with permission from Ref. [169]. Copyright 2018 *Electrochimica Acta*.

Zhang et al. produced transparent films by spin-casting colloidal solutions of $Ti_3C_2T_x$ nanosheets onto PET substrates and then annealing them at 200 °C [168] (Figure 12d). The DC conductivity of films with transmissions of 29% and 93%, respectively, is 9880 S cm^{-1} and 5736 S cm^{-1}. These transparent $Ti_3C_2T_x$ electrodes have an excellent volumetric capacitance in combination with a high response speed. Transparent solid-state asymmetric SCs have a greater energy density (0.05 µWh cm^{-2}) and capacitance (1.6 mF cm^{-2}) than SWCNT or graphene-based transparent SC devices, as well as a longer lifetime. Jiang et al. constructed an asymmetric flexible SC by coating the slurry of $MnO_2/Ti_3C_2T_x$ on PET substrates [169] (Figure 12e). This highly synergistic effect between $Ti_3C_2T_x$ and MnO_2, resulting from their chemical interaction, significantly enhances structural stability, rate stability, and the specific capacitance of the $MnO_2/Ti_3C_2T_x$ nanocomposite electrode. Furthermore, a symmetrical flexible SC based on a $MnO_2/Ti_3C_2T_x$ nanocomposite electrode exhibits good electrochemical performance, great flexibility, and excellent cycling ability.

In addition to being flexible, SCs are also expected to be miniature in order to power microdevices. The construction of flexible MSCs is also a future development trend. Laser scribing is a straightforward and cost-effective method for creating unique patterns on a variety of substrates. It has excellent flexibility in the depth of the field and the materials that can be ablated. The difficulty of laser processing is finding the correct wavelength, pulse energy, and scanning speed of the laser to achieve the proper resolution. Huang et al. described a laser processing method for fabricating freestanding MXene films, then mounted MXene films on PET substrates for flexible MSC device manufacturing, as in Figure 13a [170]. Since a cool laser is employed, less oxidation and no undesirable edge defects have been discovered during the process of laser scribing, which has improved the performance of the as-made MSC device. Moreover, the areal capacitance of these freestanding flexible MSCs is an astounding 340 mF cm^{-2} at 0.25 mA cm^{-2} when polyvinyl alcohol/sulfuric acid (PVA/H$_2$SO$_4$) gel is used as the electrolyte. As the device bends to 60°, it does not show any decrease in capacitance. In addition, MSCs also show the highest energy density and volumetric capacitance among (at the time) all unconventional SCs, reaching 12.4 mWh cm^{-3} and 183 F cm^{-3}, respectively.

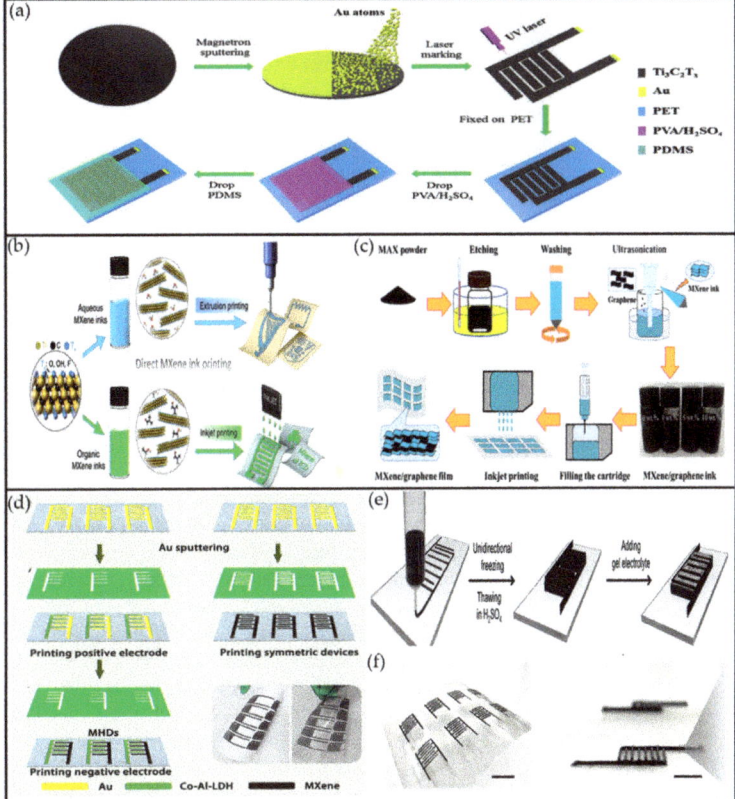

Figure 13. (a) Schematic illustration of manufacturing flexible solid-state MSCs. Reprinted with permission from Ref. [170]. Copyright 2018 WILEY-VCH Verlag GmbH & Co., KGaA. (b) Schematic illustration of direct MXene ink printing. Reprinted with permission from Ref. [170]. Copyright 2019 *Nature Communications*. (c) Schematic diagram of the inkjet printing of MXene/graphene films. Reprinted with permission from Ref. [171]. Copyright 2022 *Journal of Alloys and Compounds*. (d) Asymmetrical MXene MSCs fabricated by a modified screen-printing process. Reprinted with permission from Ref. [172]. Copyright 2022 Copyright 2018 *Nano Energy*. (e) The fabrication process of 3D printing all-MXene MSC via MSES. (f) Photographs of a 3D-printed MXene MSC; scale bar is 1 cm. Reprinted with permission from Ref. [173]. Copyright 2021 Wiley-VCH GmbH.

Inkjet printing is one of the most promising technologies for the speedy development and application of new material inks. In addition to the excellent printing precision that can be provided on a variety of substrates, its printing speed is also faster than that of screenprinting. Zhang et al. reported employing inkjet printing on an AlO_x-coated PET substrate to produce an MSC (Figure 13b) [166]. A variety of solvents have been investigated for fine printing, such as NMP, ethanol, DMSO, and DMF. Both low- and high-concentration inks show excellent printing resolution. The MSC has a volumetric capacitance of 562 F cm^{-3}, while its energy density is 0.32 µWh cm^{-2}, considered to be one of the highest among printed MSC devices. Wen et al. fabricated flexible MXene/graphene composite electrodes through inkjet printing (Figure 13c) [171]. As a result of the insertion of graphene nanosheets into composite films, the interlayer gap can be increased, thereby minimizing the self-stacking effect of MXenes. The composite electrodes exhibited high volumetric capacitance and excellent stability. Moreover, a flexible MSC based on the composite electrodes demonstrated a competitive energy density.

Due to its reproducibility and stability, screen-printing is employed for the mass production of MSCs. In this technique, a stencil is initially placed over the desired substrate, followed by the ink being pressed through a planar form stencil onto the substrate. Subsequently, the ink dries to form the desired patterns on the substrate. As shown in Figure 13d, Xu et al. fabricated a flexible coplanar asymmetric microscale hybrid device by screen-printing on PET substrates [172]. The assembled flexible device has excellent areal energy and power densities.

The 3D printing method is regarded to be a form of additive manufacturing. Extrusion-based 3D printing is the most cost-effective and versatile method for producing 3D and self-supported micro-prototypes compared with other 3D printing methods [174,175]. An extrusion-based 3D printing requires a functional ink material that is high in viscosity and exhibits the proper rheological behavior in order to achieve rapid and precise prototyping [176]. Huang et al. performed an extrusion-based 3D printing of MXene ink using a 3D printing station and then produced an MSC with interdigital patterns using a layer-by-layer printing procedure on PET substrates (Figure 13e,f) [173]. The MSC exhibits a record-high energy density of 0.1 mWh cm^{-2} at 0.38 mW cm^{-2} and excellent areal capacitance (2.0 F cm^{-2} at 1.2 mA cm^{-2}).

In addition to the fabrication techniques discussed above, Feng et al. created in-plane flexible MSCs by spray coating MXene/rGO hybrid ink onto a PET substrate [121]. The flexible MSCs have a volumetric capacitance of 33 F cm^{-3} and an area capacitance of 3.26 mF cm^{-2} at 2 mV s^{-1}. Couly et al. fabricated an asymmetric MXene-based MSC that is current-collector-free, binder-free, and flexible by spraying both $Ti_3C_2T_x$ and rGO dispersions onto a PET substrate [177]. Despite operating for 10,000 cycles, this MXene-based asymmetric MSC retains 97% of its initial capacitance. It also has a power density of 0.2 W cm^{-3} and an energy density of 8.6 mWh cm^{-3}.

An electrolyte is also an essential part of constructing flexible SCs. SCs that are flexible work in a bent state, which can potentially result in electrolyte leakage. This necessitates the development of an electrolyte with high conductivity and excellent infiltration characteristics. Moreover, in order to further expand the electrochemical voltage window of the flexible MSCs based on MXenes, Zheng et al. created ionogel-based MXene MSCs with a MXene film that was pre-intercalated by the ionic liquid. The patterned MXene-based microelectrodes were transferred onto a PET substrate to fabricate MSCs with the assistance of 20 MPa pressure [178]. Due to the pre-intercalation of ionic liquid, the interlayer spacing was enlarged to 1.45 nm, which was beneficial to the ion deintercalation and intercalation of the electrolyte. The MXene-based MSCs (M-MSCs) using $EMIMBF_4$ ionic liquid as an electrolyte showed high areal energy density and remarkably high volumetric capacitance. In addition, the solid-state M-MSCs with ionogel as an electrolyte exhibited a volumetric energy density of 41.8 mWh cm^{-3} and an excellent areal energy density of 13.3 Wh cm^{-2}, as well as long-term cyclability.

3.3. Fabric Fiber as Flexible Substrate

Because fabric fibers have stable chemical properties, high electrical conductivity, and good mechanical properties, they can be used as current collectors for flexible substrates to support or load active materials for the construction of flexible energy storage devices. Fibers have the advantage of containing 3D open-pore structures, which makes conformal coating much more effective throughout the textile network, leading to a much higher loading of active materials and, accordingly, higher energy density and power. In addition, fabric fibers are thermally stable, which expands the temperature range of flexible SCs. As a consequence, there are numerous approaches to constructing flexible electrodes based on fiber textiles, such as through chemical vapor deposition (CVD), electrodeposition, dipping, and spin coating active electrode material onto fiber textiles.

Specifically, Xia et al. presented a simple CVD method that can produce single-crystalline TiC nanowire arrays with good electrical conductivity directly on flexible carbon cloth [179]. The TiC nanowire arrays demonstrated excellent performance for flexible SCs over a wide temperature range (–25 °C to 60 °C), including high-rate characteristics and an ultra-stable cycle life. In addition, the energy density of TiC-based SCs was 18.2, which is roughly double that of commercial AC-based SCs.

Electrophoretic deposition (EPD) can be used to fabricate binder-free films with uniformity and mass-loading adjustability. Furthermore, the preparation method of EPD has unique advantages in infiltrating and depositing active material onto porous substrates, especially for the production of wearable SCs on flexible substrates. Xu et al. deposited binder-free d-$Ti_3C_2T_x$ nanoflakes on a fabric substrate in acetone solvent utilizing the EPD approach [180]. As the surface of MXene flakes that contain absorbed H^+ carries positive charges, the flakes migrate toward the cathode during the deposition of electrophoretic particles, resulting in a uniform film of MXene. In addition to great flexibility, all-solid-state SCs based on EPD film electrodes display exceptional electrochemical performance. Wang et al. employed the electrophoresis effect in depositing $Ti_3C_2T_x$/rGO composite on carbon cloth. Without adhesives, the built solid-state SCs based on the $Ti_3C_2T_x$/rGO electrode displayed outstanding cycling stability, low series resistance, high specific capacitance, and excellent mechanical flexibility [181].

In addition to deposition, dipping is a more convenient and efficient method of constructing flexible electrodes on fabric substrates [182,183]. Yan et al. fabricated conductive textile electrodes that have a specific capacitance of 182.70 F g^{-1} using dipping and drying [184]. PPy textile electrodes were electrochemically deposited on MXene textiles as a means of improving the capacitance of MXene and avoiding MXene oxidation. Furthermore, the symmetrical solid-state SCs using MXene-PPy textile electrodes also showed improved electrochemical performance and a greater degree of flexibility. Li et al. developed a synthetic technique for the construction of a high-performance, flexible SC by the in situ growth of multi-walled carbon nanotubes (MWCNTs) on MXene nanosheets placed on a CC substrate [185] (Figure 14a). Similarly, a specific concentration of MXene was loaded onto CC with multiple dipping and drying, catalyzed by nickel–aluminum-layered double hydroxide (Ni-Al-LDH), and then subjected to CVD to produce MWCNTs. The MWCNT–MXene@CC displays excellent conductivity along with an exfoliated, large surface area. Therefore, the as-manufactured electrode exhibited a large specific capacitance while retaining a high retention after 16,000 cycles at 10 mA cm^{-2} (Figure 14b,c). Recently, Li et al. developed an extremely conductive textile based on MXenes through electrostatic self-assembly [186]. In addition to providing abundant active sites, the horizontally aligned, compact MXene flakes painted on the fabric fibers may produce connected electron transport channels, as shown in Figure 14d. Thus, from 1 to 50 mA cm^{-2}, the MXene/PEI-modified fiber fabric (MXene/PMFF) delivered excellent rate performance with no reduction in capacitance. The PPy-coated MXene/PMFF electrode had a high-rate capability and areal capacitance as well as outstanding cycling stability and gravimetric capacitance (Figure 14e,f). Moreover, a solid-state symmetric SC based on the

PPy/MXene/PMFF textiles had an energy density of 40.7 Wh cm^{-2}, a maximum power density of 25 mW cm^{-2}, as well as an areal capacitance of 458 mF cm^{-2}.

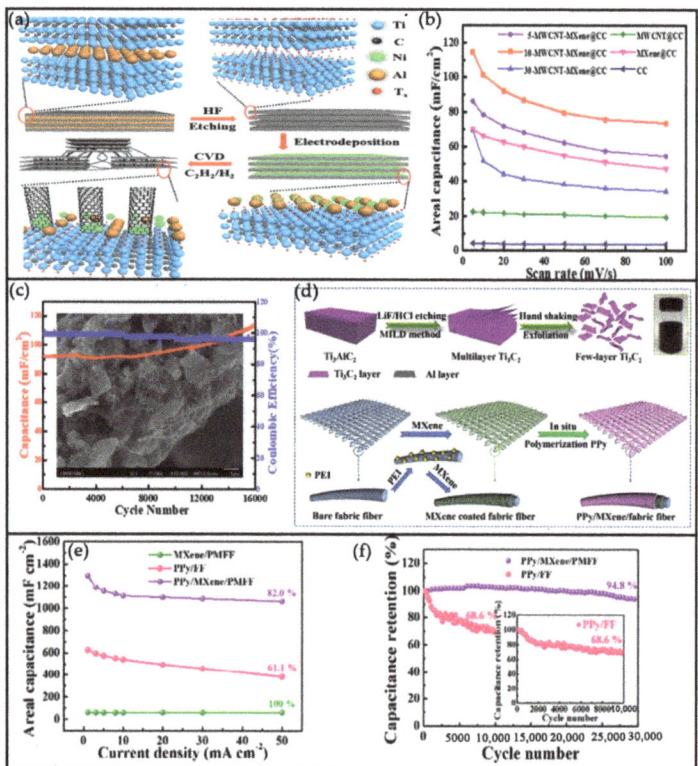

Figure 14. (a) Schematic illustration for the preparation of MXene and MWCNTs. (b) Areal specific capacitance of different samples at different scan rates. (c) Cycling stability of 10−MWCNT−MXene@CC electrode at a current density of 10 mA cm^{-2}. Reprinted with permission from Ref. [185]. Copyright 2020 WILEY-VCH Verlag GmbH & Co., KGaA. (d) Schematic illustration of the synthesis process for a MXene and PPy/MXene/PMFF textile electrode. (e) Areal capacitance comparison of the samples at different current densities. (f) Cycling performance of PPy/FF and PPy/MXene/PMFF measured at 30 mA cm^{-2}, with enlargement of cycling performance of PPy/FF in the inset. Reprinted with permission from Ref. [186]. Copyright 2020 *Energy Storage Materials*.

3.4. Other Substrates

In addition to the usual PET and fabric flexible substrates, others, such as PDMS substrates, carbon-based substrates, metal substrates, traditional paper substrates, sponge-type substrates, and cable-type substrates, can also be used as substrates of the flexible electrode. These substrates are very flexible and mechanically robust despite severe bending, enabling the employment of SCs in lightweight, wearable, and flexible electronic devices for extensive portable applications. Nevertheless, each type of flexible substrate has both advantages and disadvantages in terms of flexible SC application, which are listed in Table 1.

PDMS inherently has excellent ductility, flexibility, and mechanical strength. Therefore, flexible SCs based on PDMS substrates have better bending and electrochemical properties. Li et al. prepared stretchable MSCs on oxygen-plasma-treated PDMS substrates using 3D printing and unidirectional freezing, as in Figure 15a [187]. A nanocomposite ink consisting of MnONWs, MXene, C60, and AgNWs was constructed in a honeycomb-like porous structure. Taking advantage of the synergies between the electrode architecture and nanocomponents, the 3D-printed MSC device exhibited excellent electrochemical performance.

Table 1. Comparison of various flexible substrates for the flexible SC electrodes.

Substrate	Conductivity	Cost	Surface Area	Flexibility	Weight
metal substrate	high	moderate	low	high	high
traditional paper	low	low	moderate	high	low
carbon-based paper	high	moderate	moderate	moderate	low
sponge-type	low	low	high	high	low
cable-type	high	moderate	moderate	high	low
textile-type	low	low	high	high	low

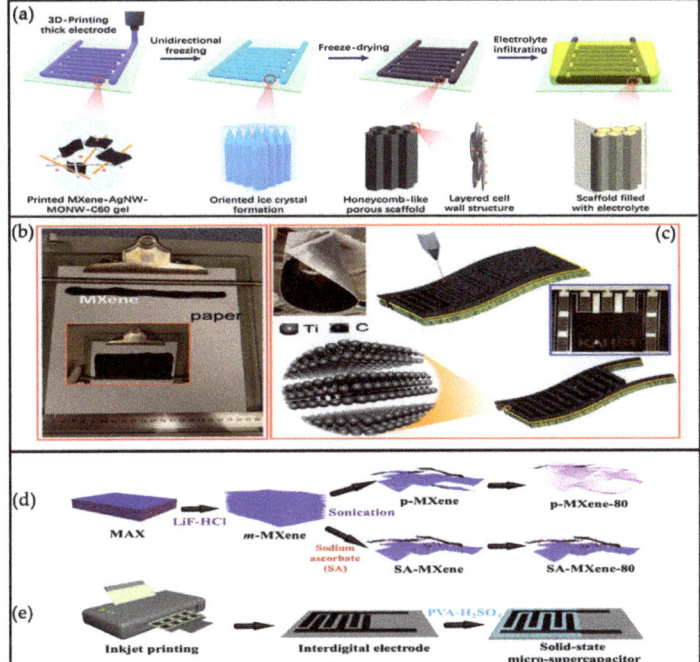

Figure 15. (a) Schematic illustration of the fabrication process of intrinsically stretchable MSCs through 3D printing and unidirectional freezing. The 3D-printed thick interdigitated electrodes possess a honeycomb-like porous structure in combination with a layered cell wall architecture. Reprinted with permission from Ref. [187]. Copyright 2020 WILEY-VCH Verlag GmbH & Co., KGaA. (b) MXene slurry on an A4 sheet of printing paper with Meyer rod; the inset shows a snapshot of the progression of the coating process. (c) Foldable MXene/paper, schematic illustration of laser patterning of MXene-coated paper to fabricate interdigitated electrodes for MSCs, and fabricated MXene-based MSC device along with the crystallographic arrangement of Ti (gray color) and C (black color) atoms in MXene sheets. Reprinted with permission from Ref. [188]. Copyright 2016 WILEY-VCH Verlag GmbH & Co., KGaA. (d) Schematic representation of the synthesis of a stable SA-MXene nanocomposite. (e) Schematic representation of a solid-state MSC fabricated through the inkjet printing of SA-MXene nanocomposites. Reprinted with permission from Ref. [189]. Copyright 2019 *Energy Storage Materials*.

It is simpler and more economical to construct flexible electrodes based on ordinary paper. The paper contains a hierarchical arrangement of cellulose fibers and can be regarded as having a rough and porous surface texture that is conducive to ink adhesion without the need for extra treatments [190]. The paper surface provides a suitable substrate for solution-processed coatings of a variety of functional materials because of the capillary nature of fibers, functional groups, and intrinsic surface charge. Kurra et al. successfully manufactured MXene-on-paper energy storage devices using Meyer rod coating and direct laser machining (Figure 15b,c) [188]. Compared to those paper-based MSCs, the Ti_3C_2 MXene-on-paper MSC produced comparable power–energy densities. Wu et al. printed interdigitated MSC electrodes on photopaper using an inkjet printer (Figure 15d,e) [189]. Adding ascorbic acid into $Ti_3C_2T_x$ MXenes can improve not only the dispersibility and oxidative stability but also enhance the spacing between the MXene layers, thereby facilitating the diffusion of the electrolyte ions. Furthermore, the manufactured solid-state MSCs displayed specific capacitance, superior mechanical flexibility, and cycle stability. Yang et al. deposited Ti_3C_2/CNTs sheets onto graphite paper for SC electrodes via electrophoretic deposition, as in Figure 16a [191]. The Ti_3C_2/CNTs electrode exhibited enhanced cycling stability and specific capacitance. (Figure 16b). Li et al. reported a flexible AMSC based on $Ti_3C_2T_x$//PPy/MnO_2 [192]. As shown in Figure 16c, the $Ti_3C_2T_x$ nanosheets were formed on graphite paper (GP) as negative electrodes. The PPy/MnO_2 materials on the GP were prepared using the same method as the positive electrodes. An AMSC based on $Ti_3C_2T_x$/PPy/MnO_2 was then constructed using a PVA/H_2SO_4 electrolyte. The maximal energy density and areal capacitance can reach 6.73 µWh cm^{-2} and 61.5 mF cm^{-2}, respectively (Figure 16d,e). In addition, the AMSC exhibited better flexibility when mechanically bent at different angles. (Figure 16f).

In summary, MXene-based films have advantages in terms of favorable metallic conductivity, high capacitance, and good flexibility, all of which are imperative for flexible energy storage devices. It is possible to assemble freestanding electrodes from the delaminated $Ti_3C_2T_x$ without the utilization of additional current collectors, polymer binders, or conductive agents. However, a significant problem associated with thin-film electrode fabrication is self-restacking for MXene nanosheets because of the van der Waals interaction between the layers, which interferes with the ability of electrolyte ions to reach the active materials, resulting in poor rate performance and sluggish redox reactions. Thus, different interlayer spacers have been introduced between $Ti_3C_2T_x$ sheets in order to alleviate the stacking problem and improve the electrochemical performance of the electrodes. The methods for preparing composite electrode materials consisting of carbon materials and MXenes include the in situ growth method, the self-assembly method, and the layer-by-layer assembly method. As a result of the increased interlayer spacing and surface area, these composite electrodes exhibit higher mechanical and electrochemical properties than pure $Ti_3C_2T_x$ electrodes. In addition, there is a strong bonding interaction between groups terminated on the surfaces of different materials, which leads to a good degree of flexibility. However, a large number of insulating groups may adversely affect the electrical conductivity of composite materials. $Ti_3C_2T_x$/polymer composites are primarily produced by polymerizing polymer monomers onto the surface of $Ti_3C_2T_x$ nanosheets. The electrodes have excellent capacitance and outstanding mechanical strength due to the superior pseudocapacitive behavior and flexibility of the polymer. Furthermore, as a result of the excellent cycling stability of $Ti_3C_2T_x$ MXene, they also have a respectable cycle life. When combined with transition metal compounds, $Ti_3C_2T_x$ MXene can also effectively improve electrochemical performance. On the one hand, the superior electrical conductivity of $Ti_3C_2T_x$ MXene can substantially facilitate electron transport. On the other hand, the theoretical capacitance of transition metal compounds is relatively high, which can greatly facilitate pseudocapacitance. However, as a result of stiffness and the poor flexibility of these transition metal compounds, a majority of $Ti_3C_2T_x$/transition metal compound composites still exhibit low mechanical strength. For the development of lightweight and flexible MXene-based films, rational methods must be developed for constructing efficient channels to facilitate ion transport. The electrochemical performance of flexible $Ti_3C_2T_x$-based composite SCs is summarized in Tables 2 and 3.

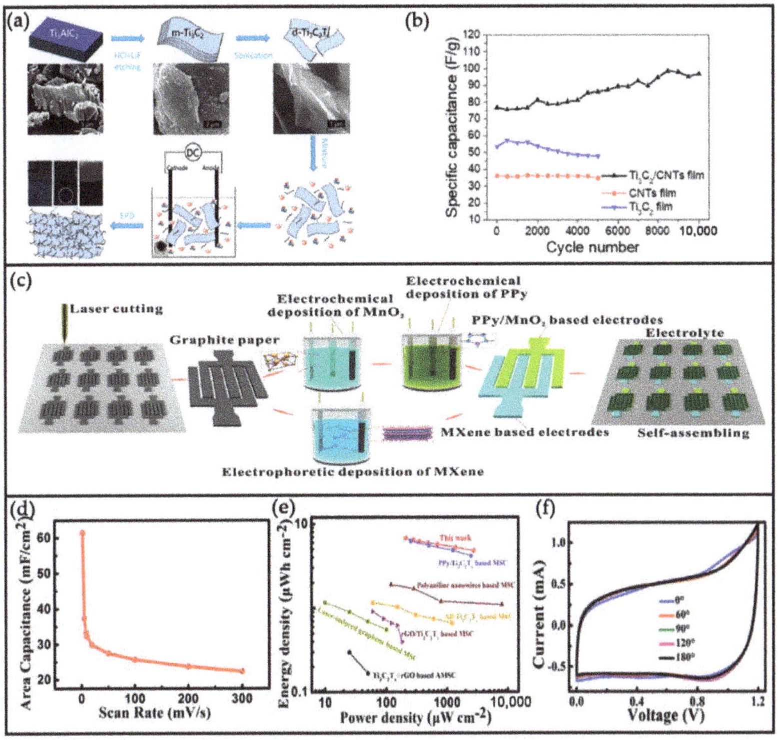

Figure 16. (a) Schematic diagram of the preparation route of electrode films using the EPD method. (b) The plots of the specific capacitance of the three electrodes after 10,000 GCD cycles at 5 A g^{-1}. Reprinted with permission from Ref. [191]. Copyright 2018 *Journal of Electroanalytical Chemistry*. (c) The schematic representation of the fabrication route of the flexible $Ti_3C_2Tx//PPy/MnO_2$-based AMSC. (d) The ACs of the $Ti_3C_2Tx//PPy/MnO_2$-based AMSC at 2–300 mV s^{-1}. (e) A Ragone plot of $Ti_3C_2T_x//PPy/MnO_2$-based AMSC compared to previously reported MSCs. (f) The CV curves of the AMSC under 0–180° bending conditions at 10 mV s^{-1}. Reprinted with permission from Ref. [192]. Copyright 2020 WILEY-VCH Verlag GmbH & Co., KG.

Table 2. Comparison of the electrochemical performance of MXene-based flexible SCs.

Substrates	Electrode	Electrolyte	Capacitance	Stability	Source
No	$Ti_3C_2T_x$ films	1 M H_2SO_4	245 F g^{-1} at 2 mV s^{-1}	100% after 10,000 cycles	[77]
	ak-$Ti_3C_2T_x$ film-A	1 M H_2SO_4	294 F g^{-1} at 1 A g^{-1}	91% after 4000 cycles	[106]
	d-Ti_3C_2 films	1 M Li_2SO_4	633 F cm^{-3} at 2 mV s^{-1}	95.3% after 10,000 cycles	[107]
	200-$Ti_3C_2T_x$ film	1 M H_2SO_4	429 F g^{-1} at 1 A g^{-1}	89% after 5000 cycles	[109]
	$Ti_3C_2T_x$ film	1 M H_2SO_4	223 F g^{-1} at 0.5 A g^{-1}		[110]
	f-MXene-10 film	3 M H_2SO_4	83 F g^{-1} at 1 A g^{-1}	89.3% after 1000 cycles	[113]
	$Ti_3C_2T_x$-Li film	1 M H_2SO_4	892 F cm^{-3} at 2 mV s^{-1}	100% after 10,000 cycles	[193]
	MXene/rHGO	3 M H_2SO_4	1445 F cm^{-3} at 2 mV s^{-1}	93% after 10,000 cycles	[128]
	MXene/rGO-5 wt%	1 M KCl	1040 F cm^{-3} at 2 mV s^{-1}	100% after 20,000 cycles	[130]
	MXene-rGO-20 film	3 M H_2SO_4	300.4 F g^{-1} at 2 A g^{-1}	90.7% after 40,000 cycles	[132]
	MXene/graphene	3 M H_2SO_4	127 F g^{-1} at 2 mV s^{-1}	95.7% after 10,000 cycles	[133]

Table 2. Cont.

Substrates	Electrode	Electrolyte	Capacitance	Stability	Source
No	$Ti_3C_2T_x$/SCNT films	1 M KOH	314 F cm^{-3} at 2 mV s^{-1}	95% after 10,000 cycles	[137]
	MXene/CNT paper	1M MgSO$_4$	390 F cm^{-3} at 2 mV s^{-1}	100% after 10,000 cycles	[142]
	$Ti_3C_2T_x$/CNTs film	3 M H$_2$SO$_4$	74.1 F g^{-1} at 5 mV s^{-1}	86.3% after 10,000 cycles	[144]
	MXene/CNT-5%	1 M H$_2$SO$_4$	300 F g^{-1} at 1 A g^{-1}	92% after 10,000 cycles	[194]
	layered Ti_3C_2/PPy	PVA/H$_2$SO$_4$	35.6 mF cm^{-2} at 0.3 mA cm^{-2}	100% after 10,000 cycles	[146]
	$Ti_3C_2T_x$/PDT	PVA/H$_2$SO$_4$	52.4 mF cm^{-2} at 0.1 mA cm^{-2}	excellent cycling stability	[147]
	$Ti_3C_2T_x$/PVA film	1 M KOH	528 F cm^{-3} at 2 mV s^{-1}		[148]
	$Ti_3C_2T_x$/PPy	1 M H$_2$SO$_4$	1000 F cm^{-3} at 5 mV s^{-1}	92% after 25,000 cycles	[149]
	$Ti_3C_2T_x$/PANI	1 M H$_2$SO$_4$	272.5 F g^{-1} at 1 A g^{-1}	71.4% after 4000 cycles	[150]
	$Ti_3C_2T_x$/PEDOT:PSS	1 M H$_2$SO$_4$	1065 F cm^{-3} at 2 mV s^{-1}	80% after 10,000 cycles	[151]
	$Ti_3C_2T_x$/MnO$_2$ = 6		205 mF cm^{-2} at 0.2 mA cm^{-2}	100% after 10,000 cycles	[158]
	Ti_3C_2/MnO$_x$	1 M Li$_2$SO$_4$	392.9 F cm^{-3} at 2 mV s^{-1}	89.8% after 10,000 cycles	[162]
	Ti_3C_2/FeOOH QDs	1 M Li$_2$SO$_4$	115 mF cm^{-2} at 2 mA cm^{-2}	82% after 3000 cycles	[163]
	MXene/MoO$_3$	1 M H$_2$SO$_4$	396 F cm^{-3} at 10 mV s^{-1}	90% after 5000 cycles	[164]
	MXene/Fe (OH)$_3$	3 M H$_2$SO$_4$	1142 F cm^{-3} at 0.5 A g^{-1}		[195]
PET	Mo$_{1.33}$C MXene/PEDOT:PSS	1 M H$_2$SO$_4$	1310 F cm^{-3} at 2 mV s^{-1}	90% after 10,000 cycles	[167]
	$Ti_3C_2T_x$/ MnO$_2$	1M Na$_2$SO$_4$	130.5 F g^{-1} at 0.2 A g^{-1}	100% after 1000 cycles	[169]
	$Ti_3C_2T_x$/rGO	PVA/H$_2$SO$_4$	80 F cm^{-3} at 2 mV s^{-1}	97% after 10,000 cycles	[177]
	$Ti_3C_2T_x$	PVA/H$_2$SO$_4$	340 mF cm^{-2} at 0.25 mA cm^{-2}	82.5% after 5000 cycles	[170]
	MXene	EMIMBF$_4$	140 F cm^{-3} at 0.1 mA cm^{-2}	92% after 1000 cycles	[178]
Fiber	TiC nanowires	EMIMBF$_4$	107.1 F cm^{-3} at 2.5 A g^{-1}	97% after 5000 cycles	[179]
	$Ti_3C_2T_x$/rGO	PVA/H$_3$PO$_4$	11.6 mF cm^{-2} at 0.1 mA g^{-1}	100% after 1000 cycles	[181]
	MXene/PPy	0.2 M NaClO$_4$	275.2 F g^{-1} at 1.0 mA cm^{-2}		[184]
	MXene/MWCNTs	PVA/H$_2$SO$_4$	994.79 mF cm^{-3} at 1 mA cm^{-2}	95.4% after 5000 cycles	[185]
	PPy/MXene/PMFF	PVA/Na$_2$SO$_4$	458 mF cm^{-2} at 1 mA cm^{-2}	93.7% after 3000 cycles	[186]
PDMS	MXene-AgNW-MnONW-C60	PVA/KOH	216.2 mF cm^{-2} at 10 mV s^{-1}	85% after 10,000 cycles	[187]
Paper	MXene	1 M H$_2$SO$_4$	25 mF cm^{-2} at 20 mV s^{-1}	80% after 1000 cycles	[188]
	sodium ascorbate–MXene	PVA/H$_2$SO$_4$	108.1 mF cm^{-2} at 1 A g^{-1}	94.7% after 4000 cycles	[189]
GP	Ti_3C_2/CNTs	6 M KOH	55.3 F g^{-1} at 0.5 A g^{-1}	Increase after 1000 cycles	[191]
	$Ti_3C_2T_x$//PPy/MnO$_2$	PVA/H$_2$SO$_4$	61.5 mF cm^{-2} at 2 mV s^{-1}	80.7% after 5000 cycles	[192]

Table 3. Comparison of the energy density and power density of MXene-based flexible SCs.

Substrates	Electrode	Energy Density	Power Density	Source
No	ak-$Ti_3C_2T_x$ film-A	45.2 Wh L^{-1}	326 W L^{-1}	[106]
	d-Ti_3C_2 films	41 Wh L^{-1}		[107]
	200-$Ti_3C_2T_x$ film	29.2 Wh kg^{-1}		[109]
	$Ti_3C_2T_x$ film	15.2 Wh L^{-1}	204.8 W L^{-1}	[110]
	f-MXene-10 film	6.1 Wh Kg^{-1}	175.0 W Kg^{-1}	[113]
	MXene/rHGO	11.5 Wh Kg^{-1}	62.4 W Kg^{-1}	[128]
	MXene/rGO-5 wt%	32.6 Wh L^{-1}	74.4 kW L^{-1}	[130]
	MXene/graphene	50.8 Wh kg^{-1}	215 W kg^{-1}	[133]
	$Ti_3C_2T_x$/CNTs film	23.9 Wh kg^{-1}	498.6 W kg^{-1}	[144]
	MXene-knotted CNT	59 Wh kg^{-1}	9.6 kW kg^{-1}	[145]
	$Ti_3C_2T_x$/PANI	31.18 Wh kg^{-1}	1079.3 W kg^{-1}	[150]
	$Ti_3C_2T_x$/PEDOT:PSS	23 mWh cm^{-3}	7659 mW cm^{-3}	[151]
	$Ti_3C_2T_x$/MnO$_2$ = 6	56.94 mWh cm^{-3}	0.5 W cm^{-3}	[158]
	Ti_3C_2/MnO$_x$	13.64 mWh cm^{-3}	3755.61 mW cm^{-3}	[162]
	Ti_3C_2/FeOOH QDs	40 mWh cm^{-2}	8.2 mW cm^{-2}	[163]
	MXene/MoO$_3$	13.4 Wh kg^{-1}	534.6 W kg^{-1}	[164]
	MXene/Fe (OH)$_3$	20.7 Wh L^{-1}	184.8 W L^{-1}	[195]

Table 3. Cont.

Substrates	Electrode	Energy Density	Power Density	Source
PET	$Mo_{1.33}C$ MXene/PEDOT:PSS	24.72 mWh cm^{-3}	19,470 mW cm^{-3}	[167]
	$Ti_3C_2T_x$ films	0.05 µWh cm^{-2}	2.4 µW cm^{-2}	[168]
	$Ti_3C_2T_x/MnO_2$	0.7 mWh cm^{-2}	80.0 mW cm^{-2}	[169]
	$Ti_3C_2T_x/rGO$	8.6 mWh cm^{-3}	0.2 W cm^{-3}	[177]
	$Ti_3C_2T_x$	43.5 mWh cm^{-2}	87.5 mW cm^{-2}	[170]
	MXene	19.2 mWh cm^{-3}	14 W cm^{-3}	[178]
Fiber	TiC nanowires	13.1 Wh kg^{-1}	20.2 kW kg^{-1}	[179]
	MXene/PPy	1.30 mWh g^{-1}	41.1 mW g^{-1}	[184]
	MXene/MWCNTs	22.11 mWh cm^{-3}	2.99 W cm^{-3}	[185]
	PPy/MXene/PMFF	29.2 µW h cm^{-2}	25 mW cm^{-2}	[186]
PDMS	MXene-AgNW-MnONW-C60	19.2 µWh cm^{-2}	58.3 mW cm^{-2}	[187]
Paper	MXene	0.77 µWh cm^{-2}	46.6 mW cm^{-2}	[188]
	sodium ascorbate–MXene	100.2 mWh cm^{-3}	1.9 W cm^{-3}	[189]
GP	Ti_3C_2/CNTs	0.56 Wh kg^{-1}	416.7 W kg^{-1}	[191]
	$Ti_3C_2T_x$//PPy/MnO$_2$	6.73 µWh cm^{-2}	204 µW cm^{-2}	[192]

4. Conclusions and Perspectives

In summary, this review article detailed recent advancements in the development of flexible electrodes based on MXenes for applications in SCs. A concise introduction of MXenes as emerging 2D materials was provided, along with the different synthesis methods of $Ti_3C_2T_x$ MXene and its influence on its electrochemical properties. The applications of MXene-based flexible electrodes in SCs, according to the different construction methods of flexible electrodes based on MXenes and their composite electrodes, which are the theme of this review, were also presented. Different construction methods such as self-supporting, PET-supported, fabric fiber-supported, and other substrate-supported MXene-based films as flexible electrodes for fSCs were discussed in detail. In recent years, researchers have achieved substantial advances in the study of MXene-based SCs, but there are still a great number of obstacles to their development. Consequently, we need to consider the following factors in subsequent research:

(1) In spite of the growing body of research on the preparation of MXenes, wet chemical etching remains the most common method for the production of MXenes. However, chemical etching usually uses corrosive solvents or gases, and etching conditions are harsh. The emission of pollutants is another problem that cannot be ignored. Additionally, MXenes are easily oxidized in humid environments, which limits not only their synthesis on a large scale but also their application areas and environmental status. The future direction of MXene synthesis should be facile, low-cost, green, and have excellent and stable properties. Lastly, surface functional groups have a considerable impact on their physicochemical properties, and the specific capacitance of $Ti_3C_2T_x$ is far from optimal, so it is still possible to improve the design and control of surface functional groups during the etching process.

(2) Due to the 2D lamellar structure, hydrophilic surface, and excellent metallic conductivity of MXenes, they are preferable as energy storage electrode materials. What's more, MXenes can achieve high volumetric capacitance due to impressive density and pseudocapacitive behavior. These properties make MXene very appealing for flexible SCs. However, the gravimetric capacitance of MXene flexible electrodes has yet to be further improved because of the aggregation and restacking of MXene nanosheets. On the one hand, the stacking of MXene lamellar structures can be prevented by loading small-sized pseudocapacitance materials such as nanodots. On the other hand, the density and additional pseudocapacitance can be increased by anchoring nanodots to pseudocapacitive material on MXene nanosheets so as to improve the mass-specific capacity and electrochemical performance.

(3) Along with the proliferation of emerging smart wearable electronic devices (SWEDs), there has also been an increase in the demand for flexible SCs to be integrated with other flexible devices or wearable devices to provide them with a power source or additional functions. For example, integration with sensors or environmental monitoring equipment can allow them to operate independently and work under special conditions. A self-powered integrated system composed of strain sensors and flexible SCs is capable of detecting stable human motion with precision, which makes smart flexible SCs more suitable for SWEDs. In recent years, research on self-repairing, self-charging, and electrochromic flexible SCs has also become one of the new research hotspots and development trends. In addition to traditional electrochemical properties, such as specific capacity and cyclic stability, we should also pay more attention to the photosensitivity, self-healing, transparency, and other properties of intelligent electrode materials. In conclusion, it is imperative to design novel fSCs with intelligent and interactive characteristics.

(4) It is crucial to employ solid or gel electrolytes with high conductivity and efficient infiltration because flexible SCs frequently need to work in a curved or bending state, which increases the risk of electrolyte leakage. Ionic liquid-based gel (ionogel) electrolytes have been demonstrated to have better thermal stability, chemical inertness, as well as non-flammability. In addition to this, ionogel electrolytes also have the characteristics of the ionic liquid itself, with a wide electrochemical potential window, high ionic conductivity, and negligible vapor pressure, making them a promising electrolyte choice for the production of all-solid flexible SCs with high energy density. However, the use of ionogel electrolytes is still limited by the low capacitance and slow speed of ion transport. Thus, more efforts have been made toward improving the ionic conductivity of ionogel electrolytes and expanding the voltage window of $Ti_3C_2T_x$ electrodes in aqueous electrolytes.

In addition, the design of the flexible current collector is important, as it is a crucial component of the electrode. The conventional current collectors, such as carbon-based, planar metal-based, and 3D metal-based, increase the whole mass of the supercapacitor and thus reduce the energy density. Therefore, considerable attempts and methods should be made to develop ultrathin 3D (to load more active materials) metallic current collectors. The manufacturing cost of flexible supercapacitors can also be reduced if components such as current collectors, adhesives, and encapsulation films are used as little as possible under the premise of ensuring performance stability. As a consequence, using self-supporting MXene film as a flexible electrode without current collectors is also a development direction for future flexible SCs.

Author Contributions: B.S. conceived, designed, wrote the introduction, conclusions and perspectives, and edited the manuscript. R.H. wrote the synthesis of $Ti_3C_2T_x$ MXene and MXene-based flexible electrode materials, and edited the manuscript. Y.H., Z.G. and X.Z. participated in the editing. All authors have read and agreed to the published version of the manuscript.

Funding: This work was financially supported by the National Key Research and Development Program of China (2021YFC1808902, 2021YFC1808903), the Natural Science Basic Research Program of Shaanxi (2020JQ-575), the Natural Science Foundation of the Shaanxi Provincial Department of Education (19JK0844) and the Key R&D Industrial Project of the Xianyang Science and Technology Bureau (2021ZDYF-GY-0032).

Conflicts of Interest: The authors declare no conflict of interest.

References

1. Lu, X.; Yu, M.; Wang, G.; Tong, Y.; Li, Y. Flexible solid-state supercapacitors: Design, fabrication and applications. *Energy Environ. Sci.* **2014**, *7*, 2160–2181. [CrossRef]
2. Niu, Z.; Liu, L.; Zhang, L.; Zhou, W.; Chen, X.; Xie, S. Programmable Nanocarbon-Based Architectures for Flexible Supercapacitors. *Adv. Energy Mater.* **2015**, *5*, 1500677. [CrossRef]
3. Masarapu, C.; Zeng, H.F.; Hung, K.H.; Wei, B. Effect of temperature on the capacitance of carbon nanotube supercapacitors. *ACS Nano* **2009**, *3*, 2199–2206. [CrossRef]
4. Zhong, C.; Deng, Y.; Hu, W.; Qiao, J.; Zhang, L.; Zhang, J. A review of electrolyte materials and compositions for electrochemical supercapacitors. *Chem. Soc. Rev.* **2015**, *44*, 7484–7539. [CrossRef]

5. Shao, Y.; El-Kady, M.F.; Wang, L.J.; Zhang, Q.; Li, Y.; Wang, H.; Mousavi, M.F.; Kaner, R.B. Graphene-based materials for flexible supercapacitors. *Chem. Soc. Rev.* **2015**, *44*, 3639–3665. [CrossRef]
6. Xiao, X.; Li, T.; Yang, P.; Gao, Y.; Jin, H.; Ni, W.; Zhan, W.; Zhang, X.; Cao, Y.; Zhong, J.; et al. Fiber-based all-solid-state flexible supercapacitors for self-powered systems. *ACS Nano* **2012**, *6*, 9200–9206. [CrossRef] [PubMed]
7. Wang, L.; Feng, X.; Ren, L.; Piao, Q.; Zhong, J.; Wang, Y.; Li, H.; Chen, Y.; Wang, B. Flexible Solid-State Supercapacitor Based on a Metal-Organic Framework Interwoven by Electrochemically-Deposited PANI. *J. Am. Chem. Soc.* **2015**, *137*, 4920–4923. [CrossRef]
8. Han, Y.; Ge, Y.; Chao, Y.; Wang, C.; Wallace, G.G. Recent progress in 2D materials for flexible supercapacitors. *J. Energy Chem.* **2018**, *27*, 57–72. [CrossRef]
9. Huang, X.; Zeng, Z.; Fan, Z.; Liu, J.; Zhang, H. Graphene-based electrodes. *Adv. Mater.* **2012**, *24*, 5979–6004. [CrossRef]
10. Lu, Q.; Chen, J.G.; Xiao, J.Q. Nanostructured electrodes for high-performance pseudocapacitors. *Angew. Chem. Int. Ed. Engl.* **2013**, *52*, 1882–1889. [CrossRef]
11. Wang, Y.; Lei, Y.; Li, J.; Gu, L.; Yuan, H.; Xiao, D. Synthesis of 3D-nanonet hollow structured Co_3O_4 for high capacity supercapacitor. *ACS Appl. Mater. Interfaces* **2014**, *6*, 6739–6747. [CrossRef] [PubMed]
12. Fan, Z.; Zhao, Q.; Li, T.; Yan, J.; Ren, Y.; Feng, J.; Wei, T. Easy synthesis of porous graphene nanosheets and their use in supercapacitors. *Carbon* **2012**, *50*, 1699–1703. [CrossRef]
13. Fan, M.; Ren, B.; Yu, L.; Liu, Q.; Wang, J.; Song, D.; Liu, J.; Jing, X.; Liu, L. Facile growth of hollow porous NiO microspheres assembled from nanosheet building blocks and their high performance as a supercapacitor electrode. *CrystEngComm* **2014**, *16*, 10389–10394. [CrossRef]
14. Yu, X.-Y.; Yu, L.; Lou, X.W.D. Metal Sulfide Hollow Nanostructures for Electrochemical Energy Storage. *Adv. Energy Mater.* **2016**, *6*, 1501333. [CrossRef]
15. Huang, Y.; Liang, J.; Chen, Y. An overview of the applications of graphene-based materials in supercapacitors. *Small* **2012**, *8*, 1805–1834. [CrossRef] [PubMed]
16. Peng, X.; Peng, L.; Wu, C.; Xie, Y. Two dimensional nanomaterials for flexible supercapacitors. *Chem. Soc. Rev.* **2014**, *43*, 3303–3323. [CrossRef]
17. Hong Ng, V.M.; Huang, H.; Zhou, K.; Lee, P.S.; Que, W.; Xu, Z.J.; Kong, L.B. Correction: Recent progress in layered transition metal carbides and/or nitrides (MXenes) and their composites: Synthesis and applications. *J. Mater. Chem. A* **2017**, *5*, 8769. [CrossRef]
18. Eklund, P.; Rosen, J.; Persson, P.O.Å. Layered ternary $M_{n+1}AX_n$ phases and their 2D derivative MXene: An overview from a thin-film perspective. *J. Phys. D Appl. Phys.* **2017**, *50*, 113001. [CrossRef]
19. Zhang, C.; Ma, Y.; Zhang, X.; Abdolhosseinzadeh, S.; Sheng, H.; Lan, W.; Pakdel, A.; Heier, J.; Nüesch, F. Two-Dimensional Transition Metal Carbides and Nitrides (MXenes): Synthesis, Properties, and Electrochemical Energy Storage Applications. *Energy Environ. Mater.* **2020**, *3*, 29–55. [CrossRef]
20. Anasori, B.; Lukatskaya, M.R.; Gogotsi, Y. 2D metal carbides and nitrides (MXenes) for energy storage. *Nat. Rev. Mater.* **2017**, *2*, 16098. [CrossRef]
21. Naguib, M.; Mochalin, V.N.; Barsoum, M.W.; Gogotsi, Y. 25th anniversary article: MXenes: A new family of two-dimensional materials. *Adv. Mater.* **2014**, *26*, 992–1005. [CrossRef] [PubMed]
22. Naguib, M.; Halim, J.; Lu, J.; Cook, K.M.; Hultman, L.; Gogotsi, Y.; Barsoum, M.W. New two-dimensional niobium and vanadium carbides as promising materials for Li-ion batteries. *J. Am. Chem. Soc.* **2013**, *135*, 15966–15969. [CrossRef] [PubMed]
23. Lukatskaya, M.R.; Mashtalir, O.; Ren, C.E.; Dall'Agnese, Y.; Rozier, P.; Taberna, P.L.; Naguib, M.; Simon, P.; Barsoum, M.W.; Gogotsi, Y. Cation intercalation and high volumetric capacitance of two-dimensional titanium carbide. *Science* **2013**, *341*, 1502–1505. [CrossRef] [PubMed]
24. Lukatskaya, M.R.; Kota, S.; Lin, Z.; Zhao, M.-Q.; Shpigel, N.; Levi, M.D.; Halim, J.; Taberna, P.-L.; Barsoum, M.W.; Simon, P. Ultra-high-rate pseudocapacitive energy storage in two-dimensional transition metal carbides. *Nat. Energy* **2017**, *2*, 17105. [CrossRef]
25. Tao, Q.; Dahlqvist, M.; Lu, J.; Kota, S.; Meshkian, R.; Halim, J.; Palisaitis, J.; Hultman, L.; Barsoum, M.W.; Persson, P.O. Two-dimensional $Mo_{1.33}C$ MXene with divacancy ordering prepared from parent 3D laminate with in-plane chemical ordering. *Nat. Commun.* **2017**, *8*, 14949. [CrossRef]
26. Halim, J.; Kota, S.; Lukatskaya, M.R.; Naguib, M.; Zhao, M.-Q.; Moon, E.J.; Pitock, J.; Nanda, J.; May, S.J.; Gogotsi, Y.; et al. Synthesis and Characterization of 2D Molybdenum Carbide (MXene). *Adv. Funct. Mater.* **2016**, *26*, 3118–3127. [CrossRef]
27. Soundiraraju, B.; George, B.K. Two-dimensional titanium nitride (Ti_2N) MXene: Synthesis, characterization, and potential application as surface-enhanced Raman scattering substrate. *ACS Nano* **2017**, *11*, 8892–8900. [CrossRef]
28. Zhou, J.; Zha, X.; Zhou, X.; Chen, F.; Gao, G.; Wang, S.; Shen, C.; Chen, T.; Zhi, C.; Eklund, P. Synthesis and electrochemical properties of two-dimensional hafnium carbide. *ACS Nano* **2017**, *11*, 3841–3850. [CrossRef]
29. Naguib, M.; Mashtalir, O.; Carle, J.; Presser, V.; Lu, J.; Hultman, L.; Gogotsi, Y.; Barsoum, M.W. Two-Dimensional Transition Metal Carbides. *ACS Nano* **2012**, *6*, 1322–1331. [CrossRef]
30. Urbankowski, P.; Anasori, B.; Makaryan, T.; Er, D.; Kota, S.; Walsh, P.L.; Zhao, M.; Shenoy, V.B.; Barsoum, M.W.; Gogotsi, Y. Synthesis of two-dimensional titanium nitride Ti_4N_3 (MXene). *Nanoscale* **2016**, *8*, 11385–11391. [CrossRef]
31. Ghidiu, M.; Naguib, M.; Shi, C.; Mashtalir, O.; Pan, L.; Zhang, B.; Yang, J.; Gogotsi, Y.; Billinge, S.J.; Barsoum, M.W. Synthesis and characterization of two-dimensional Nb_4C_3 (MXene). *Chem. Commun.* **2014**, *50*, 9517–9520. [CrossRef] [PubMed]

32. Zhou, J.; Zha, X.; Chen, F.Y.; Ye, Q.; Eklund, P.; Du, S.; Huang, Q. A two-dimensional zirconium carbide by selective etching of Al_3C_3 from nanolaminated $Zr_3Al_3C_5$. *Angew. Chem. Int. Ed.* **2016**, *55*, 5008–5013. [CrossRef] [PubMed]
33. Tran, M.H.; Schäfer, T.; Shahraei, A.; Dürrschnabel, M.; Molina-Luna, L.; Kramm, U.I.; Birkel, C.S. Adding a new member to the MXene family: Synthesis, structure, and electrocatalytic activity for the hydrogen evolution reaction of $V_4C_3T_x$. *ACS Appl. Energy Mater.* **2018**, *1*, 3908–3914. [CrossRef]
34. Yoon, Y.; Lee, M.; Kim, S.K.; Bae, G.; Song, W.; Myung, S.; Lim, J.; Lee, S.S.; Zyung, T.; An, K.S. A strategy for synthesis of carbon nitride induced chemically doped 2D MXene for high-performance supercapacitor electrodes. *Adv. Energy Mater.* **2018**, *8*, 1703173. [CrossRef]
35. Shan, Q.; Mu, X.; Alhabeb, M.; Shuck, C.E.; Pang, D.; Zhao, X.; Chu, X.-F.; Wei, Y.; Du, F.; Chen, G. Two-dimensional vanadium carbide (V_2C) MXene as electrode for supercapacitors with aqueous electrolytes. *Electrochem. Commun.* **2018**, *96*, 103–107. [CrossRef]
36. Wang, X.; Lin, S.; Tong, H.; Huang, Y.; Tong, P.; Zhao, B.; Dai, J.; Liang, C.; Wang, H.; Zhu, X.; et al. Two-dimensional V_4C_3 MXene as high performance electrode materials for supercapacitors. *Electrochim. Acta* **2019**, *307*, 414–421. [CrossRef]
37. Zou, X.; Liu, H.; Xu, H.; Wu, X.; Han, X.; Kang, J.; Reddy, K.M. A simple approach to synthesis Cr_2CT_x MXene for efficient hydrogen evolution reaction. *Mater. Today Energy* **2021**, *20*, 100668. [CrossRef]
38. Yin, T.; Li, Y.; Wang, R.; Al-Hartomy, O.A.; Al-Ghamdi, A.; Wageh, S.; Luo, X.; Tang, X.; Zhang, H. Synthesis of $Ti_3C_2F_x$ MXene with controllable fluorination by electrochemical etching for lithium-ion batteries applications. *Ceram. Int.* **2021**, *47*, 28642–28649. [CrossRef]
39. Huo, X.; Zhong, J.; Yang, Z.; Feng, J.; Li, J.; Kang, F. In Situ Preparation of MXenes in Ambient-Temperature Organic Ionic Liquid Aluminum Batteries with Ultrastable Cycle Performance. *ACS Appl. Mater. Interfaces* **2021**, *13*, 55112–55122. [CrossRef]
40. Nasrin, K.; Sudharshan, V.; Subramani, K.; Sathish, M. Insights into 2D/2D MXene Heterostructures for Improved Synergy in Structure toward Next-Generation Supercapacitors: A Review. *Adv. Funct. Mater.* **2022**, *32*, 2110267. [CrossRef]
41. Jiang, Q.; Lei, Y.; Liang, H.; Xi, K.; Xia, C.; Alshareef, H.N. Review of MXene electrochemical microsupercapacitors. *Energy Storage Mater.* **2020**, *27*, 78–95. [CrossRef]
42. Xiong, D.; Li, X.; Bai, Z.; Lu, S. Recent Advances in Layered $Ti_3C_2T_x$ MXene for Electrochemical Energy Storage. *Small* **2018**, *14*, e1703419. [CrossRef] [PubMed]
43. Hu, M.; Zhang, H.; Hu, T.; Fan, B.; Wang, X.; Li, Z. Emerging 2D MXenes for supercapacitors: Status, challenges and prospects. *Chem. Soc. Rev.* **2020**, *49*, 6666–6693. [CrossRef]
44. Zhu, Q.; Li, J.; Simon, P.; Xu, B. Two-dimensional MXenes for electrochemical capacitor applications: Progress, challenges and perspectives. *Energy Storage Mater.* **2021**, *35*, 630–660. [CrossRef]
45. Pang, J.; Mendes, R.G.; Bachmatiuk, A.; Zhao, L.; Ta, H.Q.; Gemming, T.; Liu, H.; Liu, Z.; Rummeli, M.H. Applications of 2D MXenes in energy conversion and storage systems. *Chem. Soc. Rev.* **2019**, *48*, 72–133. [CrossRef]
46. Aslam, M.K.; Xu, M. A Mini-Review: MXene composites for sodium/potassium-ion batteries. *Nanoscale* **2020**, *12*, 15993–16007. [CrossRef] [PubMed]
47. Bao, Z.; Lu, C.; Cao, X.; Zhang, P.; Yang, L.; Zhang, H.; Sha, D.; He, W.; Zhang, W.; Pan, L.; et al. Role of MXene surface terminations in electrochemical energy storage: A review. *Chin. Chem. Lett.* **2021**, *32*, 2648–2658. [CrossRef]
48. Liu, Y.T.; Zhu, X.D.; Pan, L. Hybrid Architectures based on 2D MXenes and Low-Dimensional Inorganic Nanostructures: Methods, Synergies, and Energy-Related Applications. *Small* **2018**, *14*, e1803632. [CrossRef]
49. Ma, R.; Chen, Z.; Zhao, D.; Zhang, X.; Zhuo, J.; Yin, Y.; Wang, X.; Yang, G.; Yi, F. $Ti_3C_2T_x$ MXene for electrode materials of supercapacitors. *J. Mater. Chem. A* **2021**, *9*, 11501–11529. [CrossRef]
50. Xu, J.; Peng, T.; Qin, X.; Zhang, Q.; Liu, T.; Dai, W.; Chen, B.; Yu, H.; Shi, S. Recent advances in 2D MXenes: Preparation, intercalation and applications in flexible devices. *J. Mater. Chem. A* **2021**, *9*, 14147–14171. [CrossRef]
51. Carey, M.; Barsoum, M.W. MXene polymer nanocomposites: A review. *Mater. Today Adv.* **2021**, *9*, 100120. [CrossRef]
52. Gao, L.; Li, C.; Huang, W.; Mei, S.; Lin, H.; Ou, Q.; Zhang, Y.; Guo, J.; Zhang, F.; Xu, S.; et al. MXene/Polymer Membranes: Synthesis, Properties, and Emerging Applications. *Chem. Mater.* **2020**, *32*, 1703–1747. [CrossRef]
53. Liu, Y.; Yu, J.; Guo, D.; Li, Z.; Su, Y. $Ti_3C_2T_x$ MXene/graphene nanocomposites: Synthesis and application in electrochemical energy storage. *J. Alloys Compd.* **2020**, *815*, 152403. [CrossRef]
54. Yang, J.; Bao, W.; Jaumaux, P.; Zhang, S.; Wang, C.; Wang, G. MXene-Based Composites: Synthesis and Applications in Rechargeable Batteries and Supercapacitors. *Adv. Mater. Interfaces* **2019**, *6*, 1802004. [CrossRef]
55. Baig, M.M.; Gul, I.H.; Baig, S.M.; Shahzad, F. 2D MXenes: Synthesis, properties, and electrochemical energy storage for supercapacitors–A review. *J. Electroanal. Chem.* **2022**, *904*, 115920. [CrossRef]
56. Li, X.; Huang, Z.; Shuck, C.E.; Liang, G.; Gogotsi, Y.; Zhi, C. MXene chemistry, electrochemistry and energy storage applications. *Nat. Rev. Chem.* **2022**, *6*, 389–404. [CrossRef]
57. Kumar, J.A.; Prakash, P.; Krithiga, T.; Amarnath, D.J.; Premkumar, J.; Rajamohan, N.; Vasseghian, Y.; Saravanan, P.; Rajasimman, M. Methods of synthesis, characteristics, and environmental applications of MXene: A comprehensive review. *Chemosphere* **2022**, *286*, 131607. [CrossRef] [PubMed]
58. Ma, C.; Ma, M.G.; Si, C.; Ji, X.X.; Wan, P. Flexible MXene-Based Composites for Wearable Devices. *Adv. Funct. Mater.* **2021**, *31*, 2009524. [CrossRef]

59. Huang, W.; Hu, L.; Tang, Y.; Xie, Z.; Zhang, H. Recent Advances in Functional 2D MXene-Based Nanostructures for Next-Generation Devices. *Adv. Funct. Mater.* **2020**, *30*, 2005223. [CrossRef]
60. Zhang, Y.; Mei, H.-X.; Cao, Y.; Yan, X.-H.; Yan, J.; Gao, H.-L.; Luo, H.-W.; Wang, S.-W.; Jia, X.-D.; Kachalova, L.; et al. Recent advances and challenges of electrode materials for flexible supercapacitors. *Coord. Chem. Rev.* **2021**, *438*, 213910. [CrossRef]
61. Yang, M.; Lu, H.; Liu, S. Recent Advances of MXene-Based Electrochemical Immunosensors. *Appl. Sci.* **2022**, *12*, 5630. [CrossRef]
62. Vasyukova, I.A.; Zakharova, O.V.; Kuznetsov, D.V.; Gusev, A. Synthesis, Toxicity Assessment, Environmental and Biomedical Applications of MXenes: A Review. *Nanomaterials* **2022**, *12*, 1797. [CrossRef] [PubMed]
63. Yang, R.; Chen, X.; Ke, W.; Wu, X. Recent Research Progress in the Structure, Fabrication, and Application of MXene-Based Heterostructures. *Nanomaterials* **2022**, *12*, 1907. [CrossRef] [PubMed]
64. Guo, Y.; Jin, S.; Wang, L.; He, P.; Hu, Q.; Fan, L.-Z.; Zhou, A. Synthesis of two-dimensional carbide Mo_2CT_x MXene by hydrothermal etching with fluorides and its thermal stability. *Ceram. Int.* **2020**, *46*, 19550–19556. [CrossRef]
65. Alhabeb, M.; Maleski, K.; Mathis, T.S.; Sarycheva, A.; Hatter, C.B.; Uzun, S.; Levitt, A.; Gogotsi, Y. Selective Etching of Silicon from Ti_3SiC_2 (MAX) To Obtain 2D Titanium Carbide (MXene). *Angew. Chem. Int. Ed. Engl.* **2018**, *57*, 5444–5448. [CrossRef]
66. Cheng, Y.; Wang, L.; Li, Y.; Song, Y.; Zhang, Y. Etching and Exfoliation Properties of Cr_2AlC into Cr_2CO_2 and the Electrocatalytic Performances of 2D Cr_2CO_2 MXene. *J. Phys. Chem. C* **2019**, *123*, 15629–15636. [CrossRef]
67. Liu, F.; Zhou, A.; Chen, J.; Jia, J.; Zhou, W.; Wang, L.; Hu, Q. Preparation of Ti_3C_2 and Ti_2C MXenes by fluoride salts etching and methane adsorptive properties. *Appl. Surf. Sci.* **2017**, *416*, 781–789. [CrossRef]
68. Karlsson, L.H.; Birch, J.; Halim, J.; Barsoum, M.W.; Persson, P.O. Atomically Resolved Structural and Chemical Investigation of Single MXene Sheets. *Nano Lett.* **2015**, *15*, 4955–4960. [CrossRef]
69. Cao, Q.; Yun, F.F.; Sang, L.; Xiang, F.; Liu, G.; Wang, X. Defect introduced paramagnetism and weak localization in two-dimensional metal VSe_2. *Nanotechnology* **2017**, *28*, 475703. [CrossRef]
70. Cockreham, C.B.; Zhang, X.; Li, H.; Hammond-Pereira, E.; Sun, J.; Saunders, S.R.; Wang, Y.; Xu, H.; Wu, D. Inhibition of $AlF_3 \cdot 3H_2O$ Impurity Formation in $Ti_3C_2T_x$ MXene Synthesis under a Unique CoF_x/HCl Etching Environment. *ACS Appl. Energy Mater.* **2019**, *2*, 8145–8152. [CrossRef]
71. Chaudhari, N.K.; Jin, H.; Kim, B.; San Baek, D.; Joo, S.H.; Lee, K. MXene: An emerging two-dimensional material for future energy conversion and storage applications. *J. Mater. Chem. A* **2017**, *5*, 24564–24579. [CrossRef]
72. Come, J.; Black, J.M.; Lukatskaya, M.R.; Naguib, M.; Beidaghi, M.; Rondinone, A.J.; Kalinin, S.V.; Wesolowski, D.J.; Gogotsi, Y.; Balke, N. Controlling the actuation properties of MXene paper electrodes upon cation intercalation. *Nano Energy* **2015**, *17*, 27–35. [CrossRef]
73. Coleman, J.N.; Lotya, M.; O'Neill, A.; Bergin, S.D.; King, P.J.; Khan, U.; Young, K.; Gaucher, A.; De, S.; Smith, R.J.; et al. Two-dimensional nanosheets produced by liquid exfoliation of layered materials. *Science* **2011**, *331*, 568–571. [CrossRef] [PubMed]
74. Mashtalir, O.; Naguib, M.; Dyatkin, B.; Gogotsi, Y.; Barsoum, M.W. Kinetics of aluminum extraction from Ti_3AlC_2 in hydrofluoric acid. *Mater. Chem. Phys.* **2013**, *139*, 147–152. [CrossRef]
75. Sang, X.; Xie, Y.; Lin, M.-W.; Alhabeb, M.; Van Aken, K.L.; Gogotsi, Y.; Kent, P.R.; Xiao, K.; Unocic, R.R. Atomic defects in monolayer titanium carbide ($Ti_3C_2T_x$) MXene. *ACS Nano* **2016**, *10*, 9193–9200. [CrossRef]
76. Lipatov, A.; Alhabeb, M.; Lukatskaya, M.R.; Boson, A.; Gogotsi, Y.; Sinitskii, A. Effect of synthesis on quality, electronic properties and environmental stability of individual monolayer Ti_3C_2 MXene flakes. *Adv. Electron. Mater.* **2016**, *2*, 1600255. [CrossRef]
77. Ghidiu, M.; Lukatskaya, M.R.; Zhao, M.Q.; Gogotsi, Y.; Barsoum, M.W. Conductive two-dimensional titanium carbide 'clay' with high volumetric capacitance. *Nature* **2014**, *516*, 78–81. [CrossRef]
78. Feng, A.; Yu, Y.; Wang, Y.; Jiang, F.; Yu, Y.; Mi, L.; Song, L. Two-dimensional MXene Ti_3C_2 produced by exfoliation of Ti_3AlC_2. *Mater. Des.* **2017**, *114*, 161–166. [CrossRef]
79. Zhan, C.; Naguib, M.; Lukatskaya, M.; Kent, P.R.C.; Gogotsi, Y.; Jiang, D.E. Understanding the MXene Pseudocapacitance. *J. Phys. Chem. Lett.* **2018**, *9*, 1223–1228. [CrossRef]
80. Halim, J.; Lukatskaya, M.R.; Cook, K.M.; Lu, J.; Smith, C.R.; Naslund, L.A.; May, S.J.; Hultman, L.; Gogotsi, Y.; Eklund, P.; et al. Transparent Conductive Two-Dimensional Titanium Carbide Epitaxial Thin Films. *Chem. Mater.* **2014**, *26*, 2374–2381. [CrossRef]
81. Wang, X.; Garnero, C.; Rochard, G.; Magne, D.; Morisset, S.; Hurand, S.; Chartier, P.; Rousseau, J.; Cabioc'h, T.; Coutanceau, C.; et al. A new etching environment (FeF_3/HCl) for the synthesis of two-dimensional titanium carbide MXenes: A route towards selective reactivity vs. water. *J. Mater. Chem. A* **2017**, *5*, 22012–22023. [CrossRef]
82. Mei, J.; Ayoko, G.A.; Hu, C.; Bell, J.M.; Sun, Z. Two-dimensional fluorine-free mesoporous Mo_2C MXene via UV-induced selective etching of Mo_2Ga_2C for energy storage. *Sustain. Mater. Technol.* **2020**, *25*, e00156. [CrossRef]
83. Rafieerad, A.; Amiri, A.; Sequiera, G.L.; Yan, W.; Chen, Y.; Polycarpou, A.A.; Dhingra, S. Development of Fluorine-Free Tantalum Carbide MXene Hybrid Structure as a Biocompatible Material for Supercapacitor Electrodes. *Adv. Funct. Mater.* **2021**, *31*, 2100015. [CrossRef] [PubMed]
84. Song, M.; Pang, S.Y.; Guo, F.; Wong, M.C.; Hao, J. Fluoride-Free 2D Niobium Carbide MXenes as Stable and Biocompatible Nanoplatforms for Electrochemical Biosensors with Ultrahigh Sensitivity. *Adv. Sci.* **2020**, *7*, 2001546. [CrossRef] [PubMed]
85. Xue, N.; Li, X.; Han, L.; Zhu, H.; Zhao, X.; Zhuang, J.; Gao, Z.; Tao, X. Fluorine-free synthesis of ambient-stable delaminated Ti_2CT_x (MXene). *J. Mater. Chem. A* **2022**, *10*, 7960–7967. [CrossRef]
86. Sun, Z.; Yuan, M.; Lin, L.; Yang, H.; Nan, C.; Li, H.; Sun, G.; Yang, X. Selective Lithiation–Expansion–Microexplosion Synthesis of Two-Dimensional Fluoride-Free Mxene. *ACS Mater. Lett.* **2019**, *1*, 628–632. [CrossRef]

87. Al Mayyahi, A.; Sarker, S.; Everhart, B.M.; He, X.; Amama, P.B. One-Step Fluorine-Free Synthesis of Delaminated, OH-Terminated Ti$_3$C$_2$: High Photocatalytic NO$_x$ Storage Selectivity Enabled by Coupling TiO$_2$ and Ti$_3$C$_2$-OH. *Mater. Today Commun.* **2022**, *32*, 103835. [CrossRef]
88. Thomas, T.; Pushpan, S.; Aguilar Martínez, J.A.; Torres Castro, A.; Pineda Aguilar, N.; Álvarez-Méndez, A.; Sanal, K.C. UV-assisted safe etching route for the synthesis of Mo$_2$CT$_x$ MXene from Mo–In–C non-MAX phase. *Ceram. Int.* **2021**, *47*, 35384–35387. [CrossRef]
89. Yang, S.; Zhang, P.; Wang, F.; Ricciardulli, A.G.; Lohe, M.R.; Blom, P.W.M.; Feng, X. Fluoride-Free Synthesis of Two-Dimensional Titanium Carbide (MXene) Using A Binary Aqueous System. *Angew. Chem. Int. Ed. Engl.* **2018**, *57*, 15491–15495. [CrossRef]
90. Ding, L.; Wei, Y.; Wang, Y.; Chen, H.; Caro, J.; Wang, H. A Two-Dimensional Lamellar Membrane: MXene Nanosheet Stacks. *Angew. Chem. Int. Ed. Engl.* **2017**, *56*, 1825–1829. [CrossRef]
91. Li, T.; Yao, L.; Liu, Q.; Gu, J.; Luo, R.; Li, J.; Yan, X.; Wang, W.; Liu, P.; Chen, B.; et al. Fluorine-Free Synthesis of High-Purity Ti$_3$C$_2$T$_x$ (T=OH, O) via Alkali Treatment. *Angew. Chem. Int. Ed. Engl.* **2018**, *57*, 6115–6119. [CrossRef] [PubMed]
92. Dall'Agnese, Y.; Lukatskaya, M.R.; Cook, K.M.; Taberna, P.-L.; Gogotsi, Y.; Simon, P. High capacitance of surface-modified 2D titanium carbide in acidic electrolyte. *Electrochem. Commun.* **2014**, *48*, 118–122. [CrossRef]
93. Zhang, B.; Zhu, J.; Shi, P.; Wu, W.; Wang, F. Fluoride-free synthesis and microstructure evolution of novel two-dimensional Ti$_3$C$_2$(OH)$_2$ nanoribbons as high-performance anode materials for lithium-ion batteries. *Ceram. Int.* **2019**, *45*, 8395–8405. [CrossRef]
94. Li, Y.; Shao, H.; Lin, Z.; Lu, J.; Liu, L.; Duployer, B.; Persson, P.O.A.; Eklund, P.; Hultman, L.; Li, M.; et al. A general Lewis acidic etching route for preparing MXenes with enhanced electrochemical performance in non-aqueous electrolyte. *Nat. Mater.* **2020**, *19*, 894–899. [CrossRef] [PubMed]
95. Naoi, K.; Kisu, K.; Iwama, E.; Nakashima, S.; Sakai, Y.; Orikasa, Y.; Leone, P.; Dupré, N.; Brousse, T.; Rozier, P.; et al. Ultrafast charge–discharge characteristics of a nanosized core–shell structured LiFePO$_4$ material for hybrid supercapacitor applications. *Energy Environ. Sci.* **2016**, *9*, 2143–2151. [CrossRef]
96. Lukatskaya, M.R.; Dunn, B.; Gogotsi, Y. Multidimensional materials and device architectures for future hybrid energy storage. *Nat. Commun.* **2016**, *7*, 12647. [CrossRef]
97. Wu, Z.; Zhu, S.; Bai, X.; Liang, M.; Zhang, X.; Zhao, N.; He, C. One-step in-situ synthesis of Sn-nanoconfined Ti$_3$C$_2$T$_x$ MXene composites for Li-ion battery anode. *Electrochim. Acta* **2022**, *407*, 139916. [CrossRef]
98. Vicic, D.A.; Jones, G.D. Experimental Methods and Techniques: Basic Techniques. *Compr. Organomet. Chem. III* **2007**, *1*, 197–218. [CrossRef]
99. Calvo-Flores, F.; Dobado, J.; Isac-García, J.; Martín-MartíNez, F. Structure and physicochemical properties. In *Lignin and Lignans as Renewable Raw Materials: Chemistry, Technology and Applications*; Wiley: New York, NY, USA, 2015; pp. 11–47. [CrossRef]
100. Laufersky, G.; Bradley, S.; Frecaut, F.; Lein, M.; Nann, T. Unraveling aminophosphine redox mechanisms for glovebox-free InP quantum dot syntheses. *Nanoscale* **2018**, *10*, 8752–8762. [CrossRef]
101. Anasori, B.; Xie, Y.; Beidaghi, M.; Lu, J.; Hosler, B.C.; Hultman, L.; Kent, P.R.; Gogotsi, Y.; Barsoum, M.W. Two-Dimensional, Ordered, Double Transition Metals Carbides (MXenes). *ACS Nano* **2015**, *9*, 9507–9516. [CrossRef]
102. Wu, J.; Wang, Y.; Zhang, Y.; Meng, H.; Xu, Y.; Han, Y.; Wang, Z.; Dong, Y.; Zhang, X. Highly safe and ionothermal synthesis of Ti$_3$C$_2$ MXene with expanded interlayer spacing for enhanced lithium storage. *J. Energy Chem.* **2020**, *47*, 203–209. [CrossRef]
103. Shi, H.; Zhang, P.; Liu, Z.; Park, S.; Lohe, M.R.; Wu, Y.; Shaygan Nia, A.; Yang, S.; Feng, X. Ambient-Stable Two-Dimensional Titanium Carbide (MXene) Enabled by Iodine Etching. *Angew. Chem. Int. Ed. Engl.* **2021**, *60*, 8689–8693. [CrossRef] [PubMed]
104. Sun, S.; Liao, C.; Hafez, A.M.; Zhu, H.; Wu, S. Two-dimensional MXenes for energy storage. *Chem. Eng. J.* **2018**, *338*, 27–45. [CrossRef]
105. Wang, J.; Tang, J.; Ding, B.; Malgras, V.; Chang, Z.; Hao, X.; Wang, Y.; Dou, H.; Zhang, X.; Yamauchi, Y. Hierarchical porous carbons with layer-by-layer motif architectures from confined soft-template self-assembly in layered materials. *Nat. Commun.* **2017**, *8*, 15717. [CrossRef]
106. Zhang, X.; Liu, Y.; Dong, S.; Yang, J.; Liu, X. Surface modified MXene film as flexible electrode with ultrahigh volumetric capacitance. *Electrochim. Acta* **2019**, *294*, 233–239. [CrossRef]
107. Yang, C.; Tang, Y.; Tian, Y.; Luo, Y.; He, Y.; Yin, X.; Que, W. Achieving of Flexible, Free-Standing, Ultracompact Delaminated Titanium Carbide Films for High Volumetric Performance and Heat-Resistant Symmetric Supercapacitors. *Adv. Funct. Mater.* **2018**, *28*, 1705487. [CrossRef]
108. Lu, M.; Li, H.; Han, W.; Chen, J.; Shi, W.; Wang, J.; Meng, X.-M.; Qi, J.; Li, H.; Zhang, B.; et al. 2D titanium carbide (MXene) electrodes with lower-F surface for high performance lithium-ion batteries. *J. Energy Chem.* **2019**, *31*, 148–153. [CrossRef]
109. Zhang, Z.; Yao, Z.; Zhang, X.; Jiang, Z. 2D Carbide MXene under postetch low-temperature annealing for high–performance supercapacitor electrode. *Electrochim. Acta* **2020**, *359*, 136960. [CrossRef]
110. Zhao, X.; Wang, Z.; Dong, J.; Huang, T.; Zhang, L. Annealing modification of MXene films with mechanically strong structures and high electrochemical performance for supercapacitor applications. *J. Power Sources* **2020**, *470*, 228356. [CrossRef]
111. Sun, Y.; Chen, D.; Liang, Z. Two-dimensional MXenes for energy storage and conversion applications. *Mater. Today Energy* **2017**, *5*, 22–36. [CrossRef]
112. Lukatskaya, M.R.; Bak, S.-M.; Yu, X.; Yang, X.-Q.; Barsoum, M.W.; Gogotsi, Y. Probing the Mechanism of High Capacitance in 2D Titanium Carbide Using In Situ X-Ray Absorption Spectroscopy. *Adv. Energy Mater.* **2015**, *5*, 1500589. [CrossRef]

113. Ran, F.; Wang, T.; Chen, S.; Liu, Y.; Shao, L. Constructing expanded ion transport channels in flexible MXene film for pseudocapacitive energy storage. *Appl. Surf. Sci.* **2020**, *511*, 45621–145627. [CrossRef]
114. Gutiérrez, M.C.; Ferrer, M.L.; del Monte, F. Ice-Templated Materials: Sophisticated Structures Exhibiting Enhanced Functionalities Obtained after Unidirectional Freezing and Ice-Segregation-Induced Self-Assembly. *Chem. Mater.* **2008**, *20*, 634–648. [CrossRef]
115. Xia, Y.; Mathis, T.S.; Zhao, M.Q.; Anasori, B.; Dang, A.; Zhou, Z.; Cho, H.; Gogotsi, Y.; Yang, S. Thickness-independent capacitance of vertically aligned liquid-crystalline MXenes. *Nature* **2018**, *557*, 409–412. [CrossRef] [PubMed]
116. Yan, J.; Wang, Q.; Wei, T.; Jiang, L.; Zhang, M.; Jing, X.; Fan, Z. Template-assisted low temperature synthesis of functionalized graphene for ultrahigh volumetric performance supercapacitors. *ACS Nano* **2014**, *8*, 4720–4729. [CrossRef] [PubMed]
117. Chen, C.M.; Zhang, Q.; Huang, C.H.; Zhao, X.C.; Zhang, B.S.; Kong, Q.Q.; Wang, M.Z.; Yang, Y.G.; Cai, R.; Sheng Su, D. Macroporous 'bubble' graphene film via template-directed ordered-assembly for high rate supercapacitors. *Chem. Commun.* **2012**, *48*, 7149–7151. [CrossRef]
118. Xu, S.; Wei, G.; Li, J.; Han, W.; Gogotsi, Y. Flexible MXene–graphene electrodes with high volumetric capacitance for integrated co-cathode energy conversion/storage devices. *J. Mater. Chem. A* **2017**, *5*, 17442–17451. [CrossRef]
119. Yang, L.; Zheng, W.; Zhang, P.; Chen, J.; Zhang, W.; Tian, W.B.; Sun, Z.M. Freestanding nitrogen-doped d-Ti_3C_2/reduced graphene oxide hybrid films for high performance supercapacitors. *Electrochim. Acta* **2019**, *300*, 349–356. [CrossRef]
120. Aïssa, B.; Sinopoli, A.; Ali, A.; Zakaria, Y.; Zekri, A.; Helal, M.; Nedil, M.; Rosei, F.; Mansour, S.; Mahmoud, K. Nanoelectromagnetic of a highly conductive 2D transition metal carbide (MXene)/Graphene nanoplatelets composite in the EHF M-band frequency. *Carbon* **2021**, *173*, 528–539. [CrossRef]
121. Li, H.; Hou, Y.; Wang, F.; Lohe, M.R.; Zhuang, X.; Niu, L.; Feng, X. Flexible All-Solid-State Supercapacitors with High Volumetric Capacitances Boosted by Solution Processable MXene and Electrochemically Exfoliated Graphene. *Adv. Energy Mater.* **2017**, *7*, 1601847. [CrossRef]
122. Novoselov, K.S.; Fal'ko, V.I.; Colombo, L.; Gellert, P.R.; Schwab, M.G.; Kim, K. A roadmap for graphene. *Nature* **2012**, *490*, 192–200. [CrossRef] [PubMed]
123. Zhou, Y.; Maleski, K.; Anasori, B.; Thostenson, J.O.; Pang, Y.; Feng, Y.; Zeng, K.; Parker, C.B.; Zauscher, S.; Gogotsi, Y.; et al. $Ti_3C_2T_x$ MXene-Reduced Graphene Oxide Composite Electrodes for Stretchable Supercapacitors. *ACS Nano* **2020**, *14*, 3576–3586. [CrossRef] [PubMed]
124. Wang, K.; Zheng, B.; Mackinder, M.; Baule, N.; Qiao, H.; Jin, H.; Schuelke, T.; Fan, Q.H. Graphene wrapped MXene via plasma exfoliation for all-solid-state flexible supercapacitors. *Energy Storage Mater.* **2019**, *20*, 299–306. [CrossRef]
125. Zhang, X.; Zhang, Z.; Zhou, Z. MXene-based materials for electrochemical energy storage. *J. Energy Chem.* **2018**, *27*, 73–85. [CrossRef]
126. Stankovich, S.; Dikin, D.A.; Dommett, G.H.; Kohlhaas, K.M.; Zimney, E.J.; Stach, E.A.; Piner, R.D.; Nguyen, S.T.; Ruoff, R.S. Graphene-based composite materials. *Nature* **2006**, *442*, 282–286. [CrossRef]
127. Navarro-Suárez, A.M.; Maleski, K.; Makaryan, T.; Yan, J.; Anasori, B.; Gogotsi, Y. 2D Titanium Carbide/Reduced Graphene Oxide Heterostructures for Supercapacitor Applications. *Batter. Supercaps* **2018**, *1*, 33–38. [CrossRef]
128. Fan, Z.; Wang, Y.; Xie, Z.; Wang, D.; Yuan, Y.; Kang, H.; Su, B.; Cheng, Z.; Liu, Y. Modified MXene/Holey Graphene Films for Advanced Supercapacitor Electrodes with Superior Energy Storage. *Adv. Sci.* **2018**, *5*, 1800750. [CrossRef]
129. Yang, Q.; Xu, Z.; Fang, B.; Huang, T.; Cai, S.; Chen, H.; Liu, Y.; Gopalsamy, K.; Gao, W.; Gao, C. MXene/graphene hybrid fibers for high performance flexible supercapacitors. *J. Mater. Chem. A* **2017**, *5*, 22113–22119. [CrossRef]
130. Yan, J.; Ren, C.E.; Maleski, K.; Hatter, C.B.; Anasori, B.; Urbankowski, P.; Sarycheva, A.; Gogotsi, Y. Flexible MXene/Graphene Films for Ultrafast Supercapacitors with Outstanding Volumetric Capacitance. *Adv. Funct. Mater.* **2017**, *27*, 1701264. [CrossRef]
131. Srivastava, S.; Kotov, N.A. Composite layer-by-layer (LBL) assembly with inorganic nanoparticles and nanowires. *Acc. Chem. Res.* **2008**, *41*, 1831–1841. [CrossRef]
132. Miao, J.; Zhu, Q.; Li, K.; Zhang, P.; Zhao, Q.; Xu, B. Self-propagating fabrication of 3D porous MXene-rGO film electrode for high-performance supercapacitors. *J. Energy Chem.* **2021**, *52*, 243–250. [CrossRef]
133. Yang, X.; Wang, Q.; Zhu, K.; Ye, K.; Wang, G.; Cao, D.; Yan, J. 3D Porous Oxidation-Resistant MXene/Graphene Architectures Induced by In Situ Zinc Template toward High-Performance Supercapacitors. *Adv. Funct. Mater.* **2021**, *31*, 2101087. [CrossRef]
134. Xiang, Z.; Shi, Y.; Zhu, X.; Cai, L.; Lu, W. Flexible and Waterproof $2D/1D/0D$ Construction of MXene-Based Nanocomposites for Electromagnetic Wave Absorption, EMI Shielding, and Photothermal Conversion. *Nanomicro. Lett.* **2021**, *13*, 150. [CrossRef] [PubMed]
135. Xie, X.; Zhao, M.-Q.; Anasori, B.; Maleski, K.; Ren, C.E.; Li, J.; Byles, B.W.; Pomerantseva, E.; Wang, G.; Gogotsi, Y. Porous heterostructured MXene/carbon nanotube composite paper with high volumetric capacity for sodium-based energy storage devices. *Nano Energy* **2016**, *26*, 513–523. [CrossRef]
136. Liang, X.; Rangom, Y.; Kwok, C.Y.; Pang, Q.; Nazar, L.F. Interwoven MXene Nanosheet/Carbon-Nanotube Composites as Li-S Cathode Hosts. *Adv. Mater.* **2017**, *29*, 1603040. [CrossRef]
137. Fu, Q.; Wang, X.; Zhang, N.; Wen, J.; Li, L.; Gao, H.; Zhang, X. Self-assembled $Ti_3C_2T_x$/SCNT composite electrode with improved electrochemical performance for supercapacitor. *J. Colloid. Interface Sci.* **2018**, *511*, 128–134. [CrossRef]
138. Cai, Y.-Z.; Fang, Y.-S.; Cao, W.-Q.; He, P.; Cao, M.-S. MXene-CNT/PANI ternary material with excellent supercapacitive performance driven by synergy. *J. Alloys Compd.* **2021**, *868*, 159159. [CrossRef]

139. Wang, X.; Luo, D.; Wang, J.; Sun, Z.; Cui, G.; Chen, Y.; Wang, T.; Zheng, L.; Zhao, Y.; Shui, L.; et al. Inside Cover: Strain Engineering of a MXene/CNT Hierarchical Porous Hollow Microsphere Electrocatalyst for a High-Efficiency Lithium Polysulfide Conversion Process. *Angew. Chem. Int. Ed.* **2021**, *60*, 2170. [CrossRef]
140. Kang, K.; Lee, K.H.; Han, Y.; Gao, H.; Xie, S.; Muller, D.A.; Park, J. Layer-by-layer assembly of two-dimensional materials into wafer-scale heterostructures. *Nature* **2017**, *550*, 229–233. [CrossRef]
141. Lipton, J.; Weng, G.-M.; Röhr, J.A.; Wang, H.; Taylor, A.D. Layer-by-Layer Assembly of Two-Dimensional Materials: Meticulous Control on the Nanoscale. *Matter* **2020**, *2*, 1148–1165. [CrossRef]
142. Zhao, M.Q.; Ren, C.E.; Ling, Z.; Lukatskaya, M.R.; Zhang, C.; Van Aken, K.L.; Barsoum, M.W.; Gogotsi, Y. Flexible MXene/carbon nanotube composite paper with high volumetric capacitance. *Adv. Mater.* **2015**, *27*, 339–345. [CrossRef] [PubMed]
143. Dall'Agnese, Y.; Rozier, P.; Taberna, P.-L.; Gogotsi, Y.; Simon, P. Capacitance of two-dimensional titanium carbide (MXene) and MXene/carbon nanotube composites in organic electrolytes. *J. Power Sources* **2016**, *306*, 510–515. [CrossRef]
144. Zhang, P.; Zhu, Q.; Soomro, R.A.; He, S.; Sun, N.; Qiao, N.; Xu, B. In Situ Ice Template Approach to Fabricate 3D Flexible MXene Film-Based Electrode for High Performance Supercapacitors. *Adv. Funct. Mater.* **2020**, *30*, 2000922. [CrossRef]
145. Gao, X.; Du, X.; Mathis, T.S.; Zhang, M.; Wang, X.; Shui, J.; Gogotsi, Y.; Xu, M. Maximizing ion accessibility in MXene-knotted carbon nanotube composite electrodes for high-rate electrochemical energy storage. *Nat. Commun.* **2020**, *11*, 6160. [CrossRef] [PubMed]
146. Zhu, M.; Huang, Y.; Deng, Q.; Zhou, J.; Pei, Z.; Xue, Q.; Huang, Y.; Wang, Z.; Li, H.; Huang, Q.; et al. Highly Flexible, Freestanding Supercapacitor Electrode with Enhanced Performance Obtained by Hybridizing Polypyrrole Chains with MXene. *Adv. Energy Mater.* **2016**, *6*, 1600969. [CrossRef]
147. Wu, X.; Huang, B.; Lv, R.; Wang, Q.; Wang, Y. Highly flexible and low capacitance loss supercapacitor electrode based on hybridizing decentralized conjugated polymer chains with MXene. *Chem. Eng. J.* **2019**, *378*, 122246. [CrossRef]
148. Ling, Z.; Ren, C.E.; Zhao, M.Q.; Yang, J.; Giammarco, J.M.; Qiu, J.; Barsoum, M.W.; Gogotsi, Y. Flexible and conductive MXene films and nanocomposites with high capacitance. *Proc. Natl. Acad. Sci. USA* **2014**, *111*, 16676–16681. [CrossRef]
149. Boota, M.; Anasori, B.; Voigt, C.; Zhao, M.Q.; Barsoum, M.W.; Gogotsi, Y. Pseudocapacitive Electrodes Produced by Oxidant-Free Polymerization of Pyrrole between the Layers of 2D Titanium Carbide (MXene). *Adv. Mater.* **2016**, *28*, 1517–1522. [CrossRef]
150. Luo, W.; Wei, Y.; Zhuang, Z.; Lin, Z.; Li, X.; Hou, C.; Li, T.; Ma, Y. Fabrication of $Ti_3C_2T_x$ MXene/polyaniline composite films with adjustable thickness for high-performance flexible all-solid-state symmetric supercapacitors. *Electrochim. Acta* **2022**, *406*, 139871. [CrossRef]
151. Li, L.; Zhang, N.; Zhang, M.; Zhang, X.; Zhang, Z. Flexible $Ti_3C_2T_x$/PEDOT:PSS films with outstanding volumetric capacitance for asymmetric supercapacitors. *Dalton Trans.* **2019**, *48*, 1747–1756. [CrossRef]
152. Mishra, R.K.; Manivannan, S.; Kim, K.; Kwon, H.-I.; Jin, S.H. Petal-like MoS_2 nanostructures with metallic 1 T phase for high performance supercapacitors. *Curr. Appl. Phys.* **2018**, *18*, 345–352. [CrossRef]
153. Zheng, J.; Jow, T. A new charge storage mechanism for electrochemical capacitors. *J. Electrochem. Soc.* **1995**, *142*, L6. [CrossRef]
154. Rakhi, R.B.; Ahmed, B.; Anjum, D.; Alshareef, H.N. Direct Chemical Synthesis of MnO_2 Nanowhiskers on Transition-Metal Carbide Surfaces for Supercapacitor Applications. *ACS Appl. Mater. Interfaces* **2016**, *8*, 18806–18814. [CrossRef] [PubMed]
155. Cao, J.; Li, X.; Wang, Y.; Walsh, F.C.; Ouyang, J.-H.; Jia, D.; Zhou, Y. Materials and fabrication of electrode scaffolds for deposition of MnO_2 and their true performance in supercapacitors. *J. Power Sources* **2015**, *293*, 657–674. [CrossRef]
156. Hou, C.; Yu, H. $ZnO/Ti_3C_2T_x$ monolayer electron transport layers with enhanced conductivity for highly efficient inverted polymer solar cells. *Chem. Eng. J.* **2021**, *407*, 127192. [CrossRef]
157. Zhang, M.; Zhou, J.; Yu, J.; Shi, L.; Ji, M.; Liu, H.; Li, D.; Zhu, C.; Xu, J. Mixed analogous heterostructure based on MXene and prussian blue analog derivative for high-performance flexible energy storage. *Chem. Eng. J.* **2020**, *387*, 123170. [CrossRef]
158. Zhou, J.; Yu, J.; Shi, L.; Wang, Z.; Liu, H.; Yang, B.; Li, C.; Zhu, C.; Xu, J. A Conductive and Highly Deformable All-Pseudocapacitive Composite Paper as Supercapacitor Electrode with Improved Areal and Volumetric Capacitance. *Small* **2018**, *14*, e1803786. [CrossRef]
159. Radha, N.; Kanakaraj, A.; Manohar, H.; Nidhi, M.; Mondal, D.; Nataraj, S.K.; Ghosh, D. Binder free self-standing high performance supercapacitive electrode based on graphene/titanium carbide composite aerogel. *Appl. Surf. Sci.* **2019**, *481*, 892–899. [CrossRef]
160. Huang, X.; Zhu, X.; Luo, S.; Li, R.; Rajput, N.; Chiesa, M.; Liao, K.; Chan, V. $MnO_{1.88}/R-MnO_2/Ti_3C_2(OH/F)_x$ composite electrodes for high-performance pseudo-supercapacitors prepared from reduced MXenes. *New J. Chem.* **2020**, *44*, 6583–6588. [CrossRef]
161. Zhao, R.; Wang, M.; Zhao, D.; Li, H.; Wang, C.; Yin, L. Molecular-Level Heterostructures Assembled from Titanium Carbide MXene and Ni–Co–Al Layered Double-Hydroxide Nanosheets for All-Solid-State Flexible Asymmetric High-Energy Supercapacitors. *ACS Energy Lett.* **2017**, *3*, 132–140. [CrossRef]
162. Tian, Y.; Yang, C.; Que, W.; Liu, X.; Yin, X.; Kong, L.B. Flexible and free-standing 2D titanium carbide film decorated with manganese oxide nanoparticles as a high volumetric capacity electrode for supercapacitor. *J. Power Sources* **2017**, *359*, 332–339. [CrossRef]
163. Zhao, K.; Wang, H.; Zhu, C.; Lin, S.; Xu, Z.; Zhang, X. Free-standing MXene film modified by amorphous FeOOH quantum dots for high-performance asymmetric supercapacitor. *Electrochim. Acta* **2019**, *308*, 1–8. [CrossRef]

164. Wang, Y.; Wang, X.; Li, X.; Liu, R.; Bai, Y.; Xiao, H.; Liu, Y.; Yuan, G. Intercalating Ultrathin MoO3 Nanobelts into MXene Film with Ultrahigh Volumetric Capacitance and Excellent Deformation for High-Energy-Density Devices. *Nanomicro. Lett.* **2020**, *12*, 115. [CrossRef] [PubMed]
165. Jiao, S.; Zhou, A.; Wu, M.; Hu, H. Kirigami Patterning of MXene/Bacterial Cellulose Composite Paper for All-Solid-State Stretchable Micro-Supercapacitor Arrays. *Adv. Sci.* **2019**, *6*, 1900529. [CrossRef]
166. Zhang, C.J.; McKeon, L.; Kremer, M.P.; Park, S.H.; Ronan, O.; Seral-Ascaso, A.; Barwich, S.; Coileain, C.O.; McEvoy, N.; Nerl, H.C.; et al. Additive-free MXene inks and direct printing of micro-supercapacitors. *Nat. Commun.* **2019**, *10*, 1795. [CrossRef]
167. Qin, L.; Tao, Q.; El Ghazaly, A.; Fernandez-Rodriguez, J.; Persson, P.O.Å.; Rosen, J.; Zhang, F. High-Performance Ultrathin Flexible Solid-State Supercapacitors Based on Solution Processable Mo$_{1.33}$C MXene and PEDOT:PSS. *Adv. Funct. Mater.* **2018**, *28*, 1703808. [CrossRef]
168. Zhang, C.J.; Anasori, B.; Seral-Ascaso, A.; Park, S.H.; McEvoy, N.; Shmeliov, A.; Duesberg, G.S.; Coleman, J.N.; Gogotsi, Y.; Nicolosi, V. Transparent, Flexible, and Conductive 2D Titanium Carbide (MXene) Films with High Volumetric Capacitance. *Adv. Mater.* **2017**, *29*, 1702678. [CrossRef]
169. Jiang, H.; Wang, Z.; Yang, Q.; Hanif, M.; Wang, Z.; Dong, L.; Dong, M. A novel MnO$_2$/Ti$_3$C$_2$T$_x$ MXene nanocomposite as high performance electrode materials for flexible supercapacitors. *Electrochim. Acta* **2018**, *290*, 695–703. [CrossRef]
170. Huang, H.; Su, H.; Zhang, H.; Xu, L.; Chu, X.; Hu, C.; Liu, H.; Chen, N.; Liu, F.; Deng, W.; et al. Extraordinary Areal and Volumetric Performance of Flexible Solid-State Micro-Supercapacitors Based on Highly Conductive Freestanding Ti$_3$C$_2$T$_x$ Films. *Adv. Electron. Mater.* **2018**, *4*, 1800179. [CrossRef]
171. Wen, D.; Ying, G.; Liu, L.; Li, Y.; Sun, C.; Hu, C.; Zhao, Y.; Ji, Z.; Zhang, J.; Wang, X.J.J.o.A.; et al. Direct inkjet printing of flexible MXene/graphene composite films for supercapacitor electrodes. *J. Alloys Compd.* **2022**, *900*, 163436. [CrossRef]
172. Xu, S.; Dall'Agnese, Y.; Wei, G.; Zhang, C.; Gogotsi, Y.; Han, W. Screen-printable microscale hybrid device based on MXene and layered double hydroxide electrodes for powering force sensors. *Nano Energy* **2018**, *50*, 479–488. [CrossRef]
173. Huang, X.; Huang, J.; Yang, D.; Wu, P. A Multi-Scale Structural Engineering Strategy for High-Performance MXene Hydrogel Supercapacitor Electrode. *Adv. Sci.* **2021**, *8*, e2101664. [CrossRef] [PubMed]
174. Zhang, D.; Chi, B.; Li, B.; Gao, Z.; Du, Y.; Guo, J.; Wei, J. Fabrication of highly conductive graphene flexible circuits by 3D printing. *Synth. Met.* **2016**, *217*, 79–86. [CrossRef]
175. Roh, S.; Parekh, D.P.; Bharti, B.; Stoyanov, S.D.; Velev, O.D. 3D Printing by Multiphase Silicone/Water Capillary Inks. *Adv. Mater.* **2017**, *29*, 1701554. [CrossRef]
176. Farahani, R.D.; Dube, M.; Therriault, D. Three-Dimensional Printing of Multifunctional Nanocomposites: Manufacturing Techniques and Applications. *Adv. Mater.* **2016**, *28*, 5794–5821. [CrossRef]
177. Couly, C.; Alhabeb, M.; Van Aken, K.L.; Kurra, N.; Gomes, L.; Navarro-Suárez, A.M.; Anasori, B.; Alshareef, H.N.; Gogotsi, Y. Asymmetric flexible MXene-reduced graphene oxide micro-supercapacitor. *Adv. Electron. Mater.* **2018**, *4*, 1700339. [CrossRef]
178. Zheng, S.; Zhang, C.J.; Zhou, F.; Dong, Y.; Shi, X.; Nicolosi, V.; Wu, Z.-S.; Bao, X. Ionic liquid pre-intercalated MXene films for ionogel-based flexible micro-supercapacitors with high volumetric energy density. *J. Mater. Chem. A* **2019**, *7*, 9478–9485. [CrossRef]
179. Xia, X.; Zhan, J.; Zhong, Y.; Wang, X.; Tu, J.; Fan, H.J. Single-Crystalline, Metallic TiC Nanowires for Highly Robust and Wide-Temperature Electrochemical Energy Storage. *Small* **2017**, *13*, 1602742. [CrossRef]
180. Xu, S.; Wei, G.; Li, J.; Ji, Y.; Klyui, N.; Izotov, V.; Han, W. Binder-free Ti$_3$C$_2$T$_x$ MXene electrode film for supercapacitor produced by electrophoretic deposition method. *Chem. Eng. J.* **2017**, *317*, 1026–1036. [CrossRef]
181. Wang, K.; Zheng, B.; Mackinder, M.; Baule, N.; Garratt, E.; Jin, H.; Schuelke, T.; Fan, Q.H. Efficient electrophoretic deposition of MXene/reduced graphene oxide flexible electrodes for all-solid-state supercapacitors. *J. Energy Storage* **2021**, *33*, 102070. [CrossRef]
182. Yang, L.; Lin, F.; Zabihi, F.; Yang, S.; Zhu, M. High specific capacitance cotton fiber electrode enhanced with PPy and MXene by in situ hybrid polymerization. *Int. J. Biol. Macromol.* **2021**, *181*, 1063–1071. [CrossRef]
183. Li, J.; Chen, J.; Wang, H.; Xiao, X. All-MXene Cotton-Based Supercapacitor-Powered Human Body Thermal Management System. *ChemElectroChem* **2021**, *8*, 648–655. [CrossRef]
184. Yan, J.; Ma, Y.; Zhang, C.; Li, X.; Liu, W.; Yao, X.; Yao, S.; Luo, S. Polypyrrole-MXene coated textile-based flexible energy storage device. *RSC Adv.* **2018**, *8*, 39742–39748. [CrossRef]
185. Li, H.; Chen, R.; Ali, M.; Lee, H.; Ko, M.J. In Situ Grown MWCNTs/MXenes Nanocomposites on Carbon Cloth for High-Performance Flexible Supercapacitors. *Adv. Funct. Mater.* **2020**, *30*, 2002739. [CrossRef]
186. Li, X.; Hao, J.; Liu, R.; He, H.; Wang, Y.; Liang, G.; Liu, Y.; Yuan, G.; Guo, Z. Interfacing MXene flakes on fiber fabric as an ultrafast electron transport layer for high performance textile electrodes. *Energy Storage Mater.* **2020**, *33*, 62–70. [CrossRef]
187. Li, X.; Li, H.; Fan, X.; Shi, X.; Liang, J. 3D-Printed Stretchable Micro-Supercapacitor with Remarkable Areal Performance. *Adv. Energy Mater.* **2020**, *10*, 1903794. [CrossRef]
188. Kurra, N.; Ahmed, B.; Gogotsi, Y.; Alshareef, H.N. MXene-on-Paper Coplanar Microsupercapacitors. *Adv. Energy Mater.* **2016**, *6*, 1601372. [CrossRef]
189. Wu, C.-W.; Unnikrishnan, B.; Chen, I.-W.P.; Harroun, S.G.; Chang, H.-T.; Huang, C.-C. Excellent oxidation resistive MXene aqueous ink for micro-supercapacitor application. *Energy Storage Mater.* **2020**, *25*, 563–571. [CrossRef]
190. Kurra, N.; Kulkarni, G.U. Pencil-on-paper: Electronic devices. *Lab. Chip* **2013**, *13*, 2866–2873. [CrossRef]

191. Yang, L.; Zheng, W.; Zhang, P.; Chen, J.; Tian, W.B.; Zhang, Y.M.; Sun, Z.M. MXene/CNTs films prepared by electrophoretic deposition for supercapacitor electrodes. *J. Electroanal. Chem.* **2018**, *830–831*, 1–6. [CrossRef]
192. Li, X.; Ma, Y.; Shen, P.; Zhang, C.; Cao, M.; Xiao, S.; Yan, J.; Luo, S.; Gao, Y. An Ultrahigh Energy Density Flexible Asymmetric Microsupercapacitor Based on $Ti_3C_2T_x$ and PPy/MnO_2 with Wide Voltage Window. *Adv. Mater. Technol.* **2020**, *5*, 2000272. [CrossRef]
193. Fu, Q.; Wen, J.; Zhang, N.; Wu, L.; Zhang, M.; Lin, S.; Gao, H.; Zhang, X. Free-standing Ti_3C_2Tx electrode with ultrahigh volumetric capacitance. *RSC Adv.* **2017**, *7*, 11998–12005. [CrossRef]
194. Chen, H.; Yu, L.; Lin, Z.; Zhu, Q.; Zhang, P.; Qiao, N.; Xu, B. Carbon nanotubes enhance flexible MXene films for high-rate supercapacitors. *J. Mater. Sci.* **2020**, *55*, 1148–1156. [CrossRef]
195. Fan, Z.; Wang, Y.; Xie, Z.; Xu, X.; Yuan, Y.; Cheng, Z.; Liu, Y. A nanoporous MXene film enables flexible supercapacitors with high energy storage. *Nanoscale* **2018**, *10*, 9642–9652. [CrossRef] [PubMed]

Article

Ferrocene Formic Acid Surface Modified Ni(OH)$_2$ for Highly Efficient Alkaline Oxygen Evolution

Guo-Ping Shen [1], Ruo-Yao Fan [1], Bin Dong [1,*] and Bo Chen [2,*]

[1] State Key Laboratory of Heavy Oil Processing, College of Chemistry and Chemical Engineering, University of Petroleum (East China), Qingdao 266580, China
[2] Department of Chemistry, City University of Hong Kong, Kowloon, Hong Kong, China
* Correspondence: dongbin@upc.edu.cn (B.D.); bchen005@e.ntu.edu.sg (B.C.)

Abstract: FeNi-based hybrid materials are among the most representative catalysts for alkaline oxygen evolution reaction (OER), but the modulation of their surface atoms to achieve the optimal catalytic properties is still a big challenge. Here, we report the surface modification of Ni(OH)$_2$/nickel foam (NF)-based electrocatalyst with a trace amount of ferrocene formic acid (FFA) (FFA-Ni(OH)$_2$/NF) for highly efficient OER. Owing to the strong electron interaction and synergistic effects of Fe-Ni heteroatoms, FFA-Ni(OH)$_2$/NF exhibits an overpotential of 311 mV at a current density of 100 mA cm^{-2}. Impressively, the overpotential of FFA-Ni(OH)$_2$/NF at 100 mA cm^{-2} is 108 mV less than that of bulk phase doped Ni/FFA(OH)$_2$/NF, demonstrating the surprising effect of heteroatomic surface modification. In addition, by introducing a small amount of surface modifier into the electrolyte, the weak surface reconstruction process in the electrochemical process can be fully utilized to achieve obvious modification effects. Therefore, this work fully proves the feasibility of improving catalytic activities of FeNi-based catalysts by modifying surface heterogeneous atom pairs.

Keywords: oxygen evolution reaction (OER); surface modification; Ni(OH)$_2$; FeNi-based catalysts

Citation: Shen, G.-P.; Fan, R.-Y.; Dong, B.; Chen, B. Ferrocene Formic Acid Surface Modified Ni(OH)$_2$ for Highly Efficient Alkaline Oxygen Evolution. *Crystals* **2022**, *12*, 1404. https://doi.org/10.3390/cryst12101404

Academic Editor: Leonid Kustov

Received: 29 August 2022
Accepted: 30 September 2022
Published: 4 October 2022

Publisher's Note: MDPI stays neutral with regard to jurisdictional claims in published maps and institutional affiliations.

Copyright: © 2022 by the authors. Licensee MDPI, Basel, Switzerland. This article is an open access article distributed under the terms and conditions of the Creative Commons Attribution (CC BY) license (https://creativecommons.org/licenses/by/4.0/).

1. Introduction

Developing sustainable green energy is a key way to realize a low-carbon transition [1,2]. At present, the development of main renewable green energy sources, including solar energy, tidal energy, wind energy, and geothermal energy, are severely limited by their intermittency, instability, and regionalism [3–5]. Hydrogen, by contrast, has a high energy density and good sustainability and is emerging as the latest star in the third energy revolution of the world [6–9].

Compared with traditional industrial hydrogen production, water electrolysis is a more ideal way for hydrogen production because of its advantages, such as being sustainable and pollution-free [10–12]. However, the efficiency of hydrogen production from water electrolysis is severely limited by the oxygen evolution reaction (OER) of four-electron transfer that occurs slowly at the anode [13–16]. At present, the most widely-used catalysts for OER are precious metals and their derivatives such as RuO$_2$ and IrO$_2$, which suffer from their high price and rare reserves [17–21]. It is reported that the well-designed FeNi-based hybrid catalytic materials in an alkaline environment have superior OER performance comparable to that of precious metals [22–26]. However, the key problem limiting the development of FeNi-based hybrid catalysts, especially FeNiO$_x$H$_y$, is that under the action of oxidation current, the strong oxidation and reconstruction of the surface of catalysts, lead to the loss of active sites and stability attenuation [27,28]. Therefore, the design and synthesis of FeNi-based catalytic materials with heterogeneous catalytic surfaces for the long-term OER is still a research hotspot.

Through accidental or intentional Fe doping or incorporation, the OER activity of nickel/cobalt-based electrocatalysts can be greatly increased, which is called the "Fe effect" [29–31]. Many researchers have discovered this phenomenon, but few have applied it to the synthesis and development of Fe-Ni and Fe-Co-based bimetallic composites. Furthermore, our previous work has shown that catalytic surfaces rich in nickel and low in iron are more conducive to the bonding of heterogeneous atoms to maintain the long-term stability of FeNi-based hybrid materials [32].

Herein, we synthesized trace ferrocene formic acid (FFA) modified $Ni(OH)_2$ hybrid catalyst by a simple electrochemical activation method to provide excellent OER performance in an alkaline solution. Due to the unique heteroatomic bond cooperation assisted by electrochemistry, the introduction of a small amount of ferrocene formic acid in the electrolyte can significantly improve the reaction current and oxygen evolution efficiency, indicating that it has a good modification effect on $Ni(OH)_2$ catalytic surface. Meanwhile, after ferrocene formic acid was introduced, the solid electronic interactions between Ni and Fe might change the electronic structure of $Ni(OH)_2$ to provide suitable intermediate adsorption energy. Impressively, the OER activity of $Ni/FFA(OH)_2/NF$ is far inferior to that of $FFA-Ni(OH)_2/NF$.

2. Experimental Section

2.1. Chemicals

$Ni(NO_3)_2 \cdot 6H_2O$ (AR), $C_{11}H_{10}FeO_2$ (AR), and C_3H_7NO (DMF) (AR) were purchased from Macklin (Shanghai, China). We used standard analytical pure KOH (purchased from Sinopharm Chemical Reagent Co., Ltd., Shanghai, China) to prepare the electrolyte. The content of Fe in this KOH is less than 0.001%. Nickel foams (NFs) were ultrasonically cleaned in acetone, hydrochloric acid (1 mol L^{-1}), and ethanol for 30 min. In this experiment, all electrolytes are prepared with specially purified ultrapure water to avoid any impurities. The models of electric heating constant temperature blast drying oven, muffle furnace, and tube furnace are DGG-90030G, KSL-1200X, and ZL-2011-2-0389859.8, respectively.

2.2. Synthesis of the $Ni(OH)_2/NF$

The growth solution contains ammonium fluoride (3 mmol), nickel nitrate (3 mmol), urea (10 mmol), and deionized water (40 mL) [33,34]. The prepared solution was then added to the reactor including a piece of NF (2 cm × 2 cm), and then heated at 100 °C for 12 h. Finally, a color transition of NF from silver-gray to light green can be observed.

2.3. Synthesis of the $FFA-Ni(OH)_2/NF$

$FFA-Ni(OH)_2$ was obtained by rapid electrochemical activation. First, $Ni(OH)_2/NF$ was activated to a stable state by linear sweep voltammetry (LSV) (5 mV s^{-1}, 0–0.7 V vs. SCE, 5 cycles). Then, ferrocene formic acid solution (0.05 mol L^{-1}) was added to the electrolyte drop by drop and stirred evenly. At the same time, LSV scanning was continued until stable. Finally, about 40 uL ferrocene formic acid solution was added until achieving the optimum activity. Finally, the obtained $FFA-Ni(OH)_2/NF$ catalyst was washed with deionized water.

2.4. Synthesis of the $Ni/FFA(OH)_2/NF$

The synthetic method of $Ni/FFA(OH)_2/NF$ is almost the same as that of $Ni(OH)_2/NF$, except that the growth solution is replaced with 10 mL 0.05 mol L^{-1} ferrocene formic acid solution, nickel nitrate (3 mmol), urea (10 mmol) and deionized water (40 mL).

2.5. Characterization

X-ray diffraction (XRD, Rigaku D/MAX-2500PC) with Cu Kα, λ = 1.54 Å was used to investigate the crystal phase-related information of the obtained catalysts. The instrument models of scanning electron microscopy (SEM) and transmission electron microscopy (TEM) are Hitachi S-4800 and FEI Tecni G20, respectively. The SEM-energy dispersive spectrometer (EDS) was used to analyze the detailed elemental composition of the catalyst. X-ray photoelectron spectroscopy (XPS) was tested on a Thermo Fisher Scientific II spectrometer using an Al Kα source.

2.6. Electrochemical Measurements

The electrochemical activities of catalysts were tested with a Gamry Reference 600 workstation in 1 M KOH electrolyte. In a typical three-electrode system, the working, counter, and reference electrodes are catalysts on NF, platinum sheet, and calomel electrode, respectively. Specifically, the LSV sweep is set at 5 mV s^{-1}; the test voltage and frequency ranges of electrochemical impedance spectroscopy (EIS) are 0.45 V vs. SCE and 10^5–0.1 Hz; the current density for the chronopotentiometry method is 200 mA cm^{-2}. The electrochemical active surface areas (ECSAs) of electrocatalysts were measured by cyclic voltammetry (CV). Test voltage range was set as 1.35–1.45 V vs. RHE and sweep speeds were set as 20, 40, 60, 80, and 100 mV s^{-1}. Their electrochemical double-layer capacitances (C_{dl}) were obtained from the change rate of voltage with sweep speed. According to the formula ESCA = C_{dl}/C_s (C_s = 0.04 mF), the corresponding ECSA of each sample can be obtained.

3. Results and Discussion

Figure 1 shows the detailed synthesis pathway of trace ferrocene formic acid (FFA)-modified Ni(OH)$_2$ electrocatalyst (named FFA-Ni(OH)$_2$/NF), including a one-step hydrothermal reaction and simple electrochemical activation process. It has been reported that Ni(OH)$_2$ has a special bonding effect for Fe ions, and the electrochemical reconstruction process of OER can promote the interaction of Fe-Ni heteroatomic pairs [24,25]. The XRD peaks of pure Ni foam are consistent with the characteristic peaks of standard Ni (00-003-1051) (Figure S1, see Supplementary Materials). Ni(OH)$_2$ synthesized by hydrothermal reaction has a mixed crystal phase structure. As shown in Figure 2a, XRD peaks of Ni(OH)$_2$/NF correspond to the standard peaks of α-Ni(OH)$_2$ (00-038-0715) and β-Ni(OH)$_2$ (01-074-2075). After the surface treatment with ferrocene formic acid, the original crystal structure of Ni(OH)$_2$ was basically retained, and the characteristic peaks of Fe(OH)$_3$ were also detected. This indicates that a tiny proportion of ferrocene is electro-activated to form Fe(OH)$_3$ after electrochemical action due to the particular interaction of Fe-Ni heteroatoms. In alkaline environments, OH intermediates can be adsorbed to Ni-Fe sites by terminal bonding or bridging to form heteroatomic bonds—Fe-O-Ni, which can be pivotal catalytic sites with high OER activity [35,36].

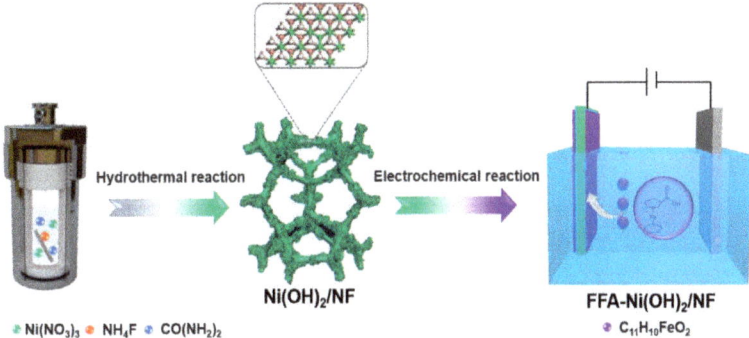

Figure 1. Schematic diagram of the preparation of FFA-Ni(OH)$_2$/NF.

In addition, the near-surface species composition and elemental valence states of all electrocatalysts were characterized by XPS. Characteristic peaks of Ni, O, and Fe appear in the full-range spectral data of FFA-Ni(OH)$_2$/NF (Figure 2b), which is consistent with the XRD data mentioned above. In detail, the Ni 2p peaks of FFA-Ni(OH)$_2$/NF could be divided into four peaks: Ni 2p$_{3/2}$ (854.83 eV), Ni 2p$_{1/2}$ (877.42 eV), Sat1 (860.51 eV) and Sat2 (878.40 eV) (Figure 2c). For the Ni 2p spectrum of Ni(OH)$_2$/NF, four similar peaks are located at 855.23, 861.03, 873.03, and 878.94 eV, respectively. A detailed comparison shows that the peak of Ni 2p$_{3/2}$ of FFA-Ni(OH)$_2$/NF is offset by 0.4 eV relative to Ni(OH)$_2$/NF, indicating an intense electron interaction between the introduced Fe and Ni.

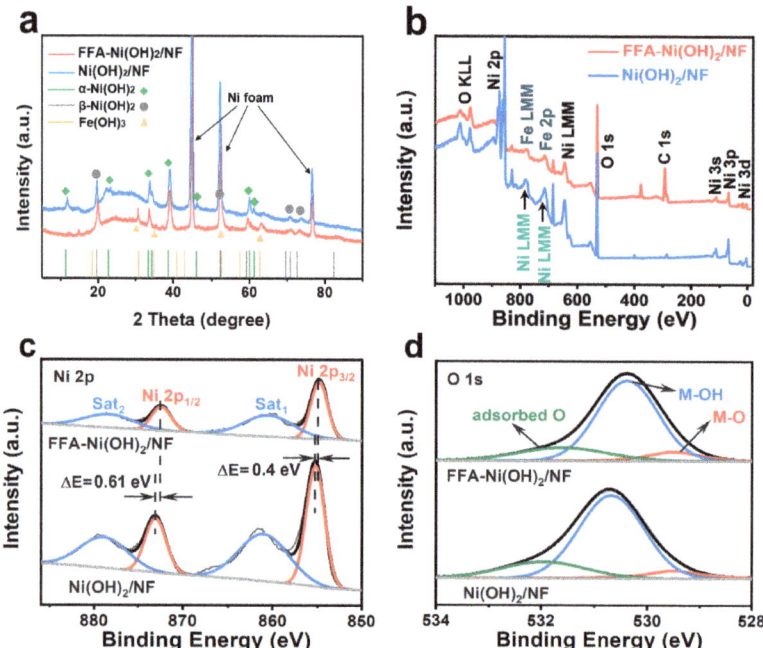

Figure 2. (a) Typical XRD patterns of Ni(OH)$_2$/NF and FFA-Ni(OH)$_2$/NF. (b) XPS spectra of FFA-Ni(OH)$_2$/NF (red) and Ni(OH)$_2$/NF (blue). (c) Ni 2p and (d) O 1s spectra of FFA-Ni(OH)$_2$/NF and Ni(OH)$_2$/NF.

In addition, the O 1s of FFA-Ni(OH)$_2$/NF and Ni(OH)$_2$/NF can be divided into three peaks corresponding to O-M (520 eV), O-H (531 eV), and adsorbed oxygen (532 eV) (Figure 2d). Compared with Ni(OH)$_2$/NF, the M-O peak of FFA-Ni(OH)$_2$/NF also exhibits a 0.3 eV shift, which is also due to the electron interaction of Fe-Ni heteroatom pairs. The 2p profile of Fe is shown in Figure S2. The peak of Fe is puzzling due to the interference of the strong Auger peak of Ni. This may also be caused by too little Fe content on the surface of FFA-Ni(OH)$_2$/NF.

The surface morphology information of electrocatalysts can be obtained from SEM. The surface of the pure NF is smooth, which is not conducive to the exposure of the catalytic active site and the direct application of the catalytic process (Figure S3). However, NF is more suitable for dispersion substrate because of its complex pore structure. In Figure 3a, the Ni(OH)$_2$/NF prepared by the hydrothermal method presents a flower-like three-dimensional structure with the aggregation of ultra-thin nanosheets. Polygonal ultra-thin nanosheets are more conducive to providing a large specific surface area and rich active sites (Figure 3d). Moreover, the interspaces formed by the cross-linking of nanosheets can provide a large number of gas transport channels to promote oxygen desorption. The surface elemental distribution of Ni(OH)$_2$/NF is demonstrated in Figure S4, and it can be found that Ni and O are distributed uniformly on the surface of the nanosheet. As we can see in Figure 3b, after ferrocene formic acid surface modification and electrochemical activation, FFA-Ni(OH)$_2$/NF can still maintain the morphology of flower-like nanosheet clusters, indicating that the main structure of the original precursor will not be damaged by FFA modification. More notably, compared with Ni(OH)$_2$/NF, FFA-Ni(OH)$_2$/NF has a thinner and more uniform nanosheet-like structure (Figure 3e), which may be more conducive to the exposure of active catalytic sites and transport of gas. In addition, FFA was added to the growth solution to prepare bulk phase doped Ni/FFA(OH)$_2$/NF, as shown in Figure 3c. It is not difficult to find that Ni/FFA(OH)$_2$/NF does not possess the flower-like morphology of nanosheet aggregation, which may be because the addition of FFA hinders the crystal growth of Ni(OH)$_2$. In Figure 3f, Ni/FFA(OH)$_2$/NF has a spongy porous structure, which is harmful to providing a large specific surface area and abundant active sites. To better reveal the influence of the morphology of electrocatalysts on the ECSA, we carried out CV tests on all samples (Figure S5). The results show that FFA-Ni(OH)$_2$/NF (37.18 mF cm^{-2}) has the largest C$_{dl}$ compared with Ni(OH)$_2$/NF (23.36 mF cm^{-2}), Ni/FFA(OH)$_2$/NF (30.55 mF cm^{-2}). The ECSAs of FFA-Ni(OH)$_2$/NF, Ni/FFA(OH)$_2$/NF, and Ni(OH)$_2$/NF are 929.50, 763.75, and 584.00 cm^{-2}, respectively (according to the formula: ESCA = C$_{dl}$/C$_s$). The results of the electrochemical test are consistent with SEM characterizations. SEM-Mapping results further showed that Fe, Ni, and O contained in FFA-Ni(OH)$_2$/NF were uniformly distributed, indicating that the electrochemical activation process could achieve uniform Fe dispersion. The content of the Fe element is really low (atomic content as low as 0.22%, Figure S6), indicating that a small amount of FFA can provide a significant modification effect.

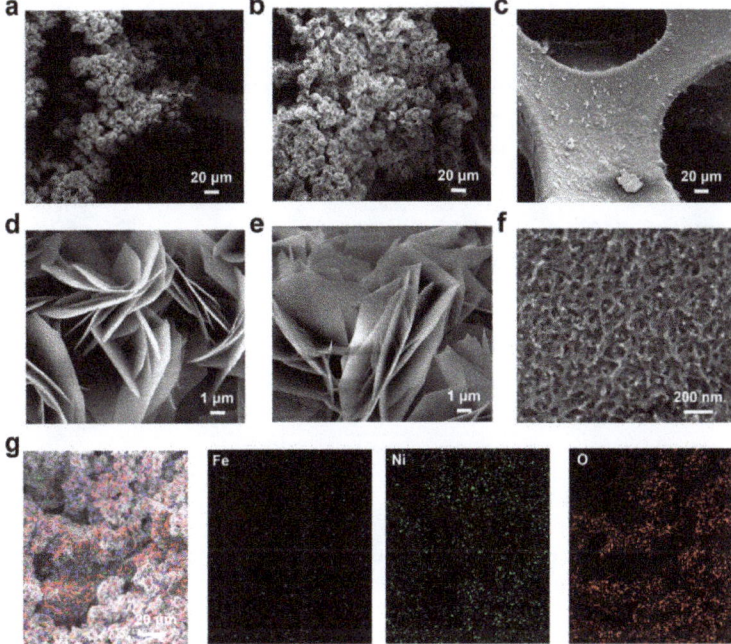

Figure 3. SEM images of Ni(OH)$_2$/NF (**a,d**), FFA-Ni(OH)$_2$/NF (**b,e**), and Ni/FFA(OH)$_2$/NF (**c,f**). (**g**) SEM image and EDX elemental mappings of FFA-Ni(OH)$_2$/NF.

To further highlight the enhanced catalytic effect of surface modification, detailed electrochemical characterizations were carried out for all electrocatalysts in 1 M KOH. In Figure 4a, compared with pure Ni(OH)$_2$/NF, the alkaline OER activity of Ni(OH)$_2$ modified by FFA was significantly improved. This may be due to the strong electron interaction and synergistic catalytic effects of Fe-Ni heteroatoms on the catalytic surface. Meanwhile, a small amount of ferrocene formic acid solution was added to the growth solution to synthesize bulk doped Ni(OH)$_2$ (Ni/FFA(OH)$_2$/NF) as the contrast sample.

As expected, the OER properties of Ni/FFA(OH)$_2$/NF are not ideal. Compared with Ni(OH)$_2$/NF, its overpotential reduces by only 30 mV, while FFA-Ni(OH)$_2$/NF reduces by 138 mV at 100 mA cm^{-2} (Figure 4b). The excellent modification effect of ferrocene formic acid surface modification can be well proved. In Figure 4g, the electrocatalysts in this work have better performances over several previously reported Fe-Ni-based OER catalysts

(Table S1). More importantly, FFA-Ni(OH)$_2$/NF exhibits an overpotential of 350 mV at a current density of 300 mA cm^{-2}, which is 240 mV less than Ni(OH)$_2$/NF. OER reaction kinetics of electrocatalysts can be reflected by the Tafel slope. As we can see in Figure 4c, FFA-Ni(OH)$_2$/NF has the smallest Tafel slope (77.68 mV dec^{-1}), indicating that it has improved OER reaction kinetics. In addition, the system resistance (R_s) and charge transfer resistance (R_{ct}) of the catalytic system were studied by electrochemical impedance spectroscopy (EIS). Our results show that the system resistances of NF, Ni(OH)$_2$/NF, Ni/FFA(OH)$_2$/NF, and FFA-Ni(OH)$_2$/NF are similar, but the R_{ct} of FFA-Ni(OH)$_2$/NF decreases nearly 100 times, which indicates that the electron transfer rate of the catalytic surface is obviously accelerated after surface modification with FFA (Figure 4d). This is probably because of the presence of conjugated electrons in the organic match, which makes the conductivity of the material significantly improved. In conclusion, FFA-Ni(OH)$_2$/NF exhibits surprising OER performances thanks to the electron interaction of Fe-Ni heteroatomic pairs, the surface modification effect of FFA, and the stereoscopic flower structure.

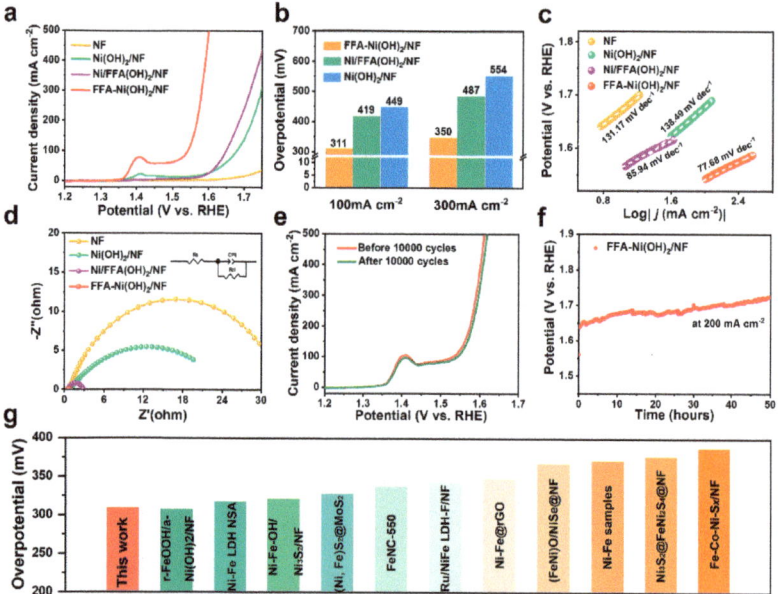

Figure 4. (**a**) LSV, (**b**) overpotential, (**c**) Tafel, and (**d**) EIS of FFA-Ni(OH)$_2$/NF, Ni(OH)$_2$/NF, Ni/FFA(OH)$_2$/NF, and NF. (**e**) LSV of FFA−Ni(OH)$_2$/NF before and after 10,000 CV cycles. (**f**) The chronopotentiometry (CP) test at 200 mA cm^{-2} of FFA-Ni(OH)$_2$/NF in 1 M KOH solution. (**g**) Comparison of the overpotentials of the catalyst in this work and the previously reported Fe-Ni-based OER catalysts.

In addition, the stability test of FFA-Ni(OH)$_2$/NF was explored in detail. As exhibited in Figure 4e, after 10,000 cycles of CV, the LSV profiles of FFA-Ni(OH)$_2$/NF basically coincide, indicating that it has excellent cycling stability. More importantly, after 50 h of stability testing, the reaction current of FFA-Ni(OH)$_2$/NF did not decay significantly, proving that it also had preeminent long-term stability (Figure 4f). Due to the interaction of abundant conjugated electron pairs, organic iron may be more conducive to Fe site stabilization than inorganic Fe salt. Importantly, the overpotential of the catalyst in this work is comparable to those of previously reported Fe-Ni-based OER catalysts (Figure 4g and Table S1) [37–47]. Therefore, the performance and stability of FFA-Ni(OH)$_2$/NF will meet the requirements of OER catalysts for future industrial applications.

4. Conclusions

In summary, FFA-Ni(OH)$_2$/NF synthesized by electrochemical activation and surface modification of FFA exhibited excellent OER performances in alkaline solution. Electrochemical tests revealed that FFA-Ni(OH)$_2$/NF only required an overpotential of 311 mV at 100 mA cm^{-2}, which was 138 mV less than that of Ni(OH)$_2$/NF. Moreover, the electrocatalytic activities of Ni(OH)$_2$/NF are stable after 10,000 cycles and up to 50 h of durability test. The excellent and comprehensive properties of FFA-Ni(OH)$_2$/NF may come from the following two aspects: (1) the strong electronic interaction of Fe-Ni heteroatom pairs positively regulates the binding energy of oxygen-containing intermediates. (2) the flower-like structure of Ni(OH)$_2$/NF composed of ultrathin nanosheets is well-maintained after micro FFA surface modification. Our research proves that surface modification is an effective way to achieve the performance enhancement of FeNi-based electrocatalysts.

Supplementary Materials: The following supporting information can be downloaded at: https://www.mdpi.com/article/10.3390/cryst12101404/s1, Figure S1: XRD of Ni foam; Figure S2: Fe 2p of FFA-Ni(OH)$_2$/NF; Figure S3: SEM images of NF; Figure S4: SEM-Mappings of Ni(OH)$_2$/NF; Figure S5: CVs of (a) FFA-Ni(OH)$_2$/NF, (b) Ni/FFA(OH)$_2$/NF, and (c) Ni(OH)$_2$/NF (at 1.35–1.45 V vs. RHE). (d) C$_{dl}$ of FFA-Ni(OH)$_2$/NF, Ni/FFA(OH)$_2$/NF and Ni(OH)$_2$/NF; Figure S6: The surface element content statistics of FFA-Ni(OH)$_2$/NF; Table S1: Comparison of the overpotential between the catalyst in this work and the previously reported Fe-Ni based OER electrocatalysts [37–47].

Author Contributions: Methodology, G.-P.S.; data curation, G.-P.S. and R.-Y.F.; formal analysis, G.-P.S. and R.-Y.F.; investigation, B.D.; resources, G.-P.S.; writing—original draft preparation, G.-P.S., R.-Y.F., B.D. and B.C.; writing—review and editing, G.-P.S., R.-Y.F., B.D. and B.C.; supervision, B.D. and B.C.; funding acquisition, B.D. All authors have read and agreed to the published version of the manuscript.

Funding: This work is financially supported by the National Natural Science Foundation of China (52174283).

Data Availability Statement: The date that support the findings of this study are available from the corresponding author upon reasonable request.

Conflicts of Interest: The authors declare no conflict of interest.

References

1. Tian, J.F.; Yu, L.G.; Xue, R.; Zhuang, S.; Shan, Y.L. Global low-carbon energy transition in the post-COVID-19 era. *Appl. Energy* **2022**, *307*, 118205. [CrossRef]
2. Kittner, N.; Lill, F.; Kammen, D.M. Energy storage deployment and innovation for the clean energy transition. *Nat. Energy* **2017**, *2*, 17125. [CrossRef]
3. Chien, F.S.; Kamran, H.W.; Albashar, G.; Iqbal, W. Dynamic planning, conversion, and management strategy of different renewable energy sources: A sustainable solution for severe energy crises in emerging economies. *Int. J. Hydrogen Energy* **2021**, *46*, 7745–7758. [CrossRef]
4. Lin, Z.M.; Zhang, B.B.; Guo, H.Y.; Wu, Z.Y.; Zou, H.Y.; Yang, J.; Wang, Z.L. Super-robust and frequency-multiplied triboelectric nanogenerator for efficient harvesting water and wind energy. *Nano Energy* **2019**, *64*, 103908. [CrossRef]
5. Sahu, B.K. A study on global solar PV energy developments and policies with special focus on the top ten solar PV power producing countries. *Renew. Sustain. Energy Rev.* **2015**, *43*, 621–634. [CrossRef]
6. Wang, X.S.; Zheng, Y.; Sheng, W.C.; Xu, Z.C.; Jaroniec, M.; Qiao, S.Z. Strategies for design of electrocatalysts for hydrogen evolution under alkaline conditions. *Mater. Today* **2020**, *36*, 125–138. [CrossRef]
7. Dawood, F.; Anda, M.; Shafiuiiah, G.M. Hydrogen production for energy: An overview. *Int. J. Hydrogen Energy* **2020**, *45*, 3847–3869. [CrossRef]
8. Zhang, X.Y.; Li, F.T.; Dong, Y.W.; Dai, F.N.; Liu, C.G.; Chai, Y.M. Dynamic anion regulation to construct S-doped FeOOH realizing 1000 mA cm(-2)-level-current-density oxygen evolution over 1000 h. *Appl. Catal. B Environ.* **2022**, *315*, 121571. [CrossRef]
9. Yue, M.L.; Lambert, H.; Pahon, E.; Roche, R.; Jemei, S.; Hissel, D. Hydrogen energy systems: A critical review of technologies, applications, trends and challenges. *Renew. Sustain. Energy Rev.* **2021**, *146*, 111180. [CrossRef]
10. Zhang, L.Y.; Zheng, Y.J.; Wang, J.C.; Geng, Y.; Zhang, B.; He, J.J.; Xue, J.M.; Frauenheim, T.; Li, M. Ni/Mo Bimetallic-oxide-derived heterointerface-rich sulfide nanosheets with Co-doping for efficient alkaline hydrogen evolution by boosting volmer reaction. *Small* **2021**, *17*, 2006730. [CrossRef] [PubMed]

11. Surendran, S.; Jesudass, S.C.; Janani, G.; Kim, J.Y.; Lim, Y.; Park, J.; Han, M.K.; Cho, I.S.; Sim, U. Sulphur assisted nitrogen-rich CNF for improving electronic interactions in Co-NiO heterostructures toward accelerated overall water splitting. *Adv. Mater. Technol.* **2022**. [CrossRef]
12. Janani, G.; Surendran, S.; Choi, H.; An, T.Y.; Han, M.K.; Song, S.J.; Park, W.; Kim, J.K.; Sim, U. Anchoring of $Ni_{12}P_5$ microbricks in nitrogen- and phosphorus-enriched carbon frameworks: Engineering bifunctional active sites for efficient water-splitting systems. *ACS Sustain. Chem. Eng.* **2022**, *10*, 1182–1194. [CrossRef]
13. Jiang, Y.Y.; Dong, K.; Lu, Y.Z.; Liu, J.W.; Chen, B.; Song, Z.Q.; Niu, L. Bimetallic oxide coupled with B-doped graphene as highly efficient electrocatalyst for oxygen evolution reaction. *Sci. China Mater.* **2020**, *63*, 1247–1256. [CrossRef]
14. Du, J.; Li, F.; Sun, L.C. Metal organic frameworks and their derivatives as electrocatalysts for the oxygen evolution reaction. *Chem. Soc. Rev.* **2021**, *50*, 2663–2695. [CrossRef]
15. Radwan, A.; Jin, H.H.; He, D.P.; Mu, S.C. Design Engineering, Synthesis Protocols, and Energy Applications of MOF-Derived Electrocatalysts. *Nano-Micro Lett.* **2021**, *13*, 132. [CrossRef]
16. Fan, R.Y.; Xie, J.Y.; Yu, N.; Chai, Y.M.; Dong, B. Interface design and composition regulation of cobalt-based electrocatalysts for oxygen evolution reaction. *Int. J. Hydrogen Energy* **2022**, *47*, 10547–10572. [CrossRef]
17. Hu, Y.D.; Luo, G.; Wang, L.G.; Liu, X.K.; Qu, Y.T.; Zhou, Y.S.; Zhou, F.Y.; Li, Z.J.; Li, Y.F.; Yao, T.; et al. Single Ru atoms stabilized by hybrid amorphous/crystalline FeCoNi layered double hydroxide for ultraefficient oxygen evolution. *Adv. Energy Mater.* **2021**, *11*, 2002816. [CrossRef]
18. Zhou, S.Z.; Jang, H.S.; Qin, Q.; Li, Z.J.; Kim, M.G.; Li, C.; Liu, X.E.; Cho, J. Three-dimensional hierarchical Co(OH)F nanosheet arrays decorated by single-atom Ru for boosting oxygen evolution reaction. *Sci. China Mater.* **2021**, *64*, 1408–1417. [CrossRef]
19. Xi, G.G.; Zuo, L.; Li, X.; Jin, Y.; Li, R.; Zhang, T. In-situ constructed Ru-rich porous framework on NiFe-based ribbon for enhanced oxygen evolution reaction in alkaline solution. *J. Mater. Sci. Technol.* **2021**, *70*, 197–204. [CrossRef]
20. Cai, C.; Wang, M.Y.; Han, S.B.; Wang, Q.; Zhang, Q.; Zhu, Y.M.; Yang, X.M.; Wu, D.J.; Zu, X.T.; Sterbinsky, G.E.; et al. Ultrahigh oxygen evolution reaction activity achieved using Ir single atoms on amorphous CoO_x nanosheets. *ACS Catal.* **2021**, *11*, 123–130. [CrossRef]
21. She, L.A.; Zhao, G.Q.; Ma, T.Y.; Chen, J.; Sun, W.P.; Pan, H.G. On the durability of Iridium-based electrocatalysts toward the oxygen evolution reaction under acid environment. *Adv. Funct. Mater.* **2022**, *32*, 2108465. [CrossRef]
22. Shi, Z.X.; Zhao, J.W.; Li, C.F.; Xu, H.; Li, G.R. Fully exposed edge/corner active sites in Fe substituted-Ni(OH)(2) tube-in-tube arrays for efficient electrocatalytic oxygen evolution. *Appl. Catal. B Environ.* **2021**, *298*, 120559. [CrossRef]
23. Trotochaud, L.; Young, S.L.; Ranney, J.K.; Boettcher, S.W. Nickel-Iron oxyhydroxide oxygen-evolution electrocatalysts: The role of intentional and incidental iron incorporation. *J. Am. Chem. Soc.* **2014**, *136*, 6744–6753. [CrossRef] [PubMed]
24. Friebel, D.; Louie, M.W.; Bajdich, M.; Sanwald, K.E.; Cai, Y.; Wise, A.M.; Cheng, M.J.; Sokaras, D.; Weng, T.C.; Alonso-Mori, R.; et al. Identification of highly active Fe sites in (Ni,Fe)OOH for electrocatalytic water splitting. *J. Am. Chem. Soc.* **2015**, *137*, 1305–1313. [CrossRef] [PubMed]
25. Anantharaj, S.; Kundu, S.; Noda, S. "The Fe Effect": A review unveiling the critical roles of Fe in enhancing OER activity of Ni and Co based catalysts. *Nano Energy* **2021**, *80*, 105514. [CrossRef]
26. Hu, C.J.; Hu, Y.F.; Fan, C.H.; Yang, L.; Zhang, Y.T.; Li, H.X.; Xie, W. Surface-enhanced raman spectroscopic evidence of key intermediate species and role of NiFe dual-catalytic center in water oxidation. *Angew. Chem. Int. Ed.* **2021**, *60*, 19774–19778. [CrossRef]
27. Wang, Y.Y.; Qiao, M.; Li, Y.F.; Wang, S.Y. Tuning surface electronic configuration of NiFe LDHs nanosheets by introducing cation vacancies (Fe or Ni) as highly efficient electrocatalysts for oxygen evolution reaction. *Small* **2018**, *14*, 1800136. [CrossRef]
28. Fan, R.Y.; Xie, J.Y.; Liu, H.J.; Wang, H.Y.; Li, M.X.; Yu, N.; Luan, R.N.; Chai, Y.M.; Dong, B. Directional regulating dynamic equilibrium to continuously update electrocatalytic interface for oxygen evolution reaction. *Chem. Eng. J.* **2022**, *431*, 134040. [CrossRef]
29. Louie, M.W.; Bell, A.T. An investigation of thin-film Ni-Fe oxide catalysts for the electrochemical evolution of oxygen. *J. Am. Chem. Soc.* **2013**, *135*, 12329–12337. [CrossRef]
30. Farhat, R.; Dhainy, J.H.; Halaoui, L.I. OER catalysis at activated and codeposited NiFe-oxo/hydroxide thin films is due to postdeposition surface-Fe and is not sustainable without Fe in solution. *ACS Catal.* **2020**, *10*, 20–35. [CrossRef]
31. Chang, J.L.; Chen, L.M.; Zang, S.Q.; Wang, Y.F.; Wu, D.P.; Xu, F.; Gao, Z.Y. The effect of Fe(III) cations in electrolyte on oxygen evolution catalytic activity of Ni(OH)(2) electrode. *J. Colloid Interface Sci.* **2020**, *569*, 50–56. [CrossRef]
32. Fan, R.Y.; Zhang, X.Y.; Yu, N.; Wang, F.G.; Zhao, H.Y.; Liu, X.; Lv, Q.X.; Liu, D.P.; Chai, Y.M.; Dong, B. Rapid "self-healing" behavior induced by chloride anions to renew the Fe–Ni(oxy)hydroxide surface for long-term alkaline seawater electrolysis. *Inorg. Chem. Front.* **2022**, *9*, 4216–4224. [CrossRef]
33. Gao, M.R.; Sheng, W.C.; Zhuang, Z.B.; Fang, Q.R.; Gu, S.; Jiang, J.; Yan, Y.S. Efficient water oxidation using nanostructured α-nickel-hydroxide as an electrocatalyst. *J. Am. Chem. Soc.* **2014**, *136*, 7077–7084. [CrossRef]
34. Xu, Q.C.; Hua, J.; Zhang, H.X.; Hu, Y.J.; Li, C.Z. Heterogeneous interface engineered atomic configuration on ultrathin $Ni(OH)_2/Ni_3S_2$ nanoforests for efficient water splitting. *Appl. Catal. B Environ.* **2019**, *242*, 60–66. [CrossRef]
35. Wang, J.; Gan, L.Y.; Zhang, W.Y.; Peng, Y.C.; Yu, H.; Yan, Q.Y.; Wang, X. In situ formation of molecular Ni-Fe active sites on heteroatom-doped graphene as a heterogeneous electrocatalyst toward oxygen evolution. *Sci. Adv.* **2018**, *4*, 7970. [CrossRef]

36. Yi, L.Y.; Niu, Y.L.; Feng, B.M.; Zhao, M.; Hu, W.H. Simultaneous phase transformation and doping via a unique photochemical–electrochemical strategy to achieve a highly active Fe-doped Ni oxyhydroxide oxygen evolution catalyst. *J. Mater. Chem. A* **2021**, *9*, 4213–4220. [CrossRef]
37. Cheng, X.D.; Yuan, J.X.; Cao, J.H.; Lei, C.J.; Yang, B.; Li, Z.J.; Zhang, X.W.; Yuan, C.; Lei, L.C.; Hou, Y. Strongly coupling of amorphous/crystalline reduced FeOOH/α-Ni(OH)$_2$ heterostructure for extremely efficient water oxidation at ultra-high current density. *J. Colloid Interface Sci.* **2020**, *579*, 340–346. [CrossRef]
38. Liu, Z.R.; Deng, Y.W.; Wang, L.; Tang, J.N. A Facile topochemical preparation of Ni-Fe LDH nanosheets array on nickel foam using in situ generated Ni^{2+} for electrochemical oxygen evolution. *J. Electrochem. Soc.* **2020**, *167*, 046502-9. [CrossRef]
39. He, W.J.; Ren, G.; Li, Y.; Jia, D.B.; Li, S.Y.; Cheng, J.N.; Liu, C.C.; Hao, Q.Y.; Zhang, J.; Liu, H. Amorphous nickel–iron hydroxide films on nickel sulfide nanoparticles for the oxygen evolution reaction. *Catal. Sci. Technol.* **2020**, *10*, 1708–1713. [CrossRef]
40. Liu, Y.K.; Jiang, S.; Li, S.J.; Zhou, L.; Li, Z.H.; Li, J.M.; Shao, M.F. Interface engineering of (Ni, Fe)S$_2$@MoS$_2$ heterostructures for synergetic electrochemical water splitting. *Appl. Catal. B* **2019**, *247*, 107–114. [CrossRef]
41. Liu, G.P.; Wang, B.; Wang, L.; Wei, W.X.; Quan, Y.; Wang, C.T.; Zhu, W.S.; Li, H.M.; Xia, J.X. MOFs derived FeNi$_3$ nanoparticles decorated hollow N-doped carbon rod for high-performance oxygen evolution reaction. *Green Energy Environ.* **2022**, *7*, 423–431. [CrossRef]
42. Wang, Y.; Zheng, P.; Li, M.X.; Li, Y.R.; Zhang, X.; Chen, J.; Fang, X.; Liu, Y.J.; Yuan, X.L.; Dai, X.P.; et al. Interfacial synergy between dispersed Ru sub-nanoclusters and porous NiFe layered double hydroxide on accelerated overall water splitting by intermediate modulation. *Nanoscale* **2020**, *12*, 9669–9679. [CrossRef]
43. Chandrasekaran, N.; Muthusamy, S. Free-standing porous interconnects of Ni–Fe alloy decorated reduced graphene oxide for oxygen evolution reaction. *Langmuir* **2017**, *33*, 2–10. [CrossRef] [PubMed]
44. Dai, W.J.; Zhu, Y.A.; Ye, Y.K.; Pan, Y.; Lu, T.; Huang, S.F. Electrochemical incorporation of heteroatom into surface reconstruction induced Ni vacancy of NixO nanosheet for enhanced water oxidation. *J. Colloid Interface Sci.* **2022**, *608*, 3030–3039. [CrossRef] [PubMed]
45. Hatami, E.; Toghraei, A.; Darband, G.B. Electrodeposition of Ni–Fe micro/nano urchin-like structure as an efficient electrocatalyst for overall water splitting. *Int. J. Hydrogen Energy* **2021**, *46*, 9394–9405. [CrossRef]
46. Yang, Y.Y.; Meng, H.X.; Kong, C.; Yan, S.H.; Ma, W.X.; Zhu, H.; Ma, F.Q.; Wang, C.J.; Hu, Z.A. Heterogeneous Ni$_3$S$_2$@FeNi$_2$S$_4$@NF nanosheet arrays directly used as high efficiency bifunctional electrocatalyst for water decomposition. *J. Colloid Interface Sci.* **2021**, *599*, 300–312. [CrossRef] [PubMed]
47. Zhu, S.S.; Lei, J.L.; Wu, S.M.; Liu, L.; Chen, T.M.; Yuan, Y.; Ding, C. Construction of Fe-Co-Ni-S-x/NF nanomaterial as bifunctional electrocatalysts for water splitting. *Mater. Lett.* **2022**, *311*, 131549. [CrossRef]

MDPI
St. Alban-Anlage 66
4052 Basel
Switzerland
Tel. +41 61 683 77 34
Fax +41 61 302 89 18
www.mdpi.com

Crystals Editorial Office
E-mail: crystals@mdpi.com
www.mdpi.com/journal/crystals

www.ingramcontent.com/pod-product-compliance
Lightning Source LLC
LaVergne TN
LVHW070628100526
838202LV00012B/753